Brewing Microbiology

Brewing Microbiology

Second edition

Edited by

F.G. Priest and I. Campbell

International Centre for Brewing and Distilling
Heriot-Watt University
Edinburgh, UK

London · Glasgow · Weinheim · New York · Tokyo · Melbourne · Madras

Published by Chapman & Hall, 2–6 Boundary Row, London SE1 8HN, UK

Chapman & Hall, 2–6 Boundary Row, London SE1 8HN, UK

Blackie Academic & Professional, Wester Cleddens Road, Bishopbriggs, Glasgow G64 2NZ, UK

Chapman & Hall GmbH, Pappelallee 3, 69469 Weinheim, Germany

Chapman & Hall USA, 115 Fifth Avenue, New York, NY 10003, USA

Chapman & Hall Japan, ITP-Japan, Kyowa Building, 3F, 2–2–1 Hirakawacho, Chiyoda-ku, Tokyo 102, Japan

Chapman & Hall Australia, 102 Dodds Street, South Melbourne, Victoria 3205, Australia

Chapman & Hall India, R. Seshadri, 32 Second Main Road, CIT East, Madras 600 035, India

---

First edition 1987

Second edition 1996

Typeset in 10/12 Palatino by WestKey Ltd, Falmouth, Cornwall

Printed in Great Britain by T.J. Press (Padstow) Ltd, Padstow, Cornwall

ISBN 0 412 59150 2

A catalogue record for this book is available from the British Library

Library of Congress Catalog Card Number: 95–70016

∞ Printed on permanent acid-free text paper, manufactured in accordance with ANSI/NISO Z39.48–1992 and ANSI/NISO Z39.48–1984 (Permanence of Paper).

Contents

List of contributors ix
Preface x

1 Systematics of yeasts 1
I. Campbell
1.1 Classification of yeasts 1
1.2 Nomenclature of yeasts 7
1.3 Properties for identification of yeasts 9
References 10

2 The biochemistry and physiology of yeast growth 13
T.W. Young
2.1 Introduction 13
2.2 Yeast nutrition 14
2.3 Yeast metabolism 15
2.4 Yeast propagation 28
2.5 Brewery fermentation 34
References 40

3 Yeast genetics 43
J.R.M. Hammond
3.1 Introduction 43
3.2 Genetic features of *Saccharomyces cerevisiae* 44
3.3 The need for new brewing yeasts 48
3.4 Genetic techniques and their application to the analysis and development of brewing yeast strains 50
3.5 The commercial use of genetically modified brewing yeasts 73
3.6 Conclusions 74
Acknowledgements 75
References 75

4 The microflora of barley and malt 83
B. Flannigan
4.1 The microflora of barley 83
4.2 The microflora of malt 96
4.3 Effects of microorganisms on malting 102
4.4 Effects of the microflora on beer and distilled spirit 107
4.5 Health hazards 111
4.6 Assessment of mould contamination 119
References 120

5 Gram-positive brewery bacteria 127
F.G. Priest
5.1 Introduction 127
5.2 Lactic acid bacteria 128
5.3 *Lactobacillus* 134
5.4 *Pediococcus* 147
5.5 *Leuconostoc* 151
5.6 Homofermentative cocci 152
5.7 *Micrococcus* and *Staphylococcus* 153
5.8 Endospore-forming bacteria 154
5.9 Identification of genera of Gram-positive bacteria of brewery origin 154
5.10 Concluding remarks 156
References 157

6 Gram-negative spoilage bacteria 163
H.J.J. Van Vuuren
6.1 Introduction 163
6.2 Acetic acid bacteria 165
6.3 Enterobacteriaceae 169
6.4 *Zymomonas* 177
6.5 Anaerobic Gram-negative rods 180
6.6 *Megasphaera* 183
6.7 Miscellaneous non-fermentative bacteria 183
6.8 Detection, enumeration and isolation 184
6.9 Conclusions 186
Acknowledgements 187
References 187

7 Wild yeasts in brewing and distilling 193
I. Campbell
7.1 Introduction 193
7.2 Detection of wild yeasts 193
7.3 Identification of wild yeasts 198
7.4 Effects of wild yeasts in the brewery 201

7.5 Elimination of wild yeasts 205
References 207

8 Rapid detection of microbial spoilage 209
I. Russell and T.M. Dowhanick
8.1 Introduction 209
8.2 Impedimetric techniques (conductance, capacitance) 211
8.3 Microcalorimetry 214
8.4 Turbidometry 214
8.5 Flow cytometry 215
8.6 ATP bioluminescence 216
8.7 Microcolony method 218
8.8 Direct epifluorescence filter technique 220
8.9 Protein fingerprinting by polyacrylamide gel electrophoresis 221
8.10 Immunoanalysis 222
8.11 Hybridization using DNA probes 224
8.12 Karyotyping (chromosome fingerprinting) 226
8.13 Polymerase chain reaction 228
8.14 Random amplified polymorphic DNA PCR 230
8.15 Summary 231
References 231

9 Methods for the rapid identification of microorganisms 237
C.S. Gutteridge and F.G. Priest
9.1 What is identification? 237
9.2 Levels of expression of the microbial genome 238
9.3 Identification at the genomic level 241
9.4 Techniques for examining proteins 244
9.5 Methods that examine aspects of cell composition 247
9.6 Developments in techniques for studying morphology and behaviour 257
9.7 Future trends in rapid identification 264
Acknowledgements 267
References 267

10 Cleaning and disinfection in the brewing industry 271
M. Singh and J. Fisher
10.1 Introduction 271
10.2 Definitions 271
10.3 Standards required within a brewery 272
10.4 Cleaning methods available 275
10.5 Soil composition 280
10.6 Process of detergency 280
10.7 Chemistry of detergents 281
10.8 Caustic and alkaline detergents 282

10.9 Sequestrants 283
10.10 Acids 286
10.11 Surface active agents 287
10.12 Disinfectants and sanitizers used in breweries 290
10.13 Oxidizing disinfectants 291
10.14 Non-oxidizing disinfectants 294
10.15 Water treatment 297
10.16 Steam 299
10.17 Summary 299
References 300

Index **301**

Contributors

I. Campbell International Centre for Brewing and Distilling, Heriot-Watt University, Edinburgh EH14 4AS, UK

T.M. Dowhanick Research Department, Labatt Breweries of Canada, London, Ontario N6A 4M3, Canada

J. Fisher Diversey (FB) Ltd Technical Centre, Greenhill Lane, Riddings, Derbyshire DL35 4LQ, UK

B. Flannigan International Centre for Brewing and Distilling, Heriot-Watt University, Edinburgh EH14 4AS, UK

C.S. Gutteridge Reading Scientific Services Ltd, Lord Zuckerman Research Centre, The University, Whiteknights, PO Box 234, Reading, Berkshire RG6 2LA, UK

J.R.M. Hammond BRF International, Lyttel Hall, Coopers Hill Road, Nutfield, Redhill, Surrey RH1 4HY, UK

F.G. Priest International Centre for Brewing and Distilling, Heriot-Watt University, Edinburgh EH14 4AS, UK

I. Russell Research Department, Labatt Breweries of Canada, London, Ontario, N6A 4M3, Canada

M. Singh Diversey (FB) Ltd Technical Centre, Greenhill Lane, Riddings, Derbyshire DL35 4LQ, UK

H.J.J. Van Vuuren Department of Microbiology, University of Stellenbosch, Stellenbosch 7600, South Africa

T.W. Young Department of Biochemistry, University of Birmingham, PO Box 363, Birmingham B15 2TT, UK

Preface

During the latter part of the last century and the early years of this century, the microbiology of beer and the brewing process played a central role in the development of modern microbiology. An important advance was Hansen's development of pure culture yeasts for brewery fermentations and the recognition of different species of brewing and wild yeasts. The discovery by Winge of the life cycles of yeasts and the possibilities of hybridization were among the first steps in yeast genetics with subsequent far-reaching consequences. Over the same period the contaminant bacteria of the fermentation industries were also studied, largely influenced by Shimwell's pioneering research and resulting in the improvement of beer quality.

Towards the end of the century, the influence of brewing microbiology within the discipline as a whole is far less important, but it retains an essential role in quality assurance in the brewing industry. Brewing microbiology has gained from advances in other aspects of microbiology and has adopted many of the techniques of biotechnology. Of particular relevance are the developments in yeast genetics and strain improvement by recombinant DNA techniques which are rapidly altering the way brewers view the most important microbiological components of the process: yeast and fermentation. Moreover, the changing emphasis in quality control from traditional plating techniques, which essentially provide a historical account of the process, to new rapid methodologies, which give up-to-the-minute microbiological status reports of the process upon which brewers can base decisions, is playing an important role in the way the modern brewery operates. Developments in other aspects of brewing, such as packaging under reduced oxygen levels, are affecting the range of spoilage organisms encountered in beer and have led to strictly anaerobic bacteria giving spoilage problems. In preparing this second edition of *Brewing Microbiology*, we have covered these and other developments in the microbiology of the brewing process and its products while retaining the comprehensive yet specialist treatment of the first edition.

Once again, we thank the authors for their contributions and Nigel Balmforth of Chapman & Hall for his help and encouragement in the preparation of the book.

Fergus G. Priest, Iain Campbell

CHAPTER 1

Systematics of yeasts

I. Campbell

Systematics includes classification, nomenclature and identification. Routine identification of culture yeasts will be discussed later (Chapter 7); this present chapter is concerned primarily with classification and nomenclature of yeasts, and the general principles of yeast identification.

1.1 CLASSIFICATION OF YEASTS

What is a yeast? No satisfactory definition exists, and such features of commonly encountered yeasts as alcoholic fermentation or growth by budding are absent from a substantial minority of species. Although yeasts are generally accepted as fungi which are predominantly unicellular, there are various borderline 'yeast-like fungi' which are difficult to classify. Also, many yeasts are capable, under appropriate cultural conditions, of growing in mycelial form, either as true mycelium or as a pseudomycelium of branched chains of elongated yeast cells.

Most mycologists accept the series of publications of the Dutch group of yeast taxonomists as definitive. The current classification is still basically that of Kreger-van Rij (1984), although updated (Barnett, Payne and Yarrow, 1990) to recognize the discovery of new species or recent changes of attitude to the significance of taxonomic tests (Kurtzman, 1984; Kreger-van Rij, 1987; Kurtzman and Phaff, 1987).

Three of the four groups of fungi include yeasts: Ascomycetes, Basidiomycetes and Deuteromycetes. Although Zygomycetes may develop yeast morphology under unusual cultural conditions, their normal existence is in the filamentous form, as non-septate hyphae. Classification of fungi (Table 1.1) is largely on the form of vegetative growth and the nature of the spores, if formed. Physiological properties, as widely used in bacteriology, are not used in identification of filamentous fungi; so far as yeasts are

Brewing Microbiology, 2nd edn. Edited by F. G. Priest and I. Campbell.
Published in 1996 by Chapman & Hall, London. ISBN 0 412 59150 2

Table 1.1 Classification of yeasts (from Kreger-van Rij, 1984)

1. Group 1: Ascomycetes. The sexual spores are formed endogenously, i.e. within the cell. In filamentous fungi a specialized spore-bearing ascus is formed; in yeasts, the spores develop within the former vegetative cell which is then correctly termed the ascus (but note the exception, *Lipomyces*).
   (a) Spermphthoraceae (needle-shaped spores; Fig. 1.1a); principal genera *Metschnikowia, Nematospora.*
   (b) Saccharomycetaceae (other forms of spore; Fig. 1.1b,c,d,e); four families, distinguished mainly by method of vegetative growth.
      (i) Schizosaccharomycoideae (growth by binary fission, Fig. 1.2a); one genus, *Schizosaccharomyces.*
      (ii) Nadsoniodeae (growth by polar budding, Fig. 1.2b); principal genera *Hanseniaspora, Nadsonia, Saccharomycodes,* distinguished by form of spores.
      (iii) Lipomycoideae (growth by multilateral budding, Fig. 1.2c, but the principal characteristic is the 'exozygotic ascus', a sac-like ascus with numerous oval spores, Fig. 1.1e); one genus, *Lipomyces.*
      (iv) Saccharomycoideae (growth by multilateral budding, Fig. 1.2c); numerous genera, distinguished mainly by details of the sporulation cycle and spore morphology. The most important genera in the brewing and distilling industries are *Debaryomyces, Dekkera, Kluyveromyces, Pichia, Saccharomyces, Schwanniomyces, Torulaspora* and *Zygosaccharomyces.*
2. Group 2: Basidiomycetes. The sexual spores are formed exogenously. Yeasts of this group are unimportant in the fermentation industries, but their non-sporing, 'imperfect' forms, especially *Sporobolomyces, Rhodotorula* and *Cryptococcus,* are common surface organisms of plant materials, including barley and malt.
3. Group 3: Blastomycetes, Deuteromycetes or Fungi Imperfecti. No sexual spores are formed.
   (a) Sporobolomycetaceae (forming exogenous asexual spores, i.e. ballistospores); two genera, *Bullera* (relatively rare) and *Sporobolomyces* (very common).
   (b) Cryptococcaceae (no exogenous asexual spores). Genera of this family, which represent those yeasts of groups 1 and 2 which have lost the ability to form sexual spores, are classified by different characteristics from the sporing yeasts and therefore do not necessarily coincide with the equivalent 'perfect' genera. Genera are distinguished by the form of vegetative growth, fermentative ability and a somewhat haphazard selection of other tests. Principal genera of importance to the fermentation industries are *Kloeckera* (which grows by polar budding), *Brettanomyces, Candida, Cryptococcus* and *Rhodotorula* (growth by multilateral budding). The genus *Trichosporon,* often a contaminant of cereal grains, grows both by multilateral budding and a form of fission. *Torulopsis,* formerly a separate genus including various species of spoilage yeasts of the brewing and related industries, is now incorporated into the genus *Candida* (Kreger-van Rij, 1984).

concerned, they are used to distinguish species. Formerly, the ability to utilize nitrate as a nitrogen source was used to distinguish genera but, following the demonstration by DNA analysis that species of *Hansenula* and

Pichia differing in utilization of NO_3^- were otherwise identical (Kurtzman, 1984), the 'nitrate-positive' genus *Hansenula* has now been abandoned. Former *Hansenula* spp. are now incorporated in the genus *Pichia*, or other related genera according to life cycles and sporulation characteristics (Barnett, Payne and Yarrow, 1990).

The definitive property of Ascomycetes is the production of endogenous sexual spores (ascospores). Further subdivision as families and genera is based on the type of spores (Fig. 1.1), and the nature of the life cycle by which they are formed. The mode of vegetative multiplication is also relevant, since growth by binary fission (Fig. 1.2a) or polar budding (Fig. 1.2b) is sufficiently different from the more common multilateral budding to be of taxonomic importance. Within the group of yeasts growing by multilateral budding, the genera *Saccharomyces, Kluyveromyces, Torulaspora* and *Zygosaccharomyces* form a closely related subgroup of vigorously fermentative yeasts. Indeed, by earlier classification (e.g. Lodder and Kreger-van Rij, 1952), species at present in these four genera were all allocated to one genus, *Saccharomyces*. Subsequently, the differences in vegetative growth cycles and sporulation were considered sufficiently different to justify four separate genera. *S. cerevisiae* is of such industrial importance that its vegetative and sporulation cycles have been carefully studied, in recent years largely from the viewpoint of applied genetics. Wild sporulating strains of *S. cerevisiae* are normally diploid, and meiosis on sporulation produces asci containing four haploid spores (Chapter 3). The typical appearance of asci of *S. cerevisiae* is illustrated in

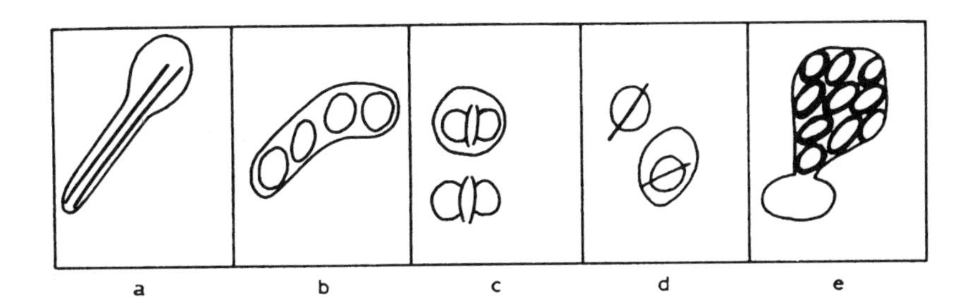

Fig. 1.1 Yeast spores. (a) Needle spores. The original vegetative cell, top right, has been distended to a club shape by development of the spores. Normally either one or two spores will be formed in *Metschnikowia* or *Nematospora* species. (b) Oval spores, as in the linear ascus of *Schizosaccharomyces pombe*. (c) 'Hat' spores in the ascus, and free. The hat appearance is created by a tangential plate or ring forming the 'brim'. *Pichia membranaefaciens* is a brewery contaminant forming this type of spore. (d) 'Saturn' spores, as in *Williopsis saturnus*. The ring is located equatorially on the spore, giving the appearance of the planet Saturn. (e) Ascus of *Lipomyces*. Although the ascospores are endogenous, i.e. within the structure of the ascus, the ascus itself is a separate structure (above) from the original vegetative cell (below). In all other ascosporogenous yeasts the spores develop within the original vegetative cell which is then, by definition, called the ascus.

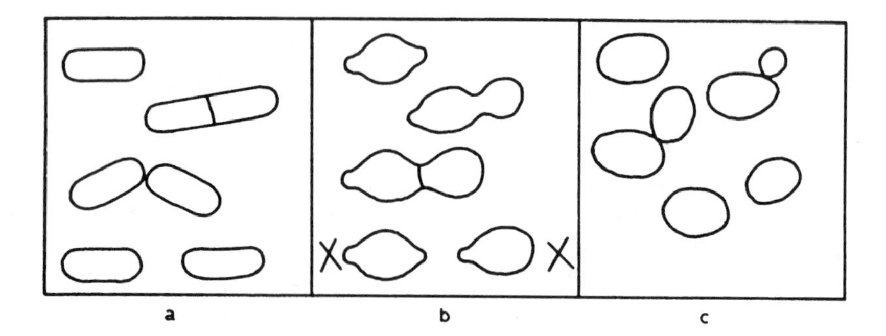

Fig. 1.2 Vegetative reproduction of yeasts. (a) Binary fission. The original cell elongates and after nuclear division a septum separates the two cells, which then complete the formation of the cell wall, and break apart. (b) Polar budding. Successive divisions are from either end (pole) of the cell, alternately. The scar tissue from budding gives the cell a 'lemon' shape. Note also the broad base to the developing bud, characteristic of this method of growth. In the next generation the two cells will bud from end X. (c) Multilateral budding, i.e. budding from many different points on the surface in successive generations. Note the typical narrow base to the developing bud.

Fig. 1.3a. Often tetrahedral asci are formed, as the most compact arrangement of the four spores. However, occasional failure of meiotic nuclear divisions can result in two- or three-spored asci. Spores of *Saccharomyces* species are not liberated immediately on maturation; spores of *Kluyveromyces* species, although superficially similar in appearance in some species, are rapidly released.

Kluyveromyces is composed of both homo- and heterothallic species with spherical or ellipsoidal spores, and homothallic species producing reniform (kidney-shaped) spores. The number of spores produced by *Kluyveromyces* species is often large: up to 60 in *K. polysporus*.

The sporulation process of *Kluyveromyces* species, and in particular the early liberation of mature spores, was judged to be sufficiently distinctive to justify a separate genus in the previous classification of yeasts (Lodder, 1970). The remaining species of actively fermenting yeasts were, somewhat illogically, retained in a single genus *Saccharomyces*, even though there were substantial differences in methods of sporulation within the genus. This was subsequently rectified in the more recent classification of Kreger-van Rij (1984), with the transfer of species to the genera *Torulaspora* and *Zygosaccharomyces*.

Torulaspora is haploid in the vegetative growth cycle, but sporulation results from homothallic fusion between the nuclei of parent cell and bud, under the cultural conditions which promote sporulation. *Zygosaccharomyces* is also haploid in the vegetative state, but fusion leading to sporulation is between independent cells, forming a diploid zygote which

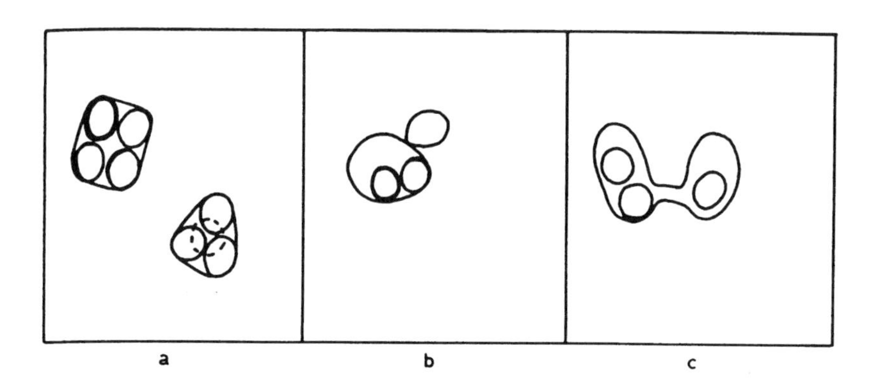

Fig. 1.3 Spores of *Saccharomyces, Torulaspora* and *Zygosaccharomyces.* (a) *Saccharomyces.* Heterozygous diploid cells undergo two successive meiotic nuclear divisions to produce four haploid spores in plane (top left) or tetrahedral (bottom right) configuration. Often only two or three spores are formed, by failure of nuclear division. (b) *Torulaspora.* Under conditions which stimulate sporulation, the haploid cell undergoes nuclear fusion with its bud, which although not yet fully developed is already, in nuclear terms, a distinct individual. In this way a transient diploid cell is formed, and then two haploid spores. (Note that in some species of *Debaryomyces,* which have a similar method of sporulation, the spores are formed in the buds. Development in the mother cell, as shown, is more normal, and invariable practice in *Torulaspora.*) (c) *Zygosaccharomyces.* Conjugation of independent haploid cells precedes ascus formation. The ascus, composed of the two former vegetative cells and the conjugation tube, contains one to four haploid spores; three, as shown, is a common occurrence.

produces two spores, one in each parent, or subsequent meiotic division may result in two spores in each former cell. Asci of *Torulaspora* and *Zygosaccharomyces* are illustrated in Figs 1.3b and 1.3c. The haploid vegetative cycle and sporulation in the manner of *Torulaspora* and *Zygosaccharomyces* spp. are also properties of the common wild yeasts of the genera *Debaryomyces, Hansenula* and *Pichia.* However, these genera ferment sugars only weakly, or not at all (Kreger-van Rij, 1984).

Yeasts of the genera included in the Basidiomycetes are unlikely to occur as contaminants of brewery or related fermentations, but are common surface contaminants of barley and malt (Chapter 4). These yeasts of malt are killed during mashing and hop boiling, and their poor growth at pH 5 or below prevents later reinfection.

It is reasonable to believe that the imperfect, non-sporing yeasts are derived from heterothallic sporing yeasts, but in the absence of the opposite mating type are unable to conjugate and form spores. Single spores germinated in the absence of the opposite mating type grow indefinitely as haploid cells, provided nutrients are available. Previously the principles of classification were applied rigidly and yeasts were allocated to perfect or

Table 1.2 Perfect and imperfect states of common yeast contaminants of the brewing industry (Barnett, Payne and Yarrow, 1990)

Spore-forming yeast	*Non-spore-forming synonym*
Dekkera bruxellensis	*Brettanomyces bruxellensis*
Dek. intermedia	*B. intermedius*
Debaryomyces hansenii	*Candida famata*
Hanseniaspora uvarum	*Kloeckera apiculata*
H. valbyensis	*K. japonica*
H. vineae	*K. africana*
Issatchenkia orientalis	*C. krusei*
Kluyveromyces marxianus	*C. kefyr*
Pichia anomala	*C. pelliculosa*
P. fabianii	*C. fabianii*
P. fermentans	*C. lambica*
P. guilliermondii *P. ohmeri*	*C. guilliermondii*
P. membranaefaciens	*C. valida*
Saccharomyces cerevisiae	*C. robusta*
S. exiguus	*C. holmii*
Torulaspora delbrueckii	*C. colliculosa*

imperfect genera according to sporulation. In the most recent classification this principle has been less rigidly applied. *Candida robusta* (Lodder, 1970) had the same properties as *S. cerevisiae*, except that no spores were formed. Sporing yeasts were obviously *S. cerevisiae*, but the logic of the classification broke down over the nomenclature of the majority of yeasts of the brewing and related industries, which, in the course of their long history of artificial cultivation, had lost the ability to form spores. To apply the name *S. cerevisiae* was technically wrong, but was nevertheless accepted practice. In the classification of Kreger-van Rij (1984) a separate species is not listed; it is now a synonym of *S. cerevisiae*. Similarly the various other examples of identical pairs of sporing and non-sporing yeasts (Table 1.2) are now recognized by the name of the sporing species.

Largely to avoid the complications arising from the use of sporulation as a fundamental property for 'classical' yeast taxonomy, and the requirement to observe spores as a first step in classical identification, various authors have provided alternative identification schemes which ignore sporulation. A simple, effective scheme introduced by Beech *et al.* (1968) has been superseded by a succession of diagnostic keys by Barnett and colleagues, who have provided a comprehensive, but unfortunately complex, system for identification by physiological properties (see Barnett, Payne and Yarrow, 1990). A simplified version of their system is the basis of the API test kit for identification of yeasts (see Chapter 7).

1.2 NOMENCLATURE OF YEASTS

Hansen, the pioneer yeast taxonomist, applied the name *Saccharomyces cerevisiae* to the traditional top-cropping ale yeast of the British and German brewing industries. The different properties of lager yeast were acknowledged by allocation to a different species, *S. carlsbergensis*. Other specific names were applied to different yeasts according to their fermentation of sugars, industrial properties or cell morphology, e.g. ale yeast (spherical, or almost so) and wine yeast *S. ellipsoideus* (ellipsoidal) were distinguished by Hansen partly on the basis of shape, and partly on different behaviour in fermentation (Lodder and Kreger-van Rij, 1952). In the reclassification of yeasts by Lodder and Kreger-van Rij (1952) the sole difference detectable in the laboratory, morphology, was judged insufficient for separation of two species and the wine yeast became *S. cerevisiae* var. *ellipsoideus*. However, another wine yeast, *S. uvarum*, and lager yeast *S. carlsbergensis*, despite their biochemical similarities, remained distinct species. In the next reclassification of yeasts (Lodder, 1970), *S. cerevisiae*, its *ellipsoideus* variety and *S. willianus*, identical in fermentation properties but of distinctively elongated cell morphology, were merged as a single species *S. cerevisiae*. An equivalent series of yeasts, comprising *S. carlsbergensis*, *S. uvarum* and *S. logos*, were united as *S. uvarum*. *S. cerevisiae* and *S. uvarum* were distinguished by the ability of only the latter to ferment melibiose. Subsequently, extensive research on yeast hybridization demonstrated interbreeding, with the production of fertile progeny, between *S. cerevisiae* and *S. carlsbergensis*, which suggested that only one species was justified. Also, the availability of these same hybrids as industrial yeasts completely blurred any distinction in physiological terms between the species. Therefore the allocation of both groups to a single species *S. cerevisiae* by Kreger-van Rij (1984) was understandable. Indeed, many other former species were also merged in the new definition of *S. cerevisiae*, as shown in Table 1.3.

It was unfortunate that 'classical' identification methods failed to distinguish the very different industrial properties of these yeasts. Gilliland's (1971) objection that 'industrial' properties were sufficient to justify separate species was confirmed by numerical analyses of data including not only the standard morphological and physiological properties, but also such industrially useful properties as tolerance to ethanol, flocculation and fining ability and optimum growth temperature (Campbell, 1972). More detailed analysis of the effect of temperature on growth of lager and non-brewing strains of *S. uvarum* (i.e. *S. carlsbergensis* and *S. uvarum*, respectively, as defined by Lodder and Kreger-van Rij in 1952) by Walsh and Martin (1977) confirmed the distinctive nature of lager yeast, however alike they may have been in the tests of 'classical' taxonomy. Unfortunately, the ability to ferment various sugars, some (galactose, melibiose, raffinose)

Table 1.3 Nomenclature of important yeast species of the brewing and related industries, 1952–1984

1952 classification (Lodder and Kreger-van Rij, 1952)	*1970 classification (Lodder, 1970)*	*1984 classification (Kreger-van Rij, 1984)*
*Saccharomyces bayanus**		
S. oviformis	*S. bayanus**	
*S. pastorianus**		
S. cerevisiae		
S. cerevisiae var. *ellipsoideus*	*S. cerevisiae*	
S. willianus		
S. carlsbergensis		*S. cerevisiae*
S. logos	*S. uvarum*	
S. uvarum		
S. chevalieri	*S. chevalieri*	
S. italicus	*S. italicus*	
	S. aceti †	
	S. diastaticus †	
S. delbrueckii	*S. delbrueckii*	
S. rosei	*S. rosei*	*Torulaspora delbrueckii*
	S. vafer †	
S. acidifaciens		
S. bailii	*S. bailii*	*Zygosaccharomyces bailii*
S. elegans		
S. fragilis		
S. marxianus	*Kluyveromyces marxianus*	*K. marxianus*
S. lactis	*K. lactis*	

**S. bayanus* and *S. pastorianus* are valid species again (Barnett, 1992).
†New species.

unlikely to occur in brewery fermentations, has arbitrarily been given more importance than industrially useful properties. However, the fundamental difference between ale, lager and non-brewing strains of *S. cerevisiae* was demonstrated again by 'DNA fingerprinting' (Pedersen, 1983; Meaden, 1990). Previous amalgamation of wine yeast species has been reversed on the basis of similar analyses (Martini and Kurtzman, 1988; Barnett, Payne and Yarrow, 1990), as shown in Table 1.3, which shows a history of the renaming over the years of *Saccharomyces* species of relevance to brewing and distilling. A more comprehensive history of the nomenclature of the genus *Saccharomyces* has recently been provided by Barnett (1992). With so many types of industrially important culture and wild yeasts all now named *S. cerevisiae*, the industrial microbiologist had to continue to distinguish these organisms, but without the assistance of distinctive specific names. Therefore the benefits of a standardized nomenclature were lost, as each microbiologist developed individual classification schemes for the different contaminant strains of the large species *S. cerevisiae*.

1.3 PROPERTIES FOR IDENTIFICATION OF YEASTS

The properties used for yeast taxonomy according to the Dutch mycologists include both morphology and physiology (Table 1.4). The large number of growth tests listed by Lodder (1970), which included over 30 carbon or nitrogen sources, has now been substantially reduced (Kreger-van Rij, 1984) to 18. Barnett, Payne and Yarrow (1983) listed 60 characters based on sugar fermentation or aerobic growth on carbon or nitrogen compounds, but suggested substantially fewer for routine identification. There is a fundamental difference between the identification procedures of the Dutch school and that of Barnett, Payne and Yarrow (1983, 1990): the former identify strictly in descending hierarchical order, i.e. family, genus, species, whereas the latter first subdivide yeasts in general into largely unrelated groups. The subdivision suggested by Barnett and co-workers was based on a small number of key tests, designed to subdivide all known yeasts into groups of approximately equal numbers of species. These groups were in turn subjected to a small number of tests chosen to identify species in that group; different tests were required for different groups. The complications largely negated the biochemical good sense of the plan, and a simpler API system, although based on the same principles, effectively identifies in a single set of tests (Anon., 1994). The three systems are compared in Table 1.4.

Numerous biochemical properties other than fermentation or aerobic growth tests have been suggested for use in classification, identification or both (Campbell, 1987). Of these methods, only the determination of the

Table 1.4 Principles of identification of yeasts

Identification by 'classical' system (Kreger-van Rij, 1984)	*Identification by physiological tests only*	
	(Barnett, Payne and Yarrow, 1983, 1990)	*API (Anon., 1994)*
Isolate	Isolate	Isolate
↓	↓	↓
Observe morphology of vegetative cells and spores (see Table 1.1)	Aerobic growth tests on various C and N compounds	Aerobic growth tests on various C compounds
↓	↓	↓
Family, genus	Physiological group	Genus, species
↓	↓	
Fermentation and aerobic growth tests (see below)	Appropriate tests for that group (usually additional growth tests)	
↓	↓	
Species	Genus, species	

Tests for identification of species (Kreger-van Rij, 1984): fermentation of mono-, di- and trisaccharides (glucose, galactose, sucrose, maltose, lactose, raffinose); aerobic utilization of above sugars, and of xylose, ethanol, glycerol, etc.

percentage of guanine + cytosine (% G + C) is still of practical value in classification (Barnett, Payne and Yarrow, 1990). However, the % G + C is of no help for identification of new isolates, since different species, and even genera, may by chance share the same value, e.g. the very different yeasts *Brettanomyces anomalus*, *Pichia quercuum* and *S. cerevisiae* are all 40.0% (Kreger-van Rij, 1984). So it is impossible to identify an unknown culture by determination of % G + C, but it is possible to recognize that an identification is wrong if % G + C values are widely different. An interesting example in the genus *Pichia* was reported by Phaff and Starmer (1987): several isolates from cacti were identified as *P. membranaefaciens* by physiological tests but named as different species because of substantially different G + C values. More recently, the reintroduction of the species *S. bayanus* and *S. pastorianus* was based partly on different G + C content from other strains of *S. cerevisiae* (Martini and Kurtzman, 1988; Barnett, Payne and Yarrow, 1990); the validity of these restored species names was subsequently confirmed by electrophoretic analysis (Martini, Martini and Cardinali, 1993). Implications of these various aspects of nucleic acid relatedness have been discussed fully by Kurtzman and Phaff (1987) and van der Walt (1987) with regard to the concept of species in yeast.

Certainly, however close % G + C values may be, only if reannealing takes place of separated DNA strands of two different yeast strains can they be regarded as the same species. DNA reassociation is possible only when the bases are in essentially the same sequence over the entire DNA molecule. At least 80% reassociation between strains is generally considered to be a requirement for allocation to the same species (e.g. Price, Fuson and Phaff, 1978; Kurtzman and Phaff, 1987), although some authors have suggested that a lower value would be acceptable (van der Walt, 1987).

The ultimate application of this principle is the determination of the entire sequence of bases of yeast DNA. While that is unlikely to be achieved for all yeasts, the method of 'fingerprinting' of strains by DNA fragments has proved useful in distinguishing industrial strains (see Chapter 3), and also has potential for classification and identification of yeasts (Meaden, 1990).

REFERENCES

Anonymous (1994) *API 20C Analytical Profile Index*, BioMérieux UK, Basingstoke.

Barnett, J.A. (1992) *Yeast*, **8**, 1.

Barnett, J.A., Payne, R.W. and Yarrow, D. (1983) *Yeasts, Characteristics and Identification*, Cambridge University Press, Cambridge.

Barnett, J.A., Payne, R.W. and Yarrow, D. (1990) *Yeasts, Characteristics and Identification*, 2nd edn, Cambridge University Press, Cambridge.

Beech, F.W., Davenport, R.R., Goswell, R.W. and Burnett, J.K. (1968) In *Identification Methods for Microbiologists*, Part B (eds B.M. Gibbs and D.A. Shapton), Academic Press, London, p. 151.

Campbell, I. (1972) *Journal of General Microbiology*, **73,** 279.
Campbell, I. (1987) In *Brewing Microbiology* (eds F.G. Priest and I. Campbell), Elsevier, London, pp. 187–205.
Gilliland, R.B. (1971) *Journal of the Institute of Brewing*, **77,** 276.
Kreger-van Rij, N.J.W. (1984) *The Yeasts, a Taxonomic Study*, 3rd edn, Elsevier/North-Holland, Amsterdam.
Kreger-van Rij, N.J.W. (1987) In *The Yeasts*, 2nd edn, Vol. 1 (eds A.H. Rose and J.S. Harrison), Academic Press, London, p. 1.
Kurtzman, C.P. (1984) *Antonie van Leeuwenhoek*, **50,** 209.
Kurtzman, C.P. and Phaff, H.J. (1987) In *The Yeasts*, 2nd edn, Vol. 1 (eds A.H. Rose and J.S. Harrison), Academic Press, London, p. 63.
Lodder, J. (1970) *The Yeasts, a Taxonomic Study*, 2nd edn, North-Holland, Amsterdam.
Lodder, J. and Kreger-van Rij (1952) *The Yeasts, a Taxonomic Study*, North-Holland, Amsterdam.
Martini, A.V. and Kurtzman, C.P. (1988) *Mycologia*, **80,** 241.
Martini, A.V., Martini, A. and Cardinali, G. (1993) *Antonie van Leeuwenhoek*, **63,** 145.
Meaden, P.G. (1990) *Journal of the Institute of Brewing*, **96,** 195.
Pedersen, M.B. (1983) *Proceedings of the 19th Congress of the European Brewery Convention, London*, HRL Press, Oxford, p. 457.
Phaff, H.J. and Starmer, W.T. (1987) In *The Yeasts*, 2nd edn, Vol. 1 (eds A.H. Rose and J.S. Harrison), Academic Press, London, p. 123.
Price, C.W., Fuson, G.B. and Phaff, H.J. (1978) *Microbiological Reviews*, **42,** 161.
van der Walt, J.P. (1987) In *The Yeasts*, 2nd edn, Vol. 1 (eds A.H. Rose and J.S. Harrison), Academic Press, London, p. 95.
Walsh, R.M. and Martin, P.A. (1977) *Journal of the Institute of Brewing*, **83,** 169.

CHAPTER 2

The biochemistry and physiology of yeast growth

T.W. Young

2.1 INTRODUCTION

When introduced into a suitable aqueous environment, containing an adequate supply of nutrients, at moderate temperature and value of pH, viable yeast cells grow. Growing yeast cells increase in volume and mass until they reach a critical size when bud formation is initiated (Pringle and Hartwell, 1981). Once initiated, provided there are sufficient nutrients to support further growth, buds increase in size and eventually separate from the parent cell. The growth of a yeast culture therefore encompasses both an increase in total cell mass (biomass) and an increase in cell number. Growth thus involves the *de novo* synthesis of new yeast cells, i.e. the synthesis of yeast cell constituent macromolecules. The mechanisms by which these syntheses occur are found in the biochemical reactions of the metabolism of the yeast cells. These reactions are also responsible for the production of the ethanol and carbon dioxide of fermentation as well as for the generation of the many compounds which contribute to the taste and aroma of beer.

In brewing, yeast growth is observed during the propagation of pure yeast cultures in specialized propagation vessels and also in the fermentation process to yield beer. Growth cannot be separated from the fermentation process and is necessary to the production of both beer and fresh yeast for use in subsequent fermentations. Control of fermentation is achieved by monitoring the changes in the fermentation medium resulting from the metabolic activities of the growing yeast cells.

Brewing Microbiology, 2nd edn. Edited by F. G. Priest and I. Campbell.
Published in 1996 by Chapman & Hall, London. ISBN 0 412 59150 2

2.2 YEAST NUTRITION

Yeasts will grow in simple media which contain fermentable carbohydrates to supply energy and 'carbon skeletons' for biosynthesis, adequate nitrogen for protein synthesis, mineral salts and one or more growth factors. It is usual for some molecular oxygen to be provided although it would seem that some strains at least of *Saccharomyces cerevisiae* do not have an absolute requirement for oxygen (Macy and Miller, 1983).

Sources of carbon include monosaccharides such as D-glucose, D-mannose, D-fructose, D-galactose and the pentose sugar D-xylulose, but not other pentoses (Wang, Johnson and Schneider, 1980). Disaccharides such as sucrose (cane sugar) and maltose are also fermented by brewer's yeast, which however will not ferment lactose (milk sugar). Trisaccharides such as maltotriose and raffinose are also fermented by brewers' yeast although in the case of raffinose some strains conduct only a partial hydrolysis. The organic compounds glycerol, ethanol and lactate are not fermented but yeast may grow aerobically by respiration using these compounds as sources of carbon and energy.

The glucose, fructose and sucrose present in brewers' wort are rapidly used by the yeast in the early stages of fermentation. Sucrose is hydrolysed by the enzyme invertase (β-D-fructofuranosidase, EC 3.2.1.26), which is located externally to the cell membrane and bound within the cell wall. The glucose and fructose produced, together with that present in wort, are then transported into the cell. Transport is facilitated possibly by a carrier molecule (permease) located in the membrane. Glucose has been estimated to enter the cell some one million times faster than can be accounted for by simple diffusion across the membrane (Heredia, Sols and De la Fuente, 1968). The uptake of maltose and maltotriose is mediated by specific inducible permeases (Sols and De la Fuente, 1961; Harris and Thompson, 1960). An α-D-glucosidase (maltase, EC 3.2.1.20) is simultaneously induced so that on entering the cell both the di- and trisaccharides are hydrolysed to glucose.

The requirement for nitrogen for protein synthesis may be met by ammonium ions although, when present, amino acids are a preferred source. Thus when mixtures of amino acids are available (e.g. in brewers' wort) growth is more rapid than when ammonium ions are the sole source of nitrogen (Thorne, 1949). Amino acids are taken up in a sequential manner during fermentation (Jones and Pierce, 1964); this is thought to reflect the properties and specificities of permeases located in the cell membrane.

Brewing yeasts show a requirement for minerals which resembles that of other living organisms and in particular a supply of potassium, iron, magnesium, manganese, calcium, copper and zinc is necessary. In addition, it is usual for strains to require one or more accessory nutrients for growth and most, if not all, need biotin. Many other nutrients are known to stimulate the growth of yeasts although they are not absolutely required for cell growth. Amongst such compounds are pantothenic acid, *meso*-inositol, nicotinic acid, thiamin, *p*-aminobenzoic acid and pyridoxine. Most of these materials are necessary cofactors of enzyme activity in yeast metabolism.

When made wholly from malt, brewers' wort contains all the nutrients needed by yeast for growth. The carbohydrate requirement is satisfied by glucose, fructose, sucrose, maltose and maltotriose: the requirement for nitrogen is met by amino acids and ample quantities of minerals and accessory nutrients are present. The addition of calcium salts to the brewing water aids yeast flocculation at the end of fermentation. Worts made using high levels of carbohydrate material (sugars, cereal starches) may be deficient in assimilable nitrogen and vitamins (particularly biotin). Such worts as well as all-malt ones may be supplemented with the addition of yeast foods based on preparations of yeast extract plus ammonium, phosphate and zinc ions. More complex formulations may also be used to ensure that the nutritional requirements of the yeast are met (Hsu, Vogt and Bernstein, 1980).

2.3 YEAST METABOLISM

2.3.1 General aspects

Metabolism describes all the enzymic reactions which occur within the cell and the organization and regulation of those reactions. Although from a biochemical standpoint individual aspects of metabolism are considered as separate pathways, such pathways do not in reality exist in isolation, but are merely parts of a whole integrated metabolic process.

Each biochemical pathway consists of a series of chemical reactions catalysed by enzymes. The catalytic moiety of an enzyme is invariably protein, although lipid and/or carbohydrate may be covalently associated with particular protein molecules. An enzyme increases the rate of a chemical reaction and enables the reaction to occur at physiological values of temperature and pH. Extremes of temperature and/or pH inactivate enzyme molecules by denaturing them. Enzymes show specificity towards their substrates and often require cofactors to participate in their catalytic activity.

Biochemical pathways may be described as catabolic, anabolic (biosynthetic), amphibolic or anaplerotic. Catabolic pathways are involved in the degradation (usually by an oxidative process) of simple organic molecules derived from the breakdown of polymers (e.g. amino acids from proteins) and retain some of the energy released in a 'biologically useful' form. Anabolic pathways consume energy and synthesize (usually by a reductive process) the simple molecules which are subsequently assembled into macromolecules. Amphibolic pathways have both catabolic and anabolic functions: they are central metabolic pathways which provide, from catabolic sequences, the intermediates which form the substrates of anabolic reactions. When during the operation of a catabolic sequence of an amphibolic pathway, intermediates are removed for biosynthetic purposes, catabolism would cease were it not for the operation of anaplerotic reactions. The function of anaplerotic reactions is to replace those intermediates which link both catabolism and anabolism, thus ensuring the continued operation of amphibolic pathways.

Biologically, energy produced from the oxidation of carbohydrate and consumed in biosynthesis is stored in the form of adenosine triphosphate (ATP). This high-energy compound on hydrolysis to adenosine diphosphate (ADP) and inorganic phosphate (P_i) liberates some 30.5 kJ mol^{-1} under standard conditions. The magnitude of the free energy of hydrolysis of ATP is a function of pH, temperature and the concentrations of ATP, ADP and P_i. In the conditions found inside the yeast cell, the hydrolysis of 1 mole of ATP may yield as much as 52 kJ.

Oxidative reactions in catabolism involve the removal of electrons from intermediates. This process is controlled by dehydrogenase enzymes and often involves the participation of the cofactor nicotinamide adenine dinucleotide (NAD^+). Electrons are transferred to NAD^+ in the form of the hydride ion [H^-] to produce reduced NAD^+ ($NADH_2$):

$$NAD^+ + [2H] \rightarrow NADH + H^+ \text{ (or } NADH_2\text{)}$$

Biosynthetic reactions are driven by ATP energy and the process of reduction is mediated by enzymes, most of which use nicotinamide adenine dinucleotide phosphate ($NADP^+$) as cofactor. The specificity of catabolic enzymes for NAD^+ and those of anabolism for $NADP^+$ is an example of 'chemical compartmentation' and enables some degree of metabolic regulation to be exerted through control of the levels of the two cofactors. The relative concentration of the oxidized and reduced forms of a particular cofactor may also serve a regulatory role.

The regulation of metabolism also involves 'biological compartmentation' whereby biochemical pathways are located within organelles or at specific points within the cell. In addition, control of biochemical pathways may occur by a variety of other mechanisms such as:

1. regulating the amount of enzyme synthesized;
2. regulating the degradation of enzymes;
3. modifying the rate of enzyme activity by allosteric inhibition or activation;
4. using isoenzymes to perform the same reaction for different purposes, e.g. alcohol dehydrogenase (ADH): ADH1 is used when cells grow on ethanol, converting ethanol to acetaldehyde, whereas ADH2 is used when cells grow on glucose and convert acetaldehyde to ethanol.

The foregoing is a very limited introduction to the subject of metabolism, and readers unfamiliar with biochemical principles are advised to consult a standard text, e.g. Stryer (1995).

The following sections will discuss only the main aspects of yeast metabolism pertinent to a consideration of brewery fermentation. Additional information may be found in Rose and Harrison (1969, 1970, 1971), Briggs *et al.* (1982) and Nykanen and Suomalainen (1983).

2.3.2 Carbohydrate metabolism during fermentation

(a) Major pathways

The main pathway for the fermentation of glucose is the Embden–Meyerhof–Parnas route (also named the EMP or glycolytic sequence). Wort-fermentable carbohydrates enter the pathway either as D-glucose or D-fructose. Similarly intracellular storage compounds such as glycogen or trehalose are hydrolysed to D-glucose.

The EMP pathway (Fig. 2.1) commences with the phosphorylation of the hexose monosaccharide glucose using the high-energy bond of ATP to produce glucose-6-phosphate (G-6-P) in a reaction mediated by the enzyme hexokinase (EC 2.7.1.1). Glycogen degradation by phosphorylase (EC 2.4.1.1) yields glucose-1-phosphate which is converted to G-6-P by the enzyme phosphoglucomutase (EC 2.7.5.1), which in yeast requires zinc for activity. The next reaction in the pathway (Fig. 2.1) is the isomerization of G-6-P to fructose-6-phosphate (F-6-P) by the enzyme glucose-phosphate isomerase (EC 5.3.1.9). This reaction is readily reversible. Intracellular fructose, derived directly from wort fructose or from the action of cell wall bound invertase, is probably phosphorylated by hexokinase to yield F-6-P. The third enzyme of the glycolytic sequence, 6-phosphofructokinase (EC 2.7.1.11), mediates the formation of fructose-1,6-bisphosphate from F-6-P and ATP. At this point in the EMP pathway, two molecules of ATP have been consumed for each molecule of glucose entering the sequence.

Fructose-bisphosphate aldolase (EC 4.1.2.13), a zinc-containing enzyme, then catalyses the reversible cleavage of the six-carbon molecule into two three-carbon molecules, triosephosphates. The triosephosphates are dihydroxyacetone phosphate and D-glyceraldehyde-3-phosphate; only the latter molecule is further processed by the EMP pathway and an equilibrium is maintained between the two triosephosphates by the enzyme triosephosphate isomerase (EC 5.3.1.1).

D-Glyceraldehyde-3-phosphate is converted into 3-phospho-D-glyceroyl phosphate by the enzyme glyceraldehyde-phosphate dehydrogenase (EC 1.2.1.12) using inorganic phosphate and transferring electrons to NAD^+. This is the first reaction of the pathway which involves the oxidation of the substrate. Furthermore, it is the first reaction which involves the formation of a high-energy bond and some of the free energy

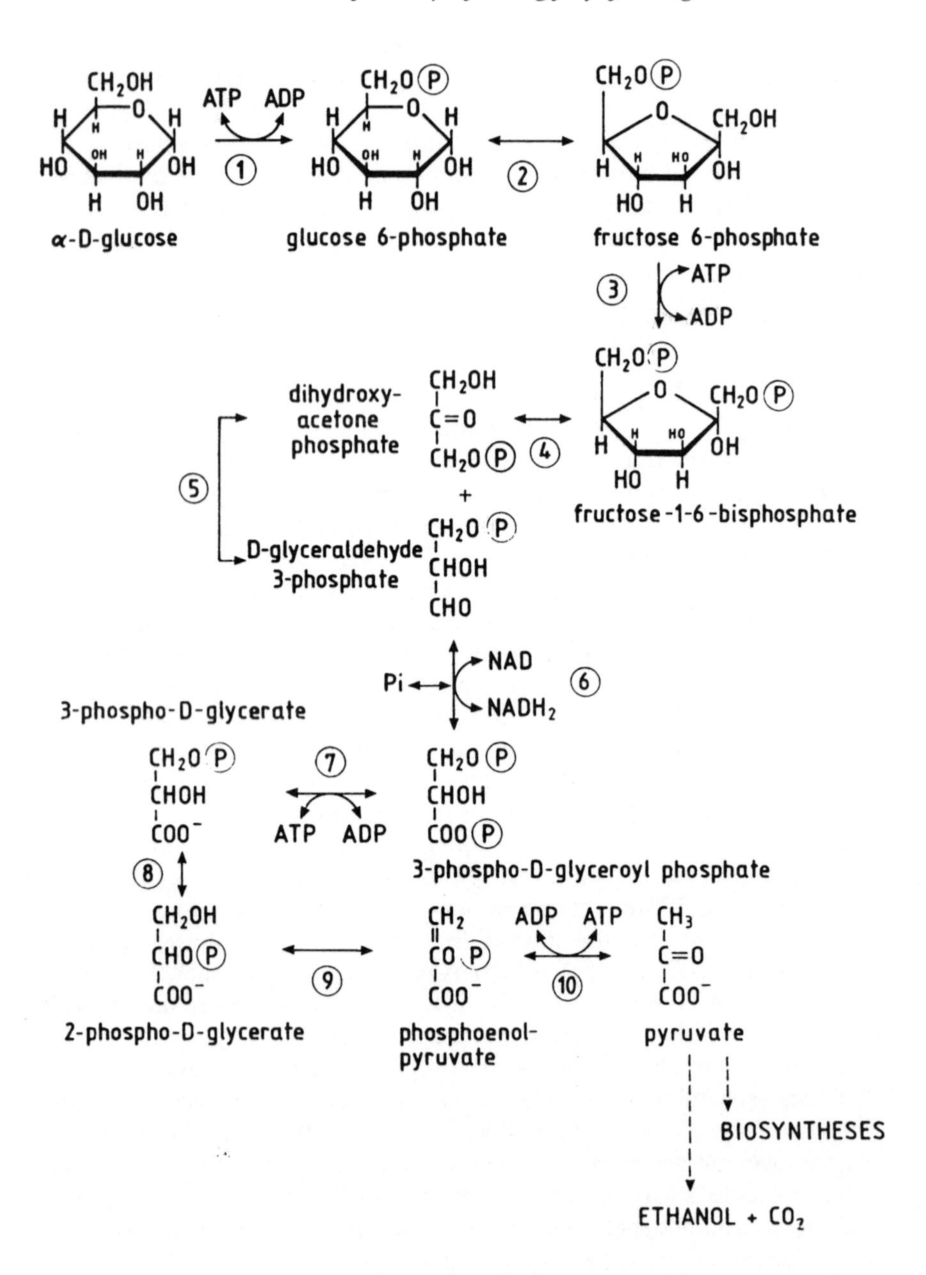

Fig. 2.1 The Embden–Meyerhof–Parnas or glycolytic pathway. 1, Hexokinase (EC 2.7.1.1); 2, glucose-phosphate isomerase (EC 5.3.1.9); 3, 6-phosphofructokinase (EC 2.7.1.11); 4, fructose-bisphosphate aldolase (EC 4.1.2.13); 5, triosephosphate isomerase (EC 5.3.1.1); 6, glyceraldehyde-phosphate dehydrogenase (EC 1.2.1.12); 7, phosphoglycerate kinase (EC 2.7.2.3); 8, phosphoglyceromutase (EC 2.7.5.3); 9, enolase (EC 4.2.1.11); 10, pyruvate decarboxylase (EC 4.1.1.1).

available from the oxidative process is trapped in this bond. Thus the phosphate bond formed at position 1 in 3-phospho-D-glyceroyl phosphate has a standard free energy of hydrolysis of -49.3 kJ mol^{-1}. The next step in the pathway is the transfer of some of the energy in this bond to ADP so forming ATP and 3-phospho-D-glycerate, a reaction catalysed by the enzyme phosphoglycerate kinase (EC 2.7.2.3). At this point in the pathway, two molecules of ATP and two of $NADH_2$ have been produced for each glucose molecule entering the sequence. Thus the expenditure of ATP earlier in the process is balanced.

3-Phospho-D-glycerate is converted to 2-phospho-D-glycerate by the enzyme phosphoglyceromutase (EC 2.7.5.3). Enolase (EC 4.2.1.11) then mediates the removal of water from 2-phospho-D-glycerate to give phosphoenolpyruvic acid. The energy-rich bond formed in phosphoenolpyruvate ($\Delta G'_0 = -61.9$ kJ mol^{-1}) is then used to phosphorylate ADP to yield ATP and pyruvic acid. At this point, for each molecule of glucose entering the pathway a net gain of two molecules of ATP is achieved. Also, two molecules of $NADH_2$ have been formed. The supply of NAD^+ is limited and therefore the continued operation of the pathway requires the oxidation of the $NADH_2$ produced to reform NAD^+. Under the conditions operating in a brewery fermentation, the yeast cell decarboxylates pyruvic acid to acetaldehyde and carbon dioxide using the enzyme pyruvate decarboxylase (EC 4.1.1.1). The carbon dioxide formed is that which is produced during fermentation and which naturally carbonates beer. The acetaldehyde produced in the reaction is subsequently reduced to ethanol, which together with carbon dioxide and yeast represent the major products of fermentation. The enzyme mediating this reaction is alcohol dehydrogenase (EC 1.1.1.1) with the participation of $NADH_2$ as cofactor so that the $NADH_2$ formed during the oxidation of glucose is finally oxidized to NAD^+, thus permitting the continued operation of the EMP pathway:

$$CH_3CO.COOH \rightarrow CO_2 + CH_3CHO \rightarrow CH_3CH_2OH$$

$$NADH_2 \quad NAD^+$$

By using acetaldehyde as the terminal electron acceptor the cell ensures the continued operation of the glycolytic pathway and therefore the continued formation of ATP for use in biosynthetic reactions.

The standard free energy change for fermentation is:

$$C_6H_{12}O_6 \rightarrow 2CO_2 + 2CH_3CH_2OH \qquad \Delta G'_0 = -234 \text{ kJ mol}^{-1}$$

The EMP pathway yields a net gain of two ATP molecules per molecule of glucose, which under standard conditions is equivalent to 61 kJ mol^{-1} free energy and the efficiency of fermentation is therefore 26%. The 74% of energy not retained by the cell is largely dissipated as heat. Thus it is mandatory to apply some degree of cooling to all but the smallest fermenters (from which heat may be readily lost to the surroundings) in order to ensure that the temperature of fermentation remains under control. The fermentation rate is not constant during the time course of fermentation and the greatest activity occurs during the first 24–36 h period. For a typical fermentation the cooling required is of the order of 3×10^5 kJ h^{-1}.

The glycolytic sequence is an example of an amphibolic pathway and the various intermediates are used by the cell in biosynthetic reactions. In particular, phosphorylated glucose is a precursor for the synthesis of cell wall carbohydrate polymers and of storage polysaccharide; triose phosphates are used in fat synthesis and phosphoenolpyruvate and pyruvate are precursors of several amino acids. However, growing cells require many more intermediates for biosynthetic reactions than can be supplied from the EMP pathway. In respiring yeast cells, i.e. those which in contrast to yeast in a brewery fermentation use molecular oxygen as a hydrogen acceptor and completely oxidize glucose, the Krebs cycle (tricarboxylic acid cycle) is used to supply additional substrates for biosynthesis. Under the anaerobic conditions of brewery fermentation, the levels of the enzymes of the Krebs cycle are greatly lowered and the extent (if any) to which the cycle operates under these circumstances is unclear. The question arises as to how the cell synthesizes essential intermediates such as succinic acid, 2-oxoglutaric acid and oxaloacetic acid. Two mechanisms have been proposed; the first envisages a limited (although complete) operation of the citric acid cycle with the formation of succinate, fumarate, malate and oxaloacetate by an oxidative process (Oura, 1977), whereas the second involves the synthesis of additional enzymes leading to the formation of succinate by a reductive pathway (Sols, Gancedo and De la Fuente, 1971). In this second mechanism it is proposed that the Krebs cycle is inoperative because of the lack of the key enzyme 2-oxoglutarate dehydrogenase (EC 1.2.4.2).

In both mechanisms, pyruvate is converted to acetyl coenzyme A (acetyl CoA) in a reaction involving the multienzyme pyruvate dehydrogenase complex. This complex contains pyruvate dehydrogenase (lipoamide; EC 1.2.4.1), which acts on pyruvic acid liberating CO_2 and forming acetyl dihydrolipoamide. Then the enzyme dihydrolipoamide acetyl transferase (EC 2.3.1.12), also part of the complex, acetylates coenzyme A (CoA), in a reaction involving thiamin diphosphate, to yield acetyl CoA and leave dihydrolipoamide. A third member of the complex, a flavoprotein, dihydrolipoamide reductase (NAD), mediates the reduction of NAD^+ to form $NADH_2$ and lipoamide. The overall reaction is therefore:

$$CH_3CO.COOH + CoASH + NAD^+ \rightarrow CO_2 + CH_3COSCoA + NADH_2$$

Also both mechanisms include the formation of oxaloacetate by the ATP-dependent carboxylation of pyruvic acid. This single reaction can be seen

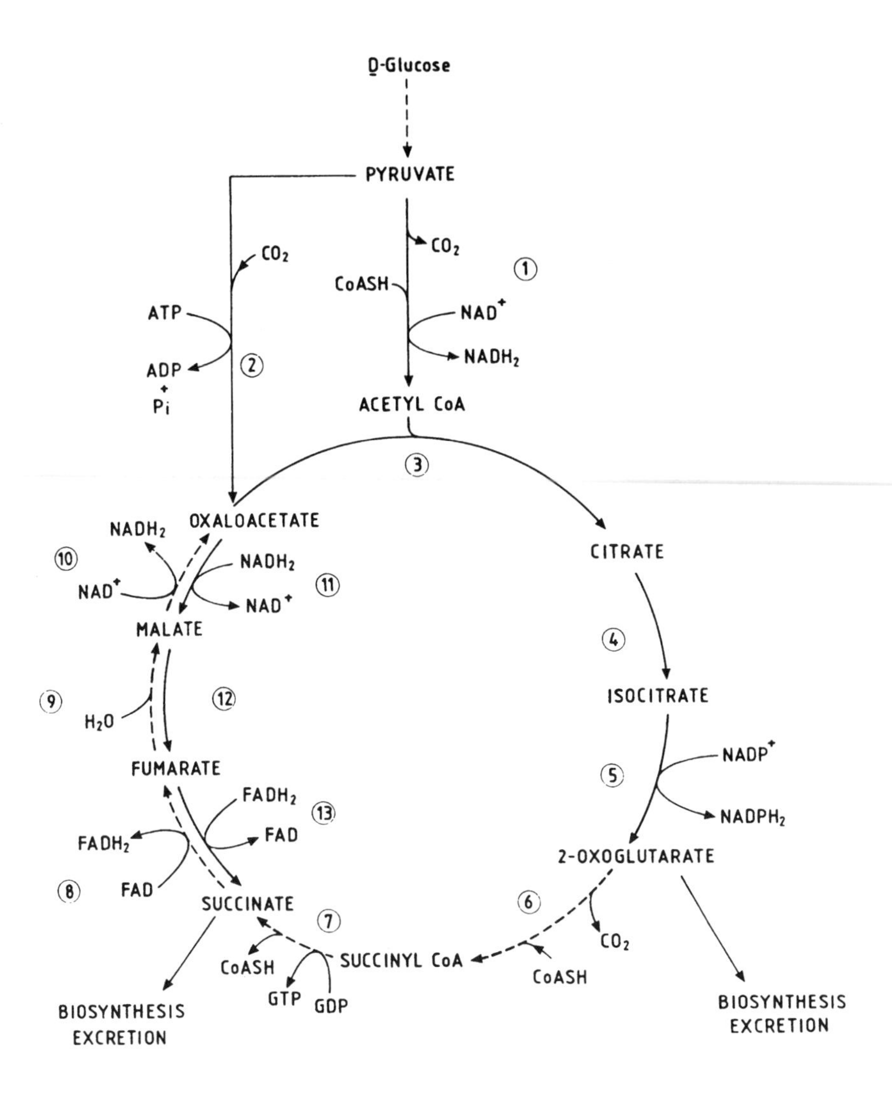

Fig. 2.2 Formation of metabolic intermediates from pyruvate by fermenting yeast. See text for details. Continuous arrows indicate the proposed reductive pathway; broken arrows show the proposed oxidative pathway. 1, Pyruvate dehydrogenase complex; 2, pyruvate carboxylase (EC 6.4.1.1); 3, citrate (*si*)-synthase (EC 4.1.3.7); 4, aconitate hydratase (EC 4.2.1.3); 5, isocitrate dehydrogenase ($NADP^+$) (EC 1.1.1.42); 6, oxoglutarate dehydrogenase (EC 1.2.4.2); 7, succinyl-CoA synthetase (GDP-forming) (EC 6.2.1.4); 8, succinate dehydrogenase (EC 1.3.99.1); 9, fumarate hydratase (EC 4.2.1.2); 10, 11, malate dehydrogenase (EC 1.1.1.37); 12, fumarate hydratase (EC 4.2.1.2); 13, fumarate reductase (induced under anaerobic conditions).

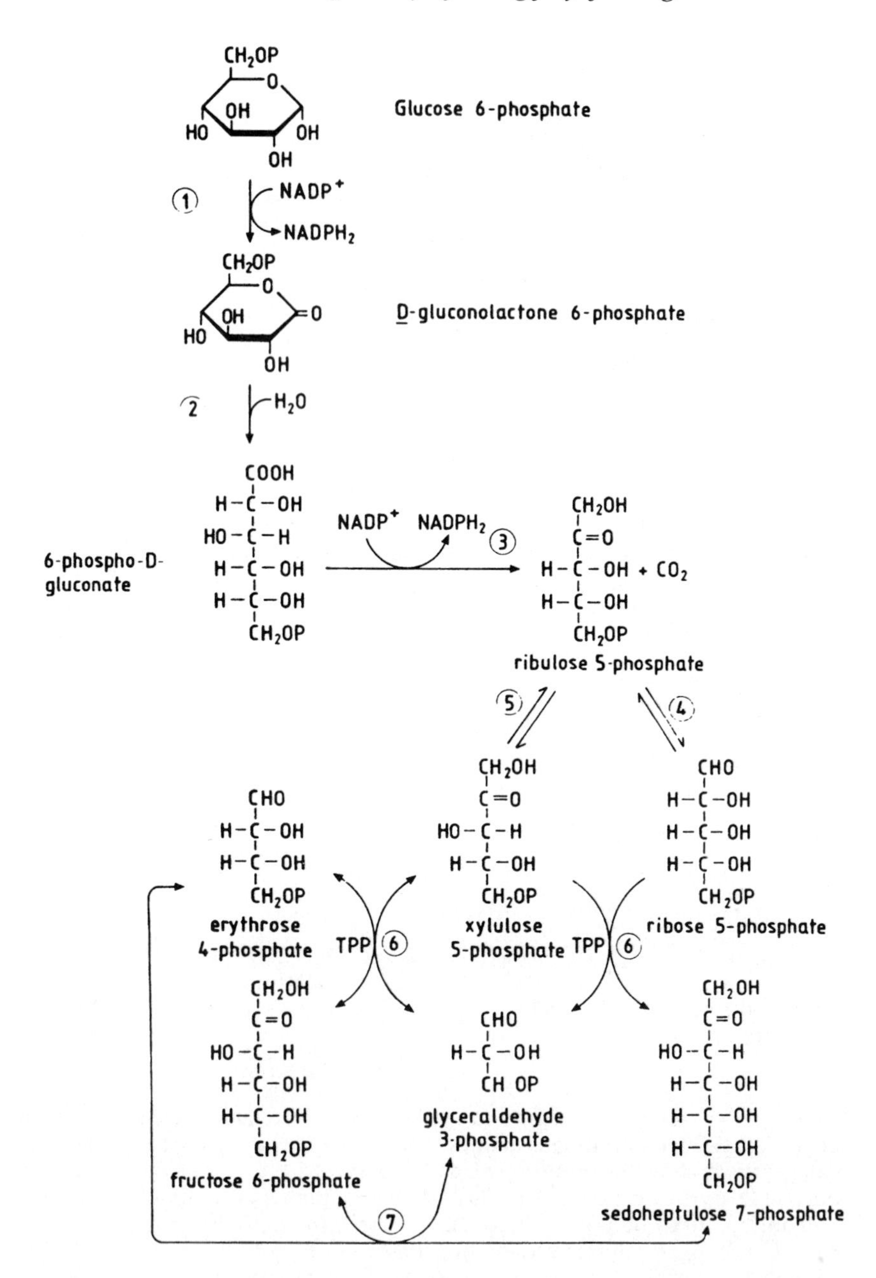

Fig. 2.3 Hexose monophosphate pathway. 1, Glucose-6-phosphate dehydrogenase (EC 1.1.1.49); 2, 6-phosphogluconolactonase (EC 3.1.1.31); 3, phosphogluconate dehydrogenase (decarboxylating) (EC 1.1.1.44); 4, ribosephosphate isomerase (EC 5.3.1.6); 5, ribulosephosphate 3-epimerase (EC 5.1.3.1); 6, transketolase (EC 2.2.1.1); 7, transaldolase (EC 2.2.1.2).

as an anaplerotic reaction for the operation of the citric acid cycle, replacing intermediates withdrawn for biosynthesis. The enzyme controlling this anaplerotic reaction is pyruvate carboxylase (EC 6.4.1.1) a zinc-containing protein. The alternative mechanisms are shown in Fig. 2.2, where the reductive pathway is depicted by the solid arrows and the oxidative pathway by the broken arrows.

The pathways of carbohydrate metabolism so far discussed do not provide for the synthesis of pentose sugars which are essential to the biosynthesis of nucleic acid precursors as well as many enzyme cofactors, e.g. NAD^+ and $NADP^+$. Pentose sugars may be synthesized from G-6-P by an oxidative pathway in which the enzymes mediating the oxidative steps use $NADP^+$ as cofactor. The pathway is depicted in Fig. 2.3 and is called the hexose monophosphate pathway (HMP) or the pentose phosphate pathway. The first reaction is the oxidation of G-6-P by the enzyme glucose-6-phosphate dehydrogenase (EC 1.1.1.49). The product of this oxidation, D-gluconolactone-6-phosphate, is acted upon by the enzyme 6-phosphogluconolactonase (EC 3.1.1.31) with the consequent formation of 6-phospho-D-gluconate; this product becomes the substrate for a second oxidative step with the NADP-enzyme phosphogluconate dehydrogenase (decarboxylating) (EC 1.1.1.44). The end-products of this reaction are CO_2 and the pentose phosphate ribulose-5-phosphate. From ribulose-5-phosphate the enzymes ribulosephosphate 3-epimerase (EC 5.1.3.1) and ribosephosphate isomerase (EC 5.3.1.6) produce xylulose-5-phosphate and ribose-5-phosphate respectively. The enzyme transketolase (EC 2.2.1.1), with thiamin diphosphate as cofactor, catalyses the transfer of the two-carbon keto-fragment of xylulose-5-phosphate to the C1 atom of ribose-5-phosphate, thereby forming sedoheptulose-7-phosphate and glyceraldehyde-3-phosphate. The enzyme transaldolase (EC 2.2.1.2) catalyses the transfer of the C1, C2 and C3 atoms of a keto-sugar to the C1 atom of an aldo sugar. Thus from sedoheptulose-7-phosphate and glyceraldehyde-3-phosphate, fructose-6-phosphate and erythrose-4-phosphate are formed.

The HMP pathway may be considered to oxidize the glucose molecule completely by the overall reaction:

$$\text{G-6-P} + 12\text{NADP}^+ \rightarrow 6\text{CO}_2 + 12\text{NADPH} + 12\text{H}^+ + \text{P}_i + 6\text{H}_2\text{O}$$

However, it is probably of greatest significance to the cell as a means of generating from glucose reduced $NADP^+$ for anabolic reactions and as a source of pentoses which yeast cannot take up from wort.

In *S. cerevisiae* during fermentation, the level of glucose-6-phosphate dehydrogenase is very low (Horecker, 1968) and it is probable that most of the yeast's requirement for pentose sugar is met by the action of transketolase on fructose-6-phosphate and glyceraldehyde-3-phosphate produced by the EMP pathway (see Fig. 2.1).

(b) Synthesis of carbohydrates

The disaccharide trehalose and the polysaccharide glycogen are both synthesized during fermentation and act as 'food' reserves for the cell. In addition cells synthesize the mannan and glucan components of the cell wall.

Both trehalose and glycogen synthesis begin with the formation of uridine diphosphate glucose (UDPG) catalysed by the enzyme glucose-1-phosphate uridylyltransferase (EC 2.7.7.9). Trehalose phosphate is synthesized from UDPG and G-6-P by the enzyme α,α-trehalose-phosphate synthase (UDP-forming; EC 2.4.1.15). Trehalose is formed by the action of trehalose-phosphatase (EC 3.1.3.12).

Glycogen synthase-D-phosphatase catalyses the sequential addition of the glucose residue of UDPG to a polysaccharide acceptor. The glucose units in the polysaccharide are linked $\alpha(1\rightarrow 4)$ in linear chains. The natural glycogen molecule is a branched structure and short $\alpha(1\rightarrow 4)$ bridging chains link longer chains together via $\alpha(1\rightarrow 6)$ bonds.

During fermentation, glycogen is accumulated by brewing yeast and accounts for the consumption of the equivalent of some 0.25% maltose. When yeast is inoculated into fresh wort, the glycogen is metabolized to yield energy (and presumably substrates) for the synthesis of sterols essential to yeast growth (Quain and Tubb, 1982). It is therefore important for yeast to contain adequate glycogen for this purpose and under normal circumstances sufficient is present. It has been shown that prolonged storage of yeast can lead to the depletion of the glycogen reserve without the concomitant formation of sterols. Under such circumstances the subsequent fermentations may be erratic both in terms of rate and final specific gravity.

(c) Regulation

Knowledge of the control of amphibolic pathways in *S. cerevisiae* is mainly derived from studies of the effects of the concentrations of glucose and oxygen on cell growth and the levels of enzymes of the pathways. Unambiguous data are obtained therefore only by using media of defined composition and under conditions of full aeration (aerobiosis) or complete lack of oxygen (anaerobiosis). These conditions are not typical of those obtaining in a brewery fermentation where the medium is complex (and glucose is not the main fermentable carbohydrate) and is air-saturated at the start but anaerobic after say the first 12 h (or less) have elapsed. Furthermore, scientific investigation of metabolic control is usually carried out with actively growing cells, whereas growth in fermentation is somewhat restricted.

S. cerevisiae belongs to the so-called glucose-sensitive yeast types. In these strains respiration is repressed in the presence of a small ($\leq$ 0.4%)

concentration of free glucose in the medium. This repression applies irrespective of the presence or absence of molecular oxygen. The effect of repressive levels of glucose is to produce changes in enzymic composition (particularly levels of TCA cycle enzymes) and influence the structure of the mitochondria of the cell (particularly levels of components of the respiratory chain). This effect of glucose is referred to as the Crabtree effect, glucose repression or catabolite repression. The consequence of catabolite repression in a brewery fermentation is not clear although since brewers' wort typically contains approximately 1% glucose it is assumed that even in air- or oxygen-saturated worts the yeast cells are repressed. When the free glucose has been consumed (probably in the first 16 h), however, exposure of cells to oxygen may result in derepression since neither maltose nor maltotriose exhibits a repressive action on respiration. Although exposure to oxygen is avoided during the later stages of fermentation, top-cropping yeast strains are obviously exposed, as is yeast stored in suspension in cold stores prior to use. The precise mechanism of catabolite repression is unknown but may be related to intracellular levels of cyclic adenosine monophosphate (cAMP) (Fiechter, Fuhrmann and Kappeli, 1981).

In yeast, regulation is also expressed through the inactivation (by enzymic modification or degradation) of enzymes. Such processes are also initiated in response to free glucose in the medium and are termed catabolite inactivation (Holzer, 1976; Hagele, Neff and Mecke, 1978).

In the absence of repressing levels of glucose and in the presence of molecular oxygen, yeast respires the carbohydrate. In this case glucose is completely oxidized to CO_2 and H_2O. This is achieved through the operation of the glycolytic pathway and tricarboxylic acid cycle. The reduced NADH produced is oxidized to NAD^+ by the electron transfer chain using molecular oxygen as hydrogen acceptor (thus forming H_2O) and in the process synthesizing ATP. It was observed by Pasteur (1867) that the uptake of glucose was lower in respiring cells than in fermenting ones. The effect has been named the Pasteur effect. It is now recognized that this effect is particularly marked when resting rather than actively growing cells are analysed (Fiechter, Fuhrmann and Kappeli, 1981). Nevertheless, the Pasteur effect points to the operation of some regulatory system on the EMP pathway. The accepted model of this regulatory system is that of Sols (Sols, Gancedo and De la Fuente, 1971) which recognizes a central role for the enzyme 6-phosphofructokinase. In this model ATP allosterically inhibits the enzyme and AMP activates it. Thus in conditions of high energy charge (respiration) the flux of glucose through the EMP pathway is lowered. In respiration, ATP synthesis also depletes the intracellular reserve of inorganic phosphate which in turn limits the operation of the EMP pathway (Lynen, 1963) and lowers the glucose flux. In both cases the overall effect is for less glucose to enter the pathway than during fermentation.

Further information on carbohydrate metabolism, particularly that of aerobic cells, and its regulation may be found in Briggs *et al.* (1982) and Sols, Gancedo and De la Fuente (1971).

2.3.3 Nitrogen metabolism

Yeast cells preferentially use the amino acids present in brewers' wort as a source of nitrogen. The nitrogen assimilated by cells is used to synthesize amino acids which in turn are used to synthesize proteins. For details of the protein-synthesizing system Stryer (1995) may be consulted.

Careful analysis of the uptake of isotopically labelled amino acids by brewers' yeast showed that negligible assimilation of intact amino acids occurred (Jones, Pragnell and Pierce, 1969). When an amino acid enters the yeast cell, a transaminase system removes the amino group and the remaining 'carbon skeleton' is metabolized. It appears that the main transamination system operating in yeast employs 2-oxoglutarate as acceptor. The glutamate formed may then be used as an amino donor to synthesize other amino acids from carbon skeletons formed by anabolic pathways. The transaminase (aminotransferase) reaction is readily reversible and the enzyme employs pyridoxal phosphate as cofactor. The reaction is shown in Fig. 2.4; the aminated cofactor (pyridoxamine phosphate) is used by the enzyme as a donor of amino groups. The oxoacid pool generated by the action of transaminases and anabolic reactions is a precursor of aldehydes and higher alcohols which contribute to beer flavour.

The assimilable nitrogen content of a brewers' wort is not specifically measured. Reliance is placed upon measurement of free amino nitrogen (FAN) with the assumption that amino acids make a greater contribution to this analysis than polypeptides. Typical worts contain 100–140 mg l^{-1} FAN and levels below these are found to be inadequate to support fermentation. Fermentable extracts made from other cereals such as sorghum (Bajomo and Young, 1992) are fermented by brewers' yeast when levels of FAN of 40 mg l^{-1} are present. Similarly wheat extracts containing FAN of 54–58 mg l^{-1} fermented successfully (Thomas and Ingledew, 1990). Clearly more research is needed to evaluate the assimilable nitrogen content and nutritional status of malt wort.

2.3.4 Lipid metabolism

Yeast lipid comprises the triacylglycerols, phospholipids and sterols. The triacylglycerols are triesters of long-chain fatty acids and glycerol. Yeast lipid contains predominantly fatty acids 16 or 18 carbon atoms long although chain lengths of 8–24 are found. The fatty acids may be saturated or unsaturated (with one or two carbon-to-carbon double bonds). Thus C_{16} (palmitic), C_{18} (stearic) are the principal saturated fatty acids and the unsaturated ones are the monoenoic $C_{16:1}$ (palmitoleic), $C_{18:1}$ (oleic) and the

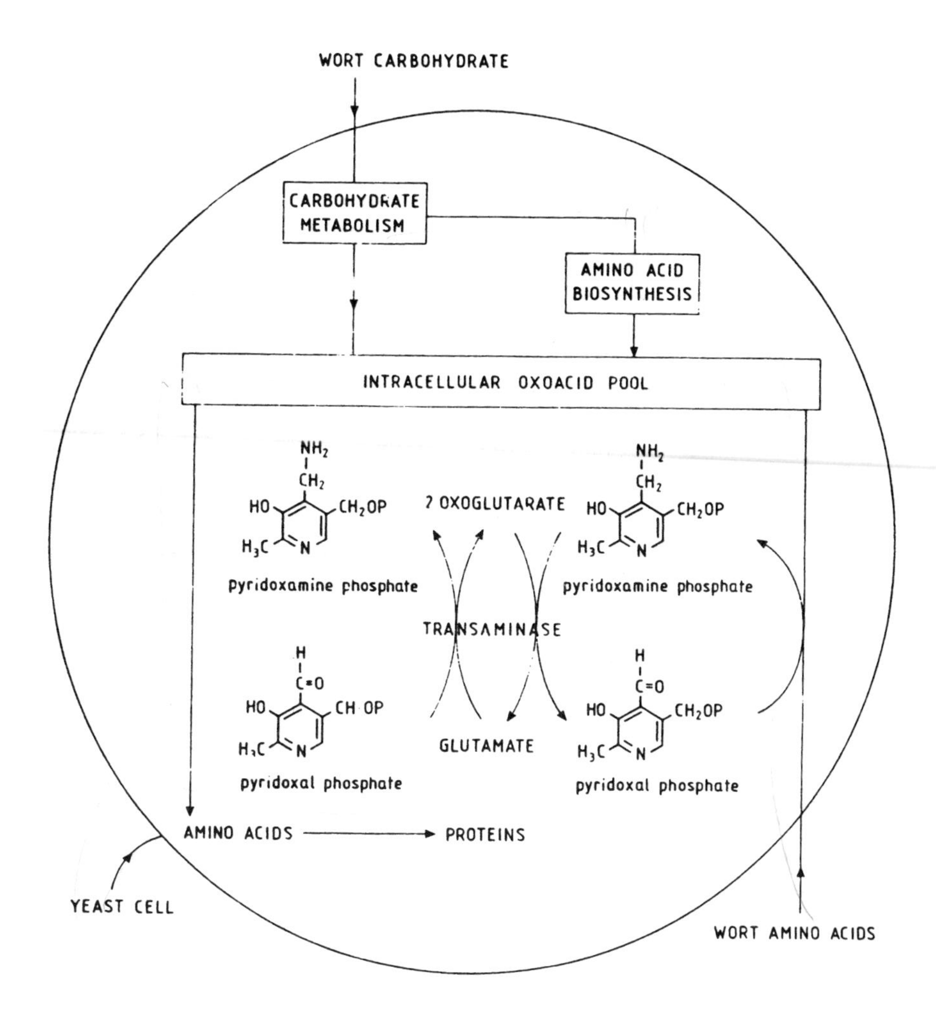

Fig. 2.4 Assimilation of wort amino acids by transamination.

dienoic $C_{18:2}$ (linoleic). Phospholipids are substituted diacylglycerophosphates with the most common substituents being choline, ethanolamine, serine or inositol. The predominant sterols are ergosterol and zymosterol; other sterols and sterol esters are also found. All three classes of lipid are important components of the yeast cell membrane. Since the cell membrane controls the entry of nutrients into, and the excretion of metabolites from, the cell its functions are essential to the growth of the cell.

Saturated fatty acids are synthesized from acetyl CoA derived from fermentable carbohydrate by the enzymes of the fatty acid synthetase complex. Unsaturated fatty acids are derived from their saturated counterparts by an enzymic reaction in which molecular oxygen acts as hydrogen acceptor. The mixed-function oxidase employed uses $NADP^+$ as a cofactor (Bloomfield and Bloch, 1960).

Sterols are also unsaturated molecules and their synthesis also involves the participation of an $NADP^+$-dependent mixed-function oxidase. The requirement of yeast for molecular oxygen is greater for sterol synthesis than for unsaturated fatty acid synthesis (David, 1974).

Failure to provide sufficient molecular oxygen to a fermentation medium restricts yeast growth and viability since the cells cannot produce unsaturated lipids for membrane biosynthesis. A secondary effect with ale yeasts is to produce elevated levels of esters in finished beer. Insufficient oxygenation (or aeration) of wort is not the only means of restricting the oxygen availability to the yeast. Since the solubility of oxygen is inversely related to the specific gravity of the wort, high-gravity worts may be unable to contain sufficient oxygen (particularly if air rather than molecular oxygen is the source) for yeast growth. Brewers therefore need to pay particular attention to satisfying the requirement of their yeast for molecular oxygen. This requirement is not fixed but varies with different yeast strains (Van den Berg, 1978).

2.4 YEAST PROPAGATION

2.4.1 Measurement of amount, viability and vitality of yeast

In brewing practice, measurement of the amount of viable yeast present is made in order to control for example the amount of yeast added to a fermenter (pitching rate) so that reproducible fermentations may be conducted, or the amount of yeast remaining in suspension after fermentation to 'condition' the beer.

Although the measurement of the dry weight of a suspension may be used to obtain a direct estimation of the biomass present, this procedure is generally too slow to be of value in practical brewing. Accordingly, less direct but more rapid methods are generally employed. Examples of these techniques are the measurement of packed cell volume following centrifugation under precisely controlled conditions, determination of wet weight after filtration and the measurement of the opacity or turbidity of a suspension. Alternatively, where pressed yeast or yeast slurry of consistent quality is available, the weight or volume of such samples may be measured. The accuracy and reliability of all these procedures is limited if a significant amount of non-yeast matter, e.g. precipitated protein (trub), is present in samples.

Cell number in suspension is most conveniently measured using an electronic particle counter. This technique does not distinguish between single cells and budding cells, chains of cells, aggregates or non-yeast matter, which are all recorded as single counts. Treatment of samples with short bursts of ultrasound prior to analysis may be used to disrupt aggregates and is of particular value when handling flocculent yeasts.

Cell counts may be obtained microscopically using a counting chamber and the numbers of cells in chains or aggregates may be estimated. Furthermore, the analyst may readily distinguish between yeast cells and particulate matter which is often dissolved by the addition of one or two drops of 20–40% sodium hydroxide or potassium hydroxide solution to the sample on a microscope slide. Counting cells in this way is considerably more time-consuming than using a particle counter or one of the indirect methods. The length of time needed is greatly increased if reproducible and statistically valid results are to be obtained (Meynell and Meynell, 1965).

The measurements of cell mass or number described cannot by themselves distinguish between living and dead cells. It is usual therefore for the brewer to combine measurement of amount of yeast with some procedure for estimating viability.

The most direct procedure for measuring viability is to prepare a suspension of yeast containing about 1000 cells ml^{-1} and spread 0.1–0.2 ml on the surface of a nutritionally rich medium solidified by the inclusion of agar. Alternatively, a larger volume of sample is mixed with molten medium at 47–49°C and the mixture poured into a Petri dish. In each case, isolated viable cells grow to produce colonies which are readily counted. In general this technique is not favoured for use with brewing yeast cultures sampled from the brewery since the estimated viability even when corrected for the presence of budding cells, chains and aggregates is invariably lower than that obtained with other methods (Richards, 1967).

Living cells contain ATP whereas this material is hydrolysed in dead cells. Detection of ATP therefore forms the basis of a technique for measuring viable cells and the procedure is claimed to be highly reproducible (Miller *et al.*, 1978) (see Chapter 8).

The most reliable procedure for obtaining accurate estimates of cell viability is the slide culture technique (Gilliland, 1959; Institute of Brewing Analysis Committee, 1971). In this method, a suitably diluted suspension of yeast is mixed with just molten wort gelatin and applied to a microscope slide or counting chamber. A coverslip is carefully lowered into position and sealed with sterile paraffin wax. After incubation for 8–16 h in a humidity chamber at 20°C, microscopic examination reveals the growth of microcolonies from viable cells. A shorter incubation time and easier visualization of microcolonies may be obtained by incorporating optical brighteners into the medium (Harrison and Webb, 1979). These compounds bind to the cell walls of the yeasts and fluoresce when illuminated with blue light. The use of a microscope equipped with an epifluorescence illumination system is therefore required. Fluorescence microscopy may also be used to estimate viability using the so-called fluorescein-diacetate technique. This compound is hydrolysed by esterases found in living but not in dead cells and the fluorescein released inside the cells causes them to fluoresce when suitably illuminated. Various fluorescent vital stains (stains which are absorbed by dead but not living cells) are available. These procedures, referred to as direct epifluorescence techniques (DEFT), often employ acridine dyes, e.g. acridine orange (Parkinnen, Oura and Suomalainen, 1976). (See Chapter 8.)

The most commonly used techniques employ non-fluorescent vital stains, and in the UK methylene blue in buffered solution (pH 5) is favoured. Dead cells stain blue whereas living cells are colourless. The reliability of the procedure is good only if population viabilities are in excess of 85% (Gilliland, 1959; Institute of Brewing Analysis Committee, 1971). Although microscopic techniques are tedious, when the methylene blue test is carried out using a counting chamber simultaneous estimates of cell number and viability are obtained.

The reliability of vital staining techniques has been questioned by many researchers. Yeast cultures after acid washing or longer than normal storage often appear viable but fail to ferment satisfactorily. In an attempt to overcome these problems, techniques aimed at measuring the physiological state of the yeast have been developed. These techniques include the rate of oxygen uptake (Kara, Daoud and Searle, 1987; Peddie *et al.*, 1991) and the rate of acidification of a medium after addition of D-glucose (Kara, Simpson and Hammond, 1988). Each method gives rapid results (minutes). These procedures are claimed to be more reliable than vital staining and are much more rapid than slide culture techniques. They are also suited to measurement in-line and hence offer great advantages for automated process control. The two methods however do not agree when used to analyse the same yeast cultures (Peddie *et al.*, 1991).

It is generally agreed that in the early stages of fermentation the oxygen consumed is used to produce unsaturated fatty acids and lipids essential to yeast viability (and vitality). Recent research has been directed at oxygenating yeast prior to fermentation in attempts to ensure adequate unsaturated lipid levels and thus eliminate wort oxygenation as a possible source of variability (Boulton, Jones and Hinchliffe, 1991; Devuyst *et al.*, 1991).

2.4.2 Handling pure yeast cultures

Hansen (1896) devised techniques for the isolation of single cells of brewing yeast. The separation of the different component strains of a brewing yeast culture was therefore feasible and each component could be analysed for its brewing properties. Because of the genetic stability of brewing yeast strains (resulting from their lack of a sexual cycle) a single isolated cell will grow to produce (excepting any chance mutation) a population of genetically identical cells, i.e. a clone. Such clones retain their brewing characteristics; thus, once selected, a fermentation strain will, in principle, produce consistent fermentations. The brewer, then, may use a pure culture if there is a means of preserving or maintaining it in a laboratory and a means of culturing it up to the large amount needed to pitch a brewery fermentation.

In breweries employing pure culture, the strain may be isolated, selected and kept at each brewery or at a central laboratory. When needed the yeast may be grown up and transported in slurry or pressed form in refrigerated containers or propagated from a laboratory culture in a yeast culture (propagation) plant in the brewery. Some companies rely on commercial laboratories to isolate and keep their yeast.

Various procedures are used to isolate pure cultures from the pitching yeast used. These include culturing from a single cell, culturing from a single colony (isolated on a medium containing agar or gelatin) and culturing from mixtures of isolated cells or colonies. Some companies isolate two or more strains which they employ in mixture to conduct fermentations.

All yeast cultures are susceptible to chance mutation; in addition, variation in mixed cultures may arise by changes in the proportions of the strains present. Such changes may be induced by use of different raw materials in the production of wort, a change in fermentation conditions or even a change in the type of fermentation vessel. On the other hand, it may be argued that a mixed culture is better able to withstand environmental changes because its increased genetic diversity enables it to adapt more readily to changed circumstances.

The successful maintenance of pure cultures requires that the isolates are given the minimum exposure to conditions which induce mutation, propagate mutant cells or encourage sporulation and must of course keep the cells in a highly viable condition. It is usual practice therefore to keep cultures in glass containers in the dark to prevent exposure to the mutagenic effects of ultraviolet light. Yeasts may be maintained on wort agar slopes (slants) at ambient temperature (to reduce the possibility of sporulation) or at 4°C, with or without an overlay of sterile mineral oil to maintain anaerobic conditions and restrict the growth of the cells. Many laboratories hold stock cultures in liquid medium such as wort or in chemically defined medium (e.g. YM medium; Haynes, Wickerham and Hesseltine, 1955) at 4°C. Whether sloped or liquid cultures are used, cell growth occurs and eventually the nutrients are consumed and starvation and cell death ensue. It is thus necessary to subculture strains regularly on to fresh slants or into fresh liquid medium every 3–6 months, to preserve the viability of the yeast. Subculture increases the chance of both mutation and contamination of the culture.

It is clearly preferable to keep the yeast in a form which removes the need for subculture, hence saving on labour and materials as well as better maintaining the integrity of the culture. One means of achieving this is to freeze-dry (lyophilize) the culture. This technique has not been generally accepted mainly because some yeasts exhibit a great loss of viability during the drying process. This phenomenon is however variable and careful control of the conditions overcomes it to some extent (Barney and

Helbert, 1976). There has also been some concern about the selection of mutants by the lyophilization process (Kirsop, 1955; Wynants, 1962); yet it has been reported that whereas routine subculture over a period of up to 20 years caused marked changes in the brewing characteristics of ale yeasts, no such changes were found when the same cultures were preserved by freeze-drying (Kirsop, 1974).

An alternative to freeze-drying is to store yeast in suspension in glycerol at the temperature of liquid nitrogen (–196°C). This technique retains the viability of cultures and is very successful (Mills, 1941; Barney and Helbert, 1976; Russell and Stewart, 1981; Kirsop, 1984).

When needed, yeasts are transferred from the laboratory culture to a small volume (approximately 10 ml) of liquid medium (wort or YM broth) and incubated at ambient temperature or 25–30°C for 16–48 h. This culture is then transferred to a larger volume, usually of sterile wort, then after further incubation this in turn is inoculated into a yet larger volume. At the end of this process the laboratory will have cultured up the yeast using stringent aseptic techniques and sterile media to a volume of approximately 20 l. The culture from each stage is used as a 5–10% v/v inoculum for the succeeding stage. The final culture will often be made in a specially designed stainless steel vessel compatible with making a good aseptic transfer of the culture to the yeast culture plant. Usually, a propagator with a working capacity some 5–10-fold the volume of the laboratory culture would be used. In some breweries however the 10–20 l laboratory culture is transferred to 4000–5000 l propagators (Van den Berg, 1978).

2.4.3 Operation of yeast culture plant

In modern breweries, where standards of general hygiene are high, re-use of yeast may often represent the major source of contaminating microorganisms. Furthermore, the performance of a brewery yeast will often decline with repeated use and the yeast is said to become 'weak'. The propagation of yeast is therefore practised routinely. In most breweries fresh yeast would be propagated every 8–10 generations (fermentation cycles) or sooner if particular problems were being experienced. The production of clean (uninfected), highly viable culture yeast is a major factor ensuring consistent fermentation performance.

In order to restrict the access of undesirable contaminating microorganisms (which would be propagated along with the culture yeast) propagation systems are designed to a high standard. Under ideal circumstances, the culture plant is placed in a room that is well separated from the rest of the brewing plant. The room should be equipped with its own filtered air supply and maintained at a slight positive pressure with respect to the surrounding areas. Access to the plant should be restricted to a few personnel and self-closing doors with disinfectant mats in the doorways are advantageous. The vessel(s) used for propagation are closed and

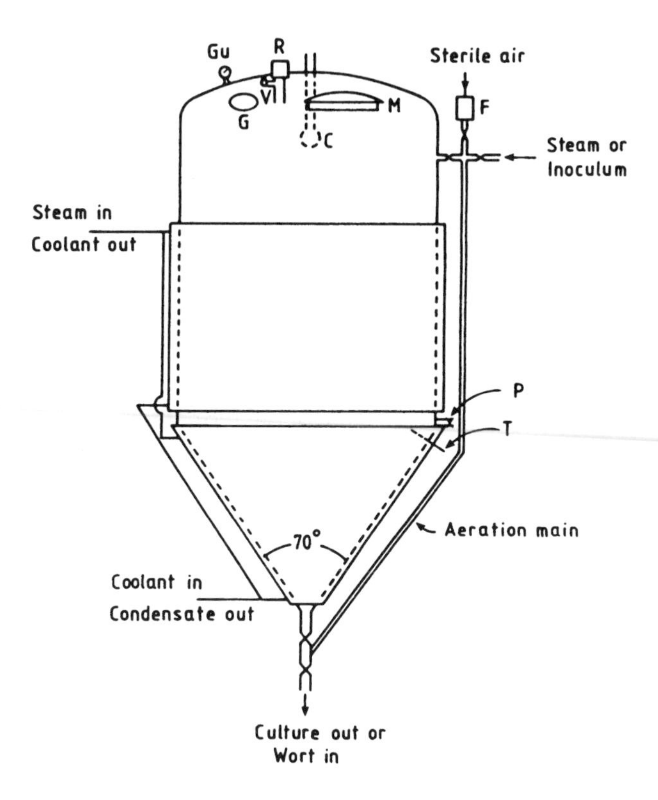

Fig. 2.5 A modern yeast propagation vessel of 49 hl (30 barrels) working volume. C, in-place cleaning spray-ball; F, secondary air filter; G, sight glass; Gu, pressure gauge; M, manway and dip-point; P, sample point; R, pressure and vacuum relief valve, fitted with sterilizing filter; T, temperature probe; V, vent with shut-off valve. (Courtesy of Shobwood Engineering Ltd, Burton-on-Trent.)

capable of being effectively cleaned either with a combined detergent sanitizer or preferably with separate caustic detergent cleaner followed by a sterilant or live steam. The wort used is either 'pasteurized' using a plate heat exchanger and passed into the sterile vessel or boiled in the vessel for 30 min or so to effect 'sterilization'; the necessary heating is obtained using pressurized steam in jackets around the vessel or by injecting live steam into the vessel. After venting air and steam, the vessel may be sealed so that sterilization may occur under pressure. After the sterilizing period, the vessel and contents are cooled and during this process air is admitted through a sterilizing filter. Most propagation vessels are fitted with cooling jackets and a system for admitting sterile air or oxygen. Agitators are not usually used, the aeration process and evolution of CO_2 during active fermentation being sufficient to maintain the yeast in suspension. A diagram of a typical propagation vessel is shown in Fig. 2.5: in this vessel the jacket serves the dual purpose of being filled with steam during sterilization and coolant during the propagation phase.

In a propagation system more than one vessel is often employed. Two vessels of different sizes, e.g. 1 hl and 20 hl, would be found and the product from the larger vessel used to recharge the smaller and pitch the first fermenter, containing (in this case) 200–400 hl wort. By retaining a proportion of propagated yeast in the smaller vessel the brewer may propagate fresh yeast from there without recourse to the laboratory culture every time. The coolant circulation around the small vessel would be maintained at such a rate and temperature as to keep the yeast below ambient temperature. In practice, recourse to the laboratory culture may be made only at intervals of 6–12 months. The sizes of the propagation vessels may be determined by the size of the first fermentor and the desire to keep the inoculum size to 5–10% of the volume of the fermentor.

In propagation the brewer wishes to obtain the maximum amount of yeast possible consistent with the flavour of the beer produced being not too far removed from the flavour of the product of a normal fermentation. In most breweries the beer from the first fermentation following propagation would in any event be blended with greater volumes of 'normal' beer. Accordingly, propagation is carried out using worts of greater specific gravity than those used in the typical fermentation, at higher temperatures and with the use of intermittent aeration throughout the process. The concentration of yeast produced would be of the order of 5×10^8 cells ml^{-1} at a viability of > 98%. The time of propagation would be of the order of 24–48 h. Yeast growth in a propagator may be described by the standard equations for microbial growth in a batch culture (Briggs *et al.*, 1982). A typical growth curve for yeast growing under conditions similar to those obtaining in a yeast propagator is shown in Fig. 2.6. In practice the lag phase would be almost non-existent (since the cells are fully adapted before inoculation) and the culture would be transferred prior to the onset of stationary phase.

2.5 BREWERY FERMENTATION

Systems of fermentation have gradually evolved around the type of brewing yeast used. It is usual to distinguish between two types of brewing yeast (and hence fermentation system): thus brewers are said to employ top yeasts (top fermentation) or bottom yeasts (bottom fermentation). The distinction arises from the practice of removing the yeast from the top of the fermentor by a skimming process or from the bottom of the vessel. It is current practice to classify all brewer's yeast as *S. cerevisiae* (Barnett, Payne and Yarrow, 1983), although in brewing practice ale yeasts are referred to as *S. cerevisiae* whereas lager strains are called *S. carlsbergensis* (or *S. uvarum*) (Lodder, 1970). It would seem useful from a practical standpoint to retain

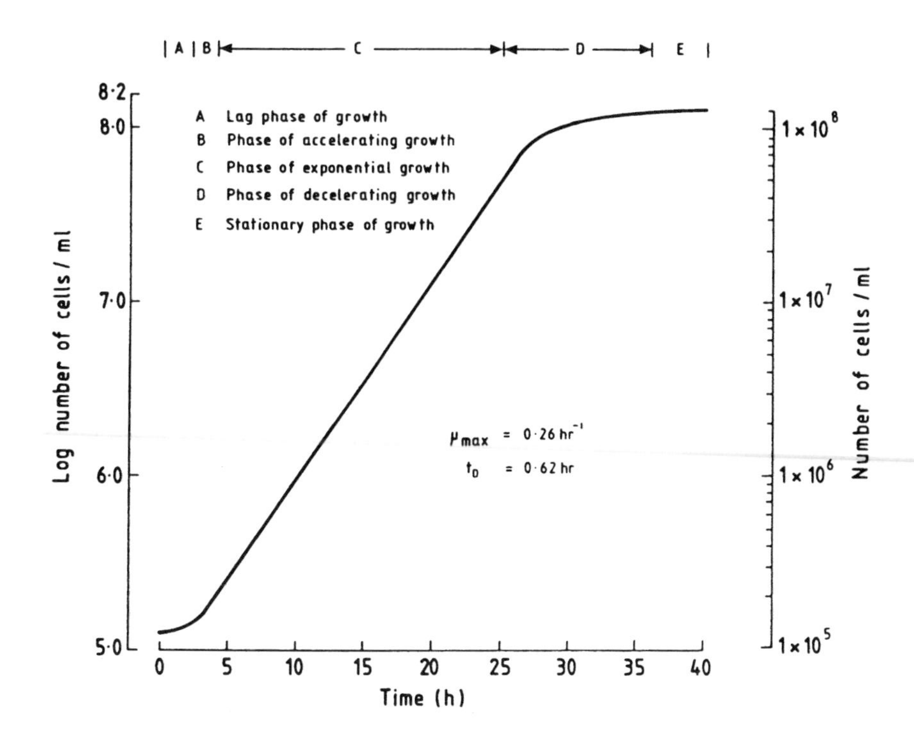

Fig. 2.6 Batch growth curve for brewing yeast culture in shake flasks at 20°C.

a distinction between the two types of brewing yeast, since they are used to produce beers under quite different conditions of fermentation. In addition the two types of organism differ in characteristics not generally included as taxonomic criteria, e.g. serotype (Campbell and Brudzynski, 1966) and maximum temperature for growth (Walsh and Martin, 1977). More recent attempts to classify brewing yeasts are based on relevant brewing characteristics (Bryant and Cowan, 1979; Cowan and Bryant, 1981), and consequently are of more value to the brewer.

Top yeasts necessarily tend to rise to the top of the fermenter and form a head, whereas bottom yeasts do not form a substantial head but tend to be sedimentary. Use of the different systems and yeasts is also typical of brewing in different geographical areas. Bottom yeasts and bottom fermentation systems were traditionally used to produce so-called lager beers in the brewing centres of mainland Europe whilst the top fermentation system is typical of the UK and was traditionally used to make the so-called ales. The production of traditional ale employed open square fermenters or one of several quite distinct types of fermentation system, e.g. Yorkshire square, Burton union, and premium quality beers are still produced in these systems. The modern trend in fermentation, however, has been to replace the open fermentation vessels (usually used where yeast is removed from the top of the vessel) with closed vessels. Closed (or covered) vessels are obviously more hygienic in operation since the fermentation is not exposed to the atmosphere and furthermore it is a relatively simple matter to arrange for them to be equipped with automatic cleaning-in-place (CIP) systems. Modern practice has also led to an increase in size of fermentation vessel and the siting of well-lagged fermenters in the open, thereby saving much on expensive building costs involved in fabricating a fermentation room (Briggs *et al.*, 1982).

Many different designs of modern batch fermenter are used; Fig. 2.7 is a diagram of a typical cylindroconical vessel used by many UK breweries for the production of both lagers and ales. These vessels were originally intended to be dual purpose and used for fermentation and conditioning (Nathan, 1930; Shardlow, 1972). However, in practice they are mainly employed for fermentation only and additional tanks are used for the conditioning process (Maule, 1977). Cylindroconical fermenters range in size from 100 to 4800 hl. Each vessel has an angled cone at its base and the most common internal angle encountered is 70°. Very large vessels tend to have much shallower slopes and angles of 105°. The cone allows for the yeast to compact at the end of fermentation and beer may be separated relatively cleanly from the sediment. Experience has shown that even top yeasts may be induced to sediment in this type of vessel although, where difficulty is experienced, antifoam is used to collapse the yeast head. Alternatively a sedimentary variant may be isolated from the yeast which routinely settles in the cone (Shardlow and Thompson, 1971). In order to allow for foam or head formation during fermentation, vessels are usually sized at 25–30% in excess of their working volume; therefore antifoam may be employed to increase the effective working volume of the fermenter (Button and Wren, 1972).

In spite of their success, cylindroconical vessels have not been used by the traditional ale brewer to produce his premium quality beer. This results from the fact that the CO_2 content of beer produced in deep fermenters is in excess of that required in a traditional ale and there is no doubt that beer produced in these vessels differs in flavour from that made in traditional fermentation systems. Nevertheless, the traditional brewer has used cylindroconical fermenters to produce lager beers most successfully.

Attemperation of fermentation is achieved by circulating coolant through cooling jackets (Maule, 1976). In the vessel shown in Fig. 2.7 three jackets are used, although many vessels do not have a cone jacket and may only have a single wall jacket two-thirds of the way up the cylindrical part of the fermentor. Cone cooling is generally accepted as advantageous when yeast for pitching is to be stored in the vessel; if a separate yeast-collecting vessel is used then cone cooling is not required. The presence of sedimented yeast effectively insulates the cone and restricts transfer of heat to and from the fermenter contents. If the vessel is to be used for

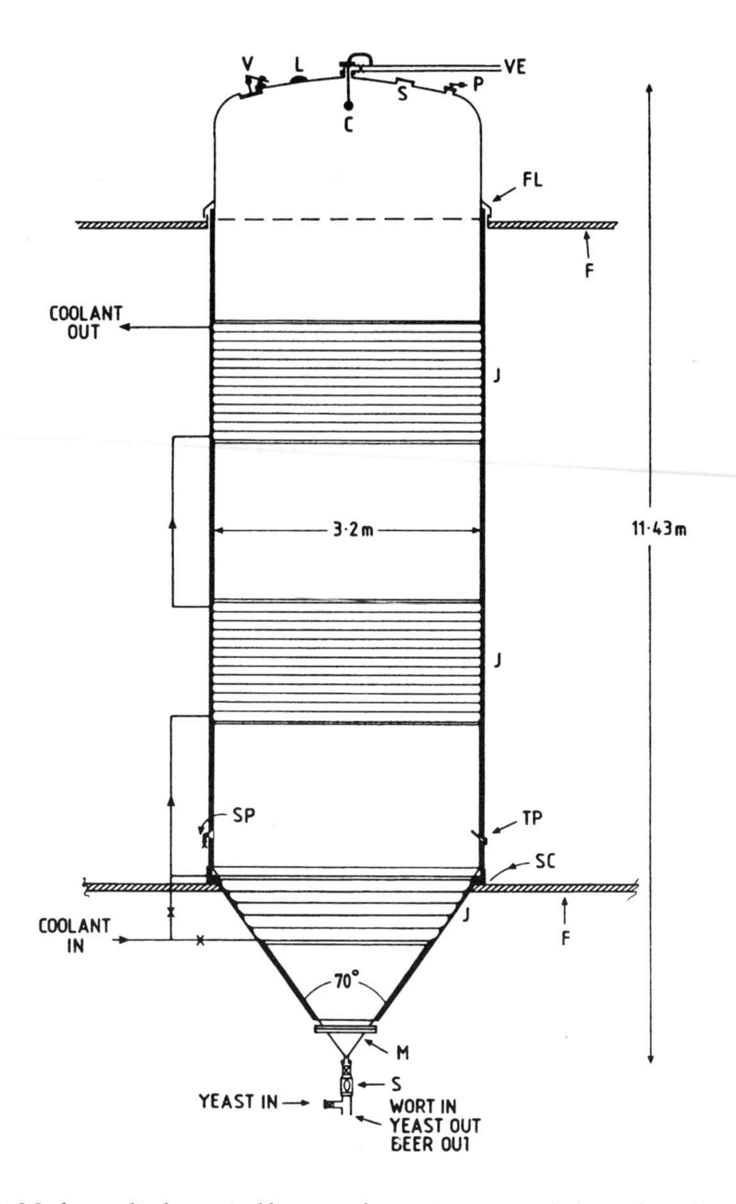

Fig. 2.7 Modern cylindroconical brewery fermentation vessel. C, in-place cleaning spray-ball; F, floors; FL, flange; J, cooling coils ('limpets'); L, inspection lamp; M, manway (detachable cone); P, pressure relief valve; S, sight glasses; SC, supporting collar; SP, sample point; TP, temperature probe; V; vacuum relief valve; VE, vent and detergent entry. (Courtesy of Gordon Smith of Bristol Ltd.)

conditioning then cone-cooling is also necessary. The positioning of cooling jackets and of temperature probes within the vessel are subjects of much research (Maule, 1976).

Cylindroconical fermentors are fitted with pressure and vacuum release valves. These relieve pressure during filling and vacuum during emptying, or cleaning with hot alkali which absorbs CO_2, thereby lowering the pressure which is further reduced as the vessel cools. The outlet from the vessel is from the apex of the cone but some fermenters have an additional outlet for beer at the junction of the cone and vertical side. Temperature probes and CIP heads are the only intrusions to the inside of the vessel, which is fabricated throughout in stainless steel and is therefore easy to clean. In some vessels, lines for admitting CO_2 for rousing and purging and air for rousing and aeration are fitted at the base of the cone. These features make the vessels suitable for use as yeast propagators.

Large fermenters have the capacity to contain more than one production volume (brew length) of wort. Such vessels are filled sequentially and the brewer may choose to aerate and pitch any or all lengths. When the first length only is pitched, the pitching rate must be increased to account for the subsequent lengths. However, as the yeast grows in the first length the amount needed is less than a *pro rata* increase based on the number of lengths in the vessel. The choice of aeration and pitching regimes influences to some extent the flavour of the final product.

A vessel taking a single brew length and used to produce ale would be filled with aerated wort, and yeast slurry would be pumped in-line with the wort. The temperature of the wort would typically be 15–18°C and the yeast concentration 2×10^7 cells ml^{-1} (corresponding to a pitching rate of 3 kg pressed yeast of 25% solids hl^{-1} wort). As fermentation proceeds the temperature will rise and at a given value (top temperature) cooling will be applied to keep the temperature steady. A typical top temperature would lie in the range 18–25°C. The modern trend is to use higher fermentation temperatures, thereby shortening the fermentation time, although this practice does affect beer flavour. Some brewers claim that flavour changes may be minimized to some extent by fermenting under pressure (5–15 psi).

Fermentation in cylindroconical vessels is faster than in traditional fermenters mainly because the rapid evolution of CO_2 from the base of the cone, coupled with attemperation of the vessel walls, results in a very thorough mixing of wort and yeast (Ladenburg, 1968; Maule, 1976). A typical fermentation would take 40–48 h and as the fermentation rate declines, heat output falls and the cooling is increased to lower the temperature of the beer and encourage the yeast to settle out. The settling period may take 2–3 days and in some operations freshly fermented beer (green beer) is centrifuged to remove yeast and passed to a separate vessel, thus releasing the fermenter for cleaning and reuse. The fermentation is monitored by taking samples for measuring the specific gravity and is controlled by varying the rate of cooling (reducing the rate allows the temperature to rise and accelerates the fermentation, and vice versa). A typical fermentation profile is shown in Fig. 2.8.

Fermentations to produce lager beers are conducted in a similar manner to those for ales except the temperatures used are lower and consequently fermentation times are longer. Typical values would be pitch at 12°C, top temperature 15°C and fermentation time 7 days.

One of the greatest operational advantages of a cylindroconical fermenter over traditional vessels is that precise control of fermentation profile may be achieved by controlling temperature (Hoggan, 1977). This flexibility has stimulated research workers to present computer simulations of the fermentation process and develop programs to enable automated fermentation to be achieved (Luckiewicz, 1978; Ruocco, Coe and Hahn, 1980).

Continuous systems of fermentation have also been used (Coutts, 1967; Ault *et al.*, 1969; Bishop, 1970; Seddon, 1975) but in the UK have been abandoned in favour of modern batch methods.

In contrast to the modern fermentation, traditional ale systems would employ a pitching temperature of 14°C, a top temperature of 17°C and take 5–7 days, and traditional German lager systems would pitch at 7°C, use a top temperature of 9°C and take 14–21 days.

In most ale fermentation systems, the yeast grows to produce a 5–8-fold increase. In lager fermentations the overall increase in yeast is generally

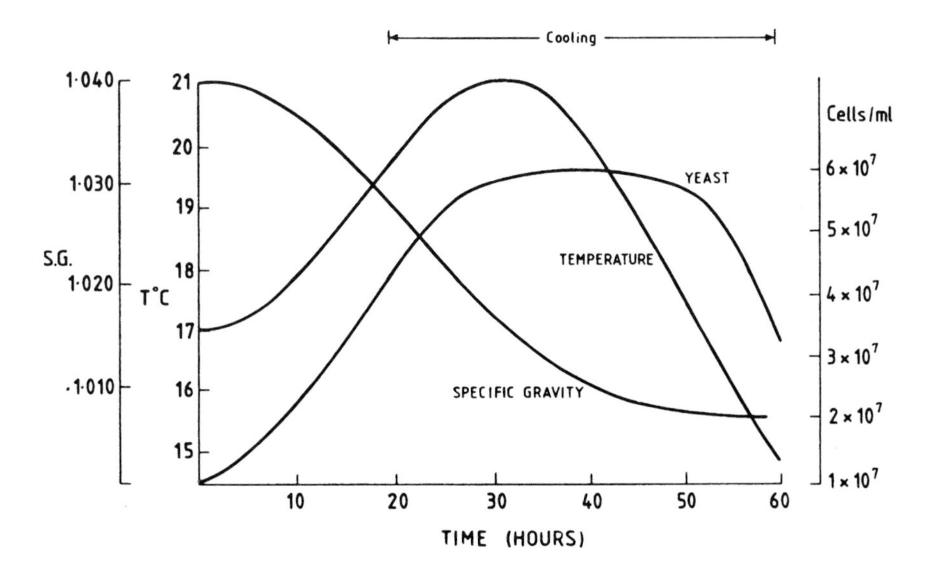

Fig. 2.8 Typical fermentation profile for modern batch fermentation to produce ale. SG, specific gravity; T, temperature.

nearer 5-fold. The amount of growth is dependent on the pitching rate, temperature regime, oxygen content and specific gravity of the wort. Excessive growth is undesirable in that wort sugars would be converted to undue amounts of yeast rather than to alcohol. Yeast is cropped from the fermentor maintained at 4°C, sometimes after washing with cold water to 'clean' it or with acidic solutions (pH 2.5) to kill any contaminating bacteria, and used to pitch subsequent fermentations. The surplus is sold to manufacturers of yeast extract and to distilleries for use in their fermentation systems.

REFERENCES

Ault, R.G., Hampton, A.N., Newton, R. and Roberts, R.H. (1969) *Journal of the Institute of Brewing*, **75**, 260.
Bajomo, M.F. and Young, T.W. (1993) *Journal of the Institute of Brewing*, **99**, 153.
Barnett, J.A., Payne, R.W. and Yarrow, D. (1983) *Yeasts: Characteristics and Identification*, Cambridge University Press, Cambridge.
Barney, M.C. and Helbert, J.R. (1976) *Journal of the American Society of Brewing Chemists*, **34**, 61.
Bishop, L.R. (1970) *Journal of the Institute of Brewing*, **76**, 173.
Bloomfield, D. and Bloch, K. (1960) *Journal of Biological Chemistry*, **235**, 337.
Boulton, C.A., Jones, A.R. and Hinchliffe, E. (1991) *Proceedings of the 23rd Congress of the European Brewery Convention, Lisbon*, IRL Press, Oxford, p. 385.
Briggs, D.E., Hough, J.S., Stevens, R. and Young, T.W. (1982) *Malting and Brewing Science*, Vol. 2, Chapman & Hall, London.
Bryant, T.N. and Cowan, W.D. (1979) *Journal of the Institute of Brewing*, **85**, 89.
Button, A.H. and Wren, J.J. (1972) *Journal of the Institute of Brewing*, **78**, 443.
Campbell, I. and Brudzynski, A. (1966) *Journal of the Institute of Brewing*, **72**, 556.
Coutts, M.W. (1967) *International Brewer and Distiller*, **1**, 33.
Cowan, W.D. and Bryant, T.N. (1981) *Journal of the Institute of Brewing*, **87**, 45.
David, M.H. (1974) *Journal of the Institute of Brewing*, **80**, 80.
Devuyst, R., Dyon, D., Ramos-Jeunehomme, C. and Masschelein, C.A. (1991) *Proceedings of the 23rd Congress of the European Brewery Convention, Lisbon*, IRL Press, Oxford, p. 377.
Fiechter, A., Fuhrmann, G.F. and Kappeli, O. (1981) *Advances in Microbial Physiology*, **22**, 123.
Gilliland, R.B. (1959) *Journal of the Institute of Brewing*, **65**, 424.
Hagele, E., Neff, J. and Mecke, D. (1978) *European Journal of Biochemistry*, **83**, 67.
Hansen, E.C. (1896) *Practical Studies in Fermentation*. Cited in Laufer, S. and Schwarz, R. (1936) *Yeast Fermentation and Pure Culture Systems*, Schwarz Laboratories Inc., New York.
Harris, G. and Thompson, C. (1960) *Journal of the Institute of Brewing*, **66**, 293.
Harrison, J. and Webb, T.J.B. (1979) *Journal of the Institute of Brewing*, **85**, 231.
Haynes, W.C., Wickerham, L.J. and Hesseltine, C.W. (1955) *Applied Microbiology*, **3**, 361.
Heredia, C.F., Sols, A. and De la Fuente, G. (1968) *European Journal of Biochemistry*, **5**, 321.
Hoggan, J. (1977) *Journal of the Institute of Brewing*, **83**, 133.
Holzer, H. (1976) *Trends in Biochemical Science*, **1**, 178.
Horecker, H. (1968) In *Aspects of Yeast Metabolism* (eds A.K. Mills and H. Krebs), Blackwell, Oxford, p. 71.

Hsu, N.P., Vogt, A. and Bernstein, L. (1980) *MBAA Technical Quarterly*, **17,** 85.
Institute of Brewing Analysis Committee (1971) *Journal of the Institute of Brewing*, **77,** 181.
Jones, M. and Pierce, J.S. (1964) *Journal of the Institute of Brewing*, **70,** 307.
Jones, M., Pragnell, M.J. and Pierce, J.S. (1969) *Journal of the Institute of Brewing*, **75,** 520.
Kara, B.V., Daoud, I. and Searle, B. (1987) *Proceedings of the 21st Congress of the European Brewery Convention, Madrid*, IRL Press, Oxford, p. 409.
Kara, B.V., Simpson, W.J. and Hammond. J.R.M. (1988) *Journal of the Institute of Brewing*, **94,** 153.
Kirsop, B. (1955) *Journal of the Institute of Brewing*, **61,** 466.
Kirsop, B. (1974) *Journal of the Institute of Brewing*, **80,** 565.
Kirsop, B.E. (1984) In *Maintenance of Microorganisms* (eds B.E. Kirsop and J.J.S. Snell), Academic Press, London, p. 109.
Ladenburg, K. (1968) *MBAA Technical Quarterly*, **5,** 81.
Lodder, J. (ed.), (1970) *The Yeasts, a Taxonomic Study*, North-Holland, Amsterdam.
Luckiewicz, E.T. (1978) *MBAA Technical Quarterly*, **15,** 190.
Lynen, F. (1963) *Control Mechanisms in Respiration and Fermentation* (ed. B. Wright). Proceedings of Annual Symposium 8, Society of General Physiologists, 1961, Woods Hole, MA, p. 290.
Macy, J.M. and Miller, M.W. (1983) *Archives of Microbiology*, **134,** 64.
Maule, D.R. (1976) *The Brewer*, **62,** 140.
Maule, D.R. (1977) *The Brewer*, **63,** 204.
Maynell, G.G. and Meynell E. (1965) *Theory and Practice in Experimental Bacteriology*, Cambridge University Press, Cambridge.
Miller, L.F., Mabee, M.S., Gress, H.S. and Jangaard, N.O. (1978) *Journal of the American Society of Brewing Chemists*, **36,** 59.
Mills, D.R. (1941) *Food Research*, **6,** 361.
Nathan, L. (1930) *Journal of the Institute of Brewing*, **36,** 538.
Nykanen, L. and Suomalainen, H. (1983) *Aroma of Beer, Wine and Distilled Beverages*, Reidel, Dordrecht.
Oura, E. (1977) *Process Biochemistry*, **12,** 19.
Parkinnen, E., Oura, E. and Suomalainen, H. (1976) *Journal of the Institute of Brewing*, **82,** 283.
Pasteur, L. (1867) *Études sur la Bière*, Imprimerie de Gauthier-Villers, Paris.
Peddie, F.L., Simpson, W.J., Kara, B.V., Robertson, S.C. and Hammond, J.R.M. (1991) *Journal of the Institute of Brewing*, **97,** 21.
Pringle, J.R. and Hartwell, L.H. (1981) In *Molecular Biology of the Yeast Saccharomyces: Life Cycle and Inheritance*, Cold Spring Harbor Laboratory, Cold Spring Harbor, NY p. 97.
Quain, D.E. and Tubb, R.S. (1982) *MBAA Technical Quarterly*, **19,** 29.
Richards, M. (1967) *Journal of the Institute of Brewing*, **73,** 162.
Rose, A.H. and Harrison, J.S. (eds) (1969) *The Yeasts*, Vol. 1, *Biology of Yeasts*, Academic Press, London.
Rose, A.H. and Harrison, J.S. (eds) (1970) *The Yeasts*, Vol. 3, *Yeast Technology*, Academic Press, London.
Rose, A.H. and Harrison, J.S. (eds) (1971) *The Yeasts*, Vol. 2, *Physiology and Biochemistry of Yeasts*, Academic Press, London.
Ruocco, J.J., Coe, R.W. and Hahn, C.W. (1980) *MBAA Technical Quarterly*, **17,** 69.
Russell, I. and Stewart, G.G. (1981) *Journal of the American Society of Brewing Chemists*, **39,** 19.
Seddon, A.W. (1975) *MBAA Technical Quarterly*, **12,** 130.
Shardlow, P.J. (1972) *MBAA Technical Quarterly*, **9,** 1.
Shardlow, P.J. and Thompson, C.C. (1971) *Brewers' Digest*, **46** (August), 76.

Sols, A. and De la Fuente, G. (1961) In *Membrane Transport and Metabolism* (eds A. Kleinzeller and A. Kotyk), Academic Press, London, p. 361.
Sols, A., Gancedo, G. and De la Fuente, G. (1971) In *The Yeasts*, Vol. 2, *Physiology and Biochemistry of Yeasts* (eds A.H. Rose and J.S. Harrison), Academic Press, London.
Stryer, L. (1995) *Biochemistry*, 4th edn, W.H. Freeman & Co., New York.
Thomas, K.C. and Ingledew, W.M. (1990) *Applied and Environmental Microbiology*, **56,** 2046.
Thorne, R.S.W. (1949) *Journal of the Institute of Brewing*, **55,** 201.
Van den Berg, R. (1978) European Brewery Convention, Fermentation and Storage Symposium, Monograph V, *Zoeterwoude*, p. 66.
Walsh, R.M. and Martin, P.A. (1977) *Journal of the Institute of Brewing*, **83,** 169.
Wang, P.Y., Johnson, B.F. and Schneider, H. (1980) *Biotechnology Letters*, **3,** 273.
Wynants, J. (1962) *Journal of the Institute of Brewing*, **68,** 350.

CHAPTER 3

Yeast genetics

J.R.M. Hammond

3.1 INTRODUCTION

Yeasts have been employed in the production of alcoholic beverages for thousands of years. Originally adventitious organisms present on raw materials and equipment were responsible for the often unpredictable fermentation which took place. More recently, strains have been selected on the basis of their fermentation behaviour and the availability of pure cultures has led to marked improvements in the reproducibility of fermentations and to beer quality.

Genetic studies with *Saccharomyces cerevisiae* were pioneered by Winge and colleagues at the Carlsberg Laboratories in Copenhagen and subsequently developed by Lindegren and co-workers. Yeasts, mainly of industrial origin, were extensively interbred to produce strains which could be mated to give healthy diploid cells capable of sporulating to produce four viable ascospores. Strain S288C and its diploid derivative X2180 provide the source of most genetically marked strains used in laboratory studies throughout the world. However, as described below, these strains differ markedly from those used for brewing and, as a consequence, studies of the genetics of brewing yeast strains have lagged behind investigations of laboratory organisms.

Against a background of fundamental knowledge of the genetics of *S. cerevisiae*, this chapter will highlight the genetic peculiarities of brewing yeasts. The limitations in their performance will be described as will the difficulties encountered in trying to breed new yeast strains using so-called 'traditional' breeding methods. Finally, the tremendous developments which the advent of recombinant DNA methods has brought about will be explained together with the new insight gained into the genetic make-up of brewing yeasts.

Brewing Microbiology, 2nd edn. Edited by F. G. Priest and I. Campbell.
Published in 1996 by Chapman & Hall, London. ISBN 0 412 59150 2

3.2 GENETIC FEATURES OF *SACCHAROMYCES CEREVISIAE*

3.2.1 Life cycle and sporulation

The life cycle of a typical laboratory strain of *S. cerevisiae* is shown in Fig. 3.1. Such a yeast has both haploid (one set of chromosomes) and diploid (two sets of chromosomes) modes of existence. Strains in which the haploid form is stable and can be maintained for many generations are termed heterothallic. The haploids from such strains exist as one of two mating

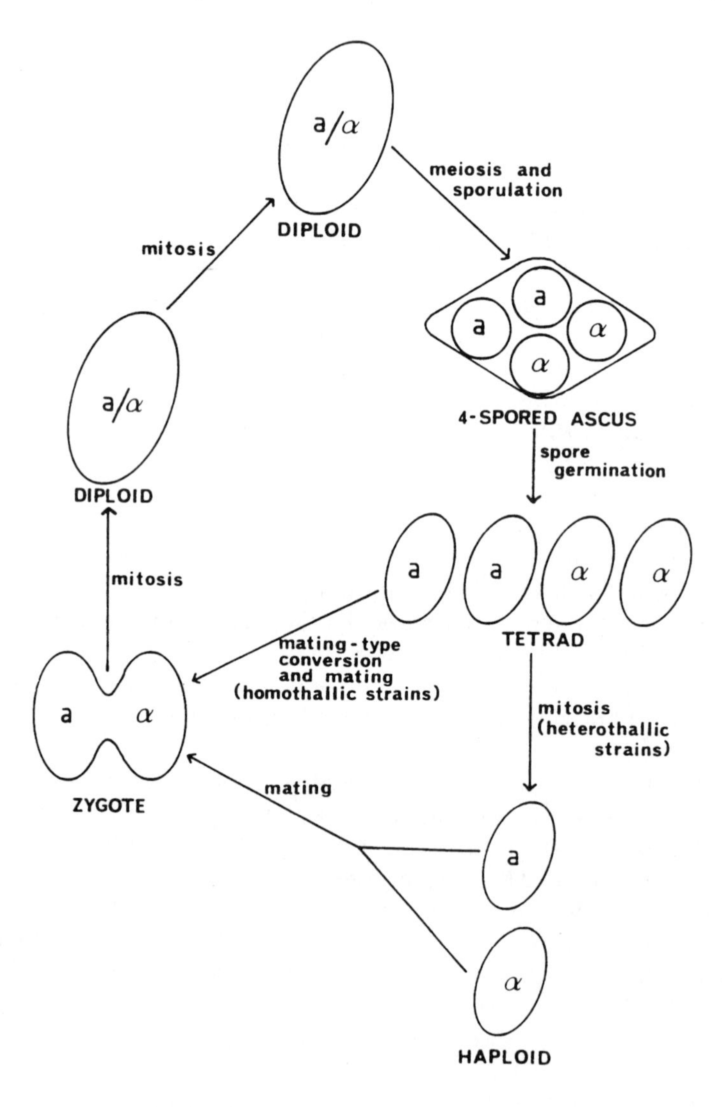

Fig. 3.1 Life cycle of *Saccharomyces cerevisiae*.

types, *a* or α, and mate to form diploids when a cell of one mating type comes into contact with a cell of the other mating type. Strains in which cell fusion and diploid formation occur among cells derived from a single spore are termed homothallic. The presence of the *HO* gene in such strains brings about a high frequency of switching between mating types during vegetative growth. Under the influence of this gene, the mating type locus, *MAT*, of such strains readily changes from *MATa* to *MAT*α or vice versa. The *MAT* gene is found on chromosome III of the yeast genome together with two silent genes, *HML*α and *HMRa*, which provide the information to allow the switch of mating type at the *MAT* locus. Cells of homothallic yeasts have to bud at least once before they are able to switch mating type, but thereafter a high frequency of switching occurs at each budding for many generations (for further details see Herskowitz and Oshima, 1981).

In both homothallic and heterothallic strains, mating takes place when cells of opposite mating type come into close proximity. Cells of α mating type produce an oligopeptide called α-factor which stops the growth of *a* cells and causes *a* and α cells to adhere to each other. Cells of mating type *a* produce *a*-factor which has similar effects on *a* cells. In the presence of these factors, the cells adhere and cytoplasmic fusion takes place to form a heterokaryon (Thorner, 1981). Nuclear fusion follows rapidly to give a zygote (Lindegren and Lindegren, 1943). By subsequent cell division this forms the diploid phase of the yeast life cycle which can be stably maintained for many generations. Meiosis and sporulation of diploid cells is triggered by nitrogen deprivation in the presence of a non-fermentable carbon source but will only occur if both *MATa* and *MAT*α genes are present. Following entry into meiosis the chromosomes in the yeast nucleus undergo DNA synthesis, pairing, recombination and segregation. Spore walls grow and envelop the four haploid genomes (two each of α and *a* mating types) forming the characteristic four-spored ascus. The spores, when placed in suitable nutrient media, germinate to form haploids and begin the whole cycle once more. Meiotic recombination is important for evolutionary change and provides the means whereby chromosomes can be mapped (for further details see Esposito and Klapholz, 1981).

Brewing strains behave very differently from laboratory-bred yeasts. They sporulate poorly, rarely form spores in tetrads and many of the spores formed are non-viable even when great efforts are made to produce ideal conditions for sporulation, such as growing cells under complete catabolite derepression and at lower temperatures than is usual for laboratory strains (Gjermansen and Sigsgaard, 1981; Bilinski, Russell and Stewart, 1986, 1987a). In one study with brewing yeasts (Anderson and Martin, 1975) ale yeasts were found to sporulate better than lager yeasts but only two- or three-spored asci were formed and many of the resulting strains were aberrant, being capable of mating with both *a* and α haploid yeasts, or were sterile.

3.2.2 Chromosomes, ploidy and genetic stability

The chromosomes of *S. cerevisiae* are located in the nucleus and make up between 80% and 85% of the total cellular DNA (Petes, 1980). In haploid cells there is sufficient DNA to account for about 15 000 genes, although to date only 769 genes have been mapped to the 16 chromosomes (Mortimer *et al.*, 1989). Each chromosome consists of a single DNA molecule of molecular size between 150 and 2500 kilobase pairs (kbp) together with basic histone protein molecules. Most genes of the haploid genome are present as single copies, the major exceptions being the ribosomal RNA genes present as about 100 copies and the approximately 15 copies of each transfer RNA gene (Fangman and Zakian, 1981). Mobile genetic elements called *Ty* also occur in *Saccharomyces* yeasts. Up to 35 copies of *Ty* can be present in haploid yeast cells and their presence, together with their ability to move from one chromosomal location to another, can cause rearrangements of the yeast genome (Philippsen *et al.*, 1983).

Whereas laboratory-bred strains of *S. cerevisiae* are either haploid or diploid most yeasts used for brewing are polyploid or aneuploid. This conclusion has been reached by measuring the DNA content of individual cells and comparing the results with those obtained from haploid strains, a notoriously unreliable procedure. Such studies have identified the presence of cells containing between two (diploid) and seven (heptaploid) times the normal haploid content of DNA as well as many aneuploids, which have a DNA content intermediate between triploid and tetraploid values (Johnston and Oberman, 1979). More recent studies involving direct genetic analysis have confirmed that brewing strains are indeed polyploid, particularly triploid, tetraploid or aneuploid.

Industrial yeasts may gain a number of benefits from being polyploid. For instance, extra copies of important genes such as those responsible for maltose utilization (*MAL*) could improve their fermentation performance. Indeed, production of α-glucosidase and hence the rate of maltose fermentation is increased with the dosage of *MAL* genes (Mowshowitz, 1979; Stewart *et al.*, 1981). It has also been argued that polyploid yeasts are more stable than haploid yeasts since multiple mutational events are required in order to change them. However, because of their very nature, polyploid yeasts can harbour non-functional recessive mutations as demonstrated by Delgado and Conde (1983). Therefore genetic stability of polyploid yeasts is likely to be a function of the frequency of segregational events leading to the expression of mutant genes, rather than the frequency of mutation itself.

3.2.3 Extrachromosomal elements

A number of extrachromosomal genetic elements are also known in yeasts, the most important being 2 μm DNA, mitochondrial DNA and double-stranded RNA (dsRNA).

Two micrometre DNA is an extrachromosomal element found in most strains of *S. cerevisiae* (Broach, 1981) including brewing strains (Tubb, 1980; Stewart, Russell and Panchal, 1981; Aigle, Erbs and Moll, 1984). When present, there are between 50 and 100 copies of 2 μm DNA per cell. Two micrometre DNA occurs as circular plasmids (Hartley and Donelson, 1980) but the only function which can be ascribed to these extrachromosomal elements is their own maintenance; there appears to be no obvious advantage for cells to possess 2 μm DNA. Although early work indicated that these plasmids were cytoplasmic, a more detailed examination now suggests they have a nuclear extrachromosomal location. Much genetic modification work carried out in recent years with brewing yeasts has involved the use of 2 μm DNA-based plasmids and for this reason these chromosomal elements have taken on a much greater importance than their lack of function *in vivo* would suggest.

Mitochondrial DNA consists of a 75 kbp circular molecule and is present as 10–40 molecules per cell in laboratory haploid yeast strains. It has a lower buoyant density than chromosomal DNA, thereby facilitating separation of the two types of molecule in density gradients (Fangman and Zakian, 1981). Mitochondrial DNA shows typical cytoplasmic inheritance and its replication is highly independent of nuclear control, taking place throughout the cell cycle (Newlon and Fangman, 1975). The mitochondrial genome carries the genetic information for only a few essential components, more than 90% of mitochondrial proteins being synthesized from nuclear-encoded genes (Dujon, 1981). Mutations in mitochondrial DNA produce petite strains which are unable to metabolize non-fermentable substrates. Such respiratory-deficient mutations can range from point mutations (mit^-) through deletion mutations (rho^-) to complete elimination of the mitochondrial DNA (rho°). Petites of *S. cerevisiae* are often deficient in uptake of sugars such as maltose and galactose (Evans and Wilkie, 1976; Mahler and Wilkie, 1978) and petite mutants of brewing yeasts have been reported which are deficient in maltotriose utilization and which therefore inadequately attenuate wort (Gyllang and Martinson, 1971). Brewing yeasts with different mitochondrial genomes produce beers of dissimilar flavour (Hammond and Eckersley, 1984) and petite mutants are very different from their parents in flocculation behaviour, lipid metabolism and higher alcohol production (Lewis, Johnston and Martin, 1976). The mitochondrial genome also has effects on diacetyl formation (Conde and Mascort, 1981; Morrison and Suggett, 1983; Debourg *et al.*, 1991) and on the production of 4-vinylguaiacol (Tubb *et al.*, 1981).

Many strains of *Saccharomyces* contain cytoplasmic linear dsRNA molecules enclosed in virus-like particles. There are two main varieties, L-dsRNA, which is present in most yeast strains including some brewing yeasts (Kreil, Kleber and Teuber, 1975; Young, 1981), and M-dsRNA, which is present only in killer strains of *Saccharomyces* (Wickner, 1983). L-dsRNA provides the code for the capsid protein of the virus particles and M-dsRNA encodes both killer toxin and the immunity factor which prevents self-killing. Cells of killer strains normally contain about 12 copies of M-dsRNA and about 100 copies of L-dsRNA. They can be cured of M-dsRNA by growth at elevated temperature or by treatment with cycloheximide (Fink and Styles, 1973). Only one brewing killer yeast has ever been found (Young and Yagiu, 1978), brewing yeasts generally being sensitive to the action of killer toxin.

3.3 THE NEED FOR NEW BREWING YEASTS

3.3.1 The nature of a 'good' brewing yeast

In general the aim of brewing, like any other industrial process, is to obtain the highest efficiency in the use of raw materials and production plant without distorting the quality of the end-product. One way of achieving this in the case of beer production is by breeding more efficient yeast strains. Breeding programmes also provide scope for the production of completely new products from brewing yeasts.

The properties which are of most importance for a 'good' brewing yeast are:

1. a rapid fermentation rate without excessive yeast growth;
2. an efficient utilization of maltose and maltotriose with good conversion to ethanol;
3. an ability to withstand the stresses imposed by the alcohol concentrations and osmotic pressures encountered in breweries;
4. a reproducible production of the correct levels of flavour and aroma compounds;
5. an ideal flocculation character for the process employed;
6. good 'handling' characteristics (e.g. retention of viability during storage, genetic stability).

3.3.2 Brewing yeast improvement

There is considerable scope for improving many of the properties of brewing yeasts listed in the preceding section. In general, such improvements will either involve extending the biochemistry of the organisms, modifying their physiology or generating completely novel strains for employment in new technologies.

(a) Extension of yeast biochemistry

There are many ways in which the biochemistry of brewing yeasts could be extended to the advantage of the brewer.

The ability of brewing yeasts to secrete amylolytic enzymes

(glucoamylases, amylases, pullulanases etc.) would considerably improve the efficiency of utilization of wort carbohydrates. Amylolytic yeasts could be used to produce well-attenuated or low-carbohydrate beers and to reduce or replace the amounts of malt and enzymes needed to convert starch to fermentable sugars. Since *S. cerevisiae* strains secrete few extracellular proteins (Lampen, 1968), amylolytic yeasts would also provide the opportunity to obtain high-purity enzyme preparations without the need for elaborate processing.

The introduction of genes for proteolytic enzymes into brewing yeasts could provide opportunities for degrading haze-forming proteins and so naturally chill-proof beer. If the proteolysis was sufficiently specific to produce 'head-positive' polypeptides this would lead to an improvement in beer foam quality.

The secretion by yeast of a suitable endo-β-glucanase would enable barley β-glucans normally present in beer to be degraded during fermentation, and so prevent the development of hazes and gels. Removal of β-glucans would also improve the filterability of the beer at the end of the process.

By engineering into yeast cells the ability to utilize carbohydrates other than those normally metabolized, more complete degradation of conventional raw materials becomes possible together with the use of new raw materials. Enzymes for the degradation of pentoses and cellulose would increase the degree of utilization of usual brewery raw materials whilst enzymes for the degradation of lactose would permit the use of whey.

Most of these modifications are technically feasible and all have already been worked on in some detail (sections 3.4.5(e) and 3.4.5(f)).

(b) *Modification of yeast physiology*

Many possibilities are available for introducing new characteristics into yeast cells in order to improve their fermentation performance. The ability to predictably change the flocculation ability of a strain would enable it to be used in a number of very different fermentation systems. Similarly, changes in cell adhesion could be useful if yeasts are to be used in immobilized cell systems. Fermentations at high specific gravities require strains with improved tolerance to ethanol and osmotic pressure. Yeasts capable of fermenting at higher temperatures, while still producing acceptable beers, would lead to reductions in cooling costs, higher productivities and a reduced risk of microbial contamination. An increased tolerance to pressure and carbon dioxide would permit fermentations to be carried out in very large, deep fermentors or in pressurized vessels. Other possibilities for the improvement of brewing yeasts include increasing fermentation rates (possibly by increasing the dosage of the genes involved in maltose utilization), improving fermentation efficiency by reducing their specific growth yields, improving control of both the type and amount of flavour compounds produced and reducing microbial contamination problems by producing yeasts with anti-contaminant properties.

Many of these potential improvements are not well defined either biochemically or genetically and so are not amenable to a direct genetic approach. Those that are better understood such as flocculation, flavour production and microbial contamination have again been worked on in some detail (sections 3.4.3(b), 3.4.5(g), 3.4.5(h) and 3.4.5(i)).

(c) Novel strains for new technologies

This area represents the most difficult one for brewing geneticists but is potentially the most exciting. Yeasts with a restricted pattern of fermentation could be engineered in order to produce a low-alcohol beer. Another worthwhile target would be to introduce into yeasts genes for the production of new flavour compounds in order to produce beers with unusual flavours. Finally there is the concept of using spent brewer's yeast to produce valuable by-products such as vitamins, enzymes, lipids etc., their formation being enhanced by suitable genetic modification. Alternatively brewing yeasts could be genetically developed to produce completely new products after they have served their role in the brewery. The possibilities here are limitless. The concept has already been tried for the production of human serum albumen (Hinchliffe, Kenny and Leaker, 1987) but because of both supply and legislative difficulties the use of spent brewing yeast has been abandoned in favour of specially grown yeast for this application. With this one exception, very little effort has so far been put into exploring the possibilities offered by the production of novel yeasts for new technologies.

3.4 GENETIC TECHNIQUES AND THEIR APPLICATION TO THE ANALYSIS AND DEVELOPMENT OF BREWING YEAST STRAINS

3.4.1 Mutation and selection

A simple and direct means for strain development is to isolate mutants, with or without prior mutagenesis. Variation is always present within a yeast population and successful isolation of mutants depends on the frequency with which they occur and the ability of the experimenter to devise appropriate selection procedures. One spontaneous mutation frequently encountered in breweries is that of changing yeast flocculation characteristics. These changes have enabled bottom-fermenting ale yeasts, for use in cylindroconical fermenting vessels, to be selected from their top-cropping predecessors. Continuous culture can be readily used to isolate variants within a yeast population without prior mutagenesis. This was demonstrated by Thorne (1970) when he grew an apparently stable lager yeast in this way for 9 months. After this time more than half of the yeasts isolated showed a difference in at least one property of brewing significance. Selection in continuous culture has also been used to obtain yeasts with improved ethanol tolerance (Brown and Oliver, 1982; Korhola, 1983).

The use of mutagens increases the proportion of mutants within a given population and has proved surprisingly successful with polyploid brewing yeasts. Molzahn (1977) mutated both ale and lager yeasts using *N*-methyl-*N*-nitro-*N*-nitrosoguanidine and ultraviolet light. By selecting for valine and methionine auxotrophs he successfully isolated a number of strains which respectively produced less diacetyl and hydrogen sulphide, both important negative flavour attributes of beer. Flocculation variants were also obtained in the same series of experiments. In other experiments, strains with a reduced ability to form esters and higher alcohols have been isolated as spontaneous mutants showing resistance to glucosamine (Hockney and Freeman, 1980). Mutants resistant to thia-isoleucine have been obtained which produce beer containing higher levels of amyl alcohol (Kielland-Brandt, Peterson and Mikkelsen, 1979). Although successful, the isolation of mutants in most of these experiments was too frequent to be explained by simultaneous inactivation of several copies of wild-type genes in polyploid brewing yeasts. Other factors such as mitotic recombination and chromosome loss must also be involved (see also section 3.2.2).

The selection of mutants resistant to the glucose analogue 2-deoxyglucose (DOG) has proved extremely valuable when applied to brewing yeasts (Jones, Russell and Stewart, 1986). In normal fermentations, due to catabolite repression, maltose is only taken up by yeast when about half of the glucose present in the wort has been metabolized. Mutants resistant to the action of DOG are derepressed, utilizing maltose and glucose simultaneously, and have improved fermentation characteristics.

Mutation methods can also be used to produce genetically marked yeast strains for further analysis. Gjermansen (1983) used ultraviolet light and ethylmethane sulphonate mutagenesis to induce auxotrophic mutations in meiotic segregants of a production lager yeast. These mutants were then used as part of a breeding programme to produce new production strains (section 3.4.2) and for detailed genetic analysis of the lager yeast (section 3.4.3).

3.4.2 Hybridization

Classically, hybridization of strains of *S. cerevisiae* involves mating haploids of opposite mating type to give a heterozygous diploid. Recombinant meiotic progeny are recovered by sporulating the diploid and recovering the individual haploid spores. This is the basis for the classical approach to genetic mapping, called tetrad analysis because of the four (tetrad) spores found in each ascus formed by a sporulating diploid cell. Detailed analysis of the results of crosses between haploid yeasts enables individual genes to be mapped to particular chromosomes and to particular parts of these chromosomes. A more detailed description of these and other mapping procedures can be found in Sherman and Wakem (1991).

As already mentioned in sections 3.2.2 and 3.2.3, brewing yeasts are polyploid and sporulate very poorly. Consequently, the use of hybridization techniques for strain development has proved difficult. Several workers have persevered and managed to isolate meiotic segregants from brewing yeasts which can then be used in breeding programmes.

Gjermansen and Sigsgaard (1981) carried out pairwise crosses between segregants obtained from one production yeast. Although most hybrids were inferior to the parent yeast, one with good brewing properties was obtained. This strain has been used at full production scale where it fermented faster, flocculated better and produced a better flavoured beer than did the normal production yeast (von Wettstein, 1983). The segregants were originally thought to be haploid (Gjermansen, 1983) but more detailed genetic analysis (section 3.4.3) has revealed that, despite their lower ploidy, they are almost as genetically complex as their polyploid parent.

An alternative approach was taken by Bilinski, Russell and Stewart (1987b) who obtained meiotic segregants from both ale and lager yeasts. By mating these, a species hybrid was produced which had both ale and lager yeast characteristics since it could both grow at 37°C and utilize the sugar melibiose. The fermentation performance of the hybrid was as good as the parent lager yeast and the ethanol yield was better. The beer produced lacked the sulphury flavour characteristic of the lager yeast parent but had an estery note more typical of beers produced by the ale yeast.

Although such hybridization methods are less specific than recombinant DNA techniques (section 3.4.5), they may provide the best approach for improving yeast characteristics which are poorly defined genetically or which result from the complex interaction of many genes.

Because of their poor sporulating ability, it is hopeless to attempt to carry out tetrad analysis with brewing yeasts or their progeny. Consequently genetic analysis of the genes of importance in brewing yeasts has been carried out in model systems based on laboratory strains of *S. cerevisiae*. Genes controlling flocculation have been defined and mapped in this way (Lewis, Johnston and Martin, 1976; Russell *et al.*, 1980; Stewart and Russell, 1986) and gene dosage effects of the genes controlling maltose metabolism have been described in constructed polyploid strains (Mowshowitz, 1979; Stewart, Goring and Russell, 1977). Further, hybridization and tetrad analysis have been used to assign the ability of yeasts to decarboxylate cinnamic acid to a single dominant gene (*POF1*). This is important because cells containing this gene impart a phenolic flavour to beer if used for fermentation (Goodey and Tubb, 1982). The *POF1* gene has been cloned and partially characterized (Meaden and Taylor, 1991). The introduction of this cloned gene into brewing yeasts leads to the production of an aroma characteristic of phenolic off-flavour when the transformed strain is used to ferment wort. This strongly suggests that brewing yeasts do not normally produce a phenolic off-flavour when used to produce beer because of their lack of a functional *POF1* allele.

3.4.3 Rare mating

(a) Rare mating techniques

Brewing yeasts which fail to show a mating type can be mated with haploid strains by a procedure known as rare mating. A large number of cells of the parental strains are mixed together and a strong positive selection procedure is applied to obtain the rare hybrids formed (Gunge and Nakatomi, 1972). These hybrids are usually selected as respiratory-sufficient prototrophs from crosses between a respiratory-deficient mutant of the industrial strain and an auxotrophic haploid strain (Fig. 3.2).

The technique has been used to construct brewing strains with the ability to ferment dextrins (Tubb *et al.*, 1981). The beers produced by these hybrids had an unacceptable phenolic off-flavour because of the presence of the *POF1* gene in the diastatic parent yeast. Only after the gene had been eliminated from this latter yeast by conventional hybridization techniques were satisfactory diastatic hybrids produced (Goodey and Tubb, 1982). Using this approach, followed by classical genetic improvement procedures, mitotically stable brewing yeasts capable of superattenuation have been constructed (Guinova-Stoyanova and Lahtchev, 1991).

(b) Cytoduction

Rare mating can also be used to generate progeny (cytoductants or heteroplasmons; Fig. 3.2) which receive cytoplasm from both parents but retain the nucleus of only one. By using haploid maters which carry the *kar1* mutation (karyogamy-deficient; i.e. deficient in nuclear fusion) cytoduction can be favoured over hybrid formation during rare mating experiments. This technique has been used extensively to transfer the dsRNA determinants for yeast killer factors from laboratory strains into suitable recipient strains in order to produce brewing yeasts with anti-contaminant properties (Young, 1981, 1983; Hammond and Eckersley, 1984). In some cases, construction of killer yeasts by cytoduction impaired the brewing characteristics of the recipient strain probably as a consequence of introducing a 'foreign' mitochondrial genome (Hammond and Eckersley, 1984). It is clear that the source of mitochondria in a modified brewing yeast can have marked effects on fermentation rate, flocculation behaviour and beer flavour (Conde and Mascort, 1981; Russell, Jones and Stewart, 1985). Although in principle the manipulation of the mitochondrial genome by cytoduction provides a means for strain improvement, an attempt to correct a sugar fermentation problem using this approach was not successful (Hammond and Wenn, 1985).

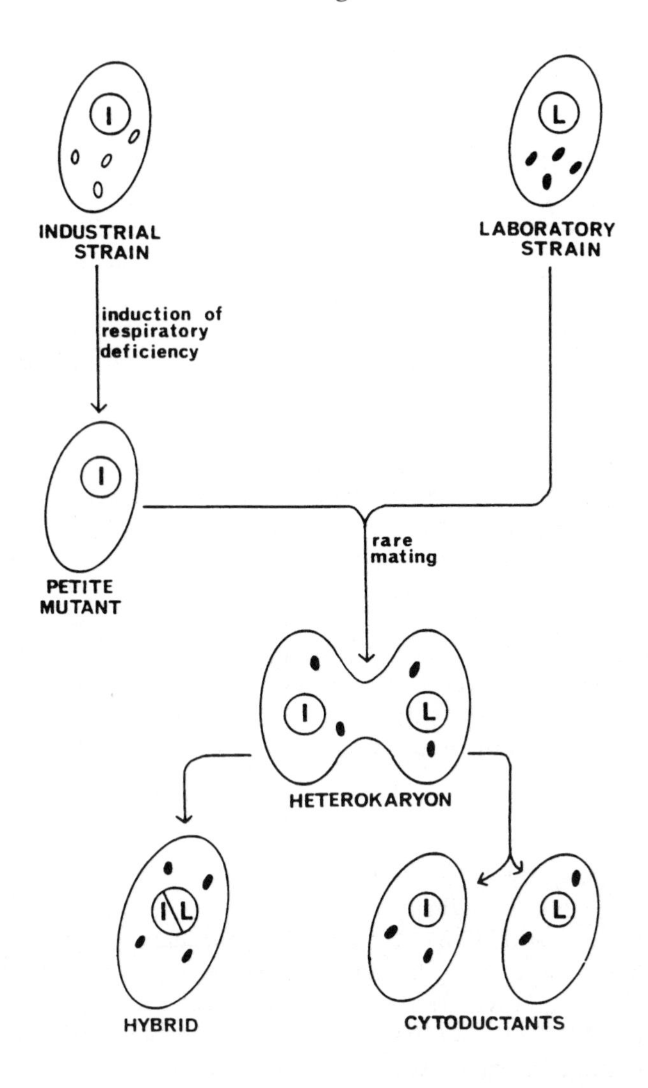

Fig. 3.2 Rare mating between industrial and laboratory strains of yeast.

(c) Single chromosome transfer

During matings between strains, one of which carries the *kar1* allele, occasional rare progeny are formed which contain the nuclear genome of one parent together with an additional chromosome from the other (Nilsson-Tillgren *et al.*, 1980; Dutcher, 1981, Goodey *et al.*, 1981). This phenomenon has been used to transfer single chromosomes from meiotic segregants of the lager brewing yeast strain M244 into laboratory yeasts, thereby allowing detailed genetic analysis of the brewing yeast chromosomes to be carried out. Chromosomes III, V, X, XII and XIII have been studied in this way.

Two forms of chromosome III have been identified. One is functionally and structurally identical to the chromosome III found in laboratory strains. The other chromosome type appears to be functionally homologous to the laboratory strain chromosome but is structurally different, recombination between the two chromosome types only occurring in certain regions (Nilsson-Tillgren *et al.*, 1981; Holmberg, 1982). The presence of two forms of chromosome III has been confirmed by an examination of the structure of the *LEU2* gene, which is found on this chromosome. Two different *LEU2* genes are present in the lager strain (Pedersen, 1985b) and one *LEU2* gene cloned from the brewing yeast has a restriction endonuclease map considerably different from that of classical *S. cerevisiae* strains (Casey and Pedersen, 1988.)

In a similar manner two forms of chromosome V have been identified, one of which is both structurally and functionally homologous to the chromosome found in laboratory strains of *S. cerevisiae*. The other fails to recombine at all despite being functionally homologous (a so-called homeologous chromosome; Nilsson-Tillgren *et al.*, 1986). Chromosome X has been examined using pulsed-field electrophoresis (section 3.4.5(j)) and three independently migrating structures identified, two of which are not homologous to chromosome X from *S. cerevisiae*, recombination occurring along only part of the chromosome (Casey, 1986a,b). The third chromosome, although the same length as that from laboratory strains, has not been examined in any detail. Petersen *et al.* (1987) analysed both chromosomes XII and XIII. They were able to isolate two different types of chromosome XII, one homologous and the other homeologous to the equivalent chromosome from laboratory strains. In contrast three different chromosomes XIII were isolated, one showing homology to that of *S. cerevisiae*, a non-recombining homeologous chromosome and a mosaic-type chromosome exhibiting partial recombination.

Thus there appear to be three types of chromosomes present in lager brewing yeast M244: chromosomes which are functionally and structurally homologous to the equivalent *S. cerevisiae* chromosome, represented by one form of chromosomes III, V, XII and XIII; homeologous chromosomes which are functionally but not structurally homologous to the equivalent *S. cerevisiae* chromosomes, represented by one form of chromosomes V, XII and XIII; mosaic chromosomes where recombination with *S. cerevisiae* chromosomes is limited to only parts of the chromosomes, represented by one form of chromosomes III and XII and both identified forms of chromosome X. Clearly lager yeast M244 has two recombination-incompatible genomes. The question of the copy number of each version of the various chromosomes has been examined by means of the *ILV2* locus found on chromosome XIII. *ILV2* gene copies were replaced by *ilv2* deletion mutants. It was found that the brewing yeast contained two copies of each of the two versions of the *ILV2* region (Gjermansen *et al.*, 1988). If this can be applied generally, then the brewing yeast would appear to be allotetraploid, although the alloploidy is clearly irregular with many mosaic chromosomes occurring. In addition, some chromosomes apparently lack certain genes (Nilsson-Tillgren *et al.*, 1986; Casey, 1986b). If all brewing yeasts are allotetraploids with irregular chromosome structures and missing or defective genes this may help to explain their low sporulation frequencies and low spore viabilities.

3.4.4 Spheroplast fusion

Spheroplast fusion provides another way of overcoming the mating/sporulation barrier found in industrial yeasts but, unlike rare mating, it is a direct asexual technique. It can be used to produce both hybrids and cytoductants. The procedure, first described by van Solingen and van der Plaat (1977), is outlined in Fig. 3.3. Spheroplasts are formed by removing the cell wall with a suitable lytic enzyme, in a medium containing an osmotic stabilizer (often 1.0 M sorbitol) to prevent cell lysis. Spheroplasts from different strains are mixed together in the presence of polyethylene glycol and calcium ions and then, after fusion has taken place, allowed to regenerate their cell walls in an osmotically stabilized agar medium. Cells can also be induced to fuse by the application of a high-intensity electric field as described by Halfmann *et al.* (1982).

The major problem with spheroplast fusion is that a wide range of cell types is produced which are often mitotically unstable, especially in the case of inter-species and inter-generic fusants. Nevertheless the technique has been applied to the development of brewing yeasts. Dextrin-fermenting strains have been constructed by fusing together *S. cerevisiae* var. *diastaticus* and brewing yeasts (Barney, Jansen and Helbert, 1980; Freeman, 1981; Russell, Hancock and Stewart, 1983; de Figueroa and de van Broock, 1985; Hansen, Rocken and Emeis, 1990). As with rare mating studies, the production of phenolic off-flavours was a problem until the *POF1* gene was eliminated by preliminary hybridization of the diastatic strain. Spheroplast fusion has also been used to create yeast strains with enhanced tolerances to ethanol, temperature and osmotic pressure (Panchal, Peacock and Stewart, 1982; Crumplen *et al.*, 1990). The fusion of a brewing yeast with a *FLO5* flocculent laboratory strain has been reported (Urano, Nishikawa and Kamimura, 1990). The fusants were intermediate between the two parents with respect to carbohydrate assimilation and flocculation. A further application of the technique has been to produce lager brewing yeasts with anti-contaminant properties; either producing yeast killer factor (Rocken, 1984) or possessing both anti-yeast and anti-bacterial properties (Sasaki *et al.*, 1984). The latter was achieved by mating a laboratory killer strain of yeast with a strain having anti-bacterial activity and then fusing the hybrid with a brewing yeast. The constructed yeast produced poor quality beer,

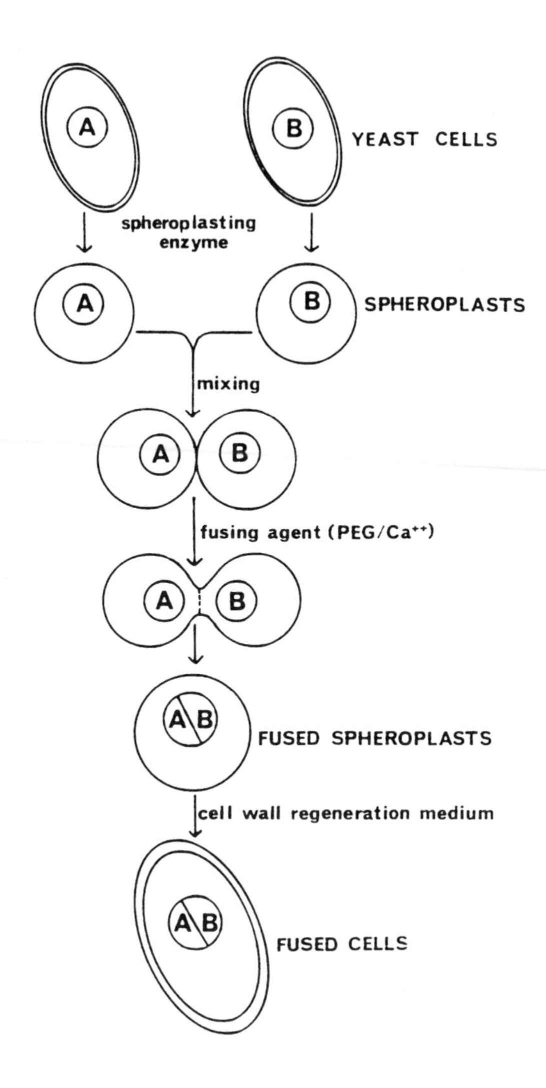

Fig. 3.3 Spheroplast fusion applied to yeasts.

probably because of the large contribution to its genome from the non-brewing strains, always a problem with this type of experimental system.

Of course, the production of inter-generic hybrids can only be carried out using spheroplast fusion. Stable fusion hybrids between *S. cerevisiae* and *Yarrowia lipolytica* (de van Broock, Sierra and de Figueroa, 1981) or *Candida utilis* (Perez, Vallin and Benitez, 1984) have been reported whereas fusion products obtained between *S. cerevisiae* and *Kluyveromyces lactis* degenerate rapidly into strains showing parental characteristics (Stewart, Russell and Panchal, 1981).

3.4.5 Transformation and recombinant DNA methods

(a) Transformation methods

Transformation involves the use of DNA molecules to bring about a change in a recipient organism. Because the mating system is bypassed the technique can readily be applied to industrial yeasts. Following early unconfirmed reports of transformation using native DNA (Oppenoorth, 1959; Khan and Sen, 1974), an unequivocal demonstration of yeast transformation awaited the availability of recombinant DNA techniques. Using this approach (Fig. 3.4), DNA is fragmented using an endonuclease (restriction enzyme). The fragments are then annealed to plasmid DNA which has been linearized, usually with the same restriction enzyme, after which recombinant plasmids are generated by 'stitching up' the joins using a DNA ligase. In this way a pool of recombinant plasmids is produced, each containing a small piece of the original donor DNA. A sufficiently large random pool in which there is a high probability of finding any single gene from the donor strain is called a gene 'bank' or 'library' (Clarke and Carbon, 1979). A number of approaches can be used to isolate or 'clone' individual genes from such a bank. Details of the

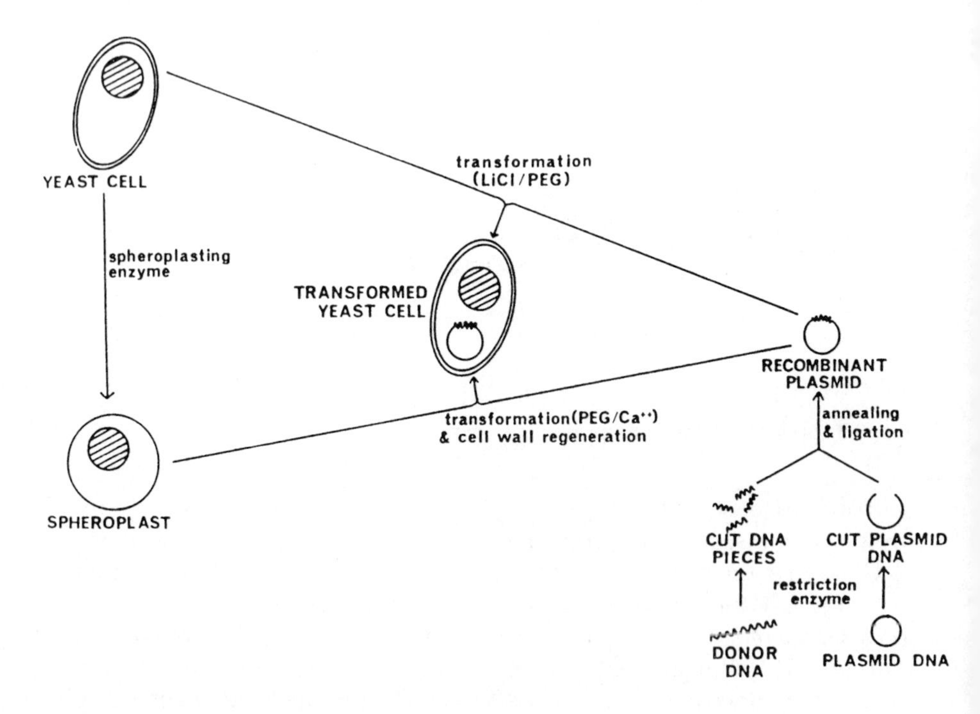

Fig. 3.4 The methods used for yeast transformation.

procedures used in recombinant DNA research can be found in Guthrie and Fink (1991).

There are various ways in which yeast cells can be induced to take up DNA from the medium in which they are suspended. One technique involves removing the cell walls using a spheroplasting enzyme as described in section 3.4.4 (Hinnen, Hicks and Fink, 1978), whilst another involves treatment of the cells with lithium or caesium salts (Ito *et al.*, 1983). In both cases the yeasts and DNA are mixed under conditions similar to those used for spheroplast fusion. Alternatively DNA can be induced to enter yeast cells by the application of an electric current, a technique known as electroporation (Becker and Guarente, 1991).

The single chromosome transfer procedure (section 3.4.3(c) has also been used to transfer plasmids from laboratory strains of yeast carrying the *kar1* mutation into brewing yeast strains. This provides an alternative transformation method to those involving isolated DNA molecules (Navas, Esteban and Delgado, 1991).

(b) Dominant selection markers

Since the first demonstration of transformation of yeast, a wide range of plasmid vector systems have become available and many have been applied to the production of new industrial yeast strains. Since brewing yeasts are polyploid, it is extremely difficult to generate the auxotrophic mutants normally used for selection purposes in transformation experiments involving laboratory strains. Consequently the plasmids used for transformation experiments with brewing yeast strains have to carry a dominant marker which is selectable against the wild-type polyploid background. The most widely used dominant selection markers are described below. Other selectable markers with potential for use with industrial yeasts have been discussed elsewhere (Knowles and Tubb, 1986; Esser and Kamper, 1988).

Copper resistance, coded by the *CUP1* gene, is probably the most popular selection method used in transformation experiments with industrial yeast strains (Henderson, Cox and Tubb, 1985; Meaden and Tubb, 1985; Hinchliffe and Daubney, 1986). The *CUP1* gene is derived from *S. cerevisiae* and is found in many wild and laboratory strains. Selection is possible since almost all brewing yeast strains are sensitive to copper. Brewing yeasts transformed with plasmids containing the *CUP1* gene ferment at the same rate as untransformed yeasts and produce beer which is indistinguishable from that normally produced (Meaden and Tubb, 1985).

Sacchararomyces yeasts are sensitive to geneticin (G418), an antibiotic which is structurally related to gentamycin and kanamycin. A bacterial gene which encodes an inactivating enzyme (aminoglycoside phosphotransferase) is expressed in yeast, thus resistance to G418 can be used as a selectable marker (Jimenez and Davies, 1980). A disadvantage of this system is that the resistance factor is a bacterial gene which must be eliminated before the transformed strains can be used for beer production. Various strategies enable this to be done (section 3.4.5(e)).

A more recently described dominant selection marker is that for resistance to the herbicide sulphometuron-methyl (*SMR1*; Casey, Xiao and Rank, 1988a). Like *CUP1*, it is derived from yeast and has no significant effect on fermentation behaviour. The gene is in fact a mutant allele of the *ILV2* gene of *S. cerevisiae* which encodes the enzyme acetolactate synthase. It is particularly useful where integration of a new gene into the yeast chromosome is required since it is not only a dominant selectable marker but also provides a site of integration in the *ILV2* gene (Casey, Xiao and Rank, 1988c; Xiao and Rank, 1989).

Resistance to chloramphenicol has also been used for selecting transformants of industrial yeasts (Hadfield, Cashmore and Meacock, 1986). Since this resistance is a mitochondrial function it can only be used when yeast cells are grown on non-fermentable substrates. However, with this proviso, it is very effective and has been successfully used for transformation of brewing yeasts. Like G418 resistance, chloramphenicol resistance is coded by a bacterial gene and so must be eliminated prior to commercial use of the transformed yeast.

Since ale yeast strains cannot use melibiose as a carbon source it is possible to use the acquisition of this ability as a selection system in such yeasts (Tubb *et al.*, 1986). However, the melibiose utilization gene, *MEL1*, encodes an extracellular enzyme and so there is a problem of Mel^+ strains cross-feeding adjacent Mel^- strains, making selection more difficult. Direct selection of dextrin-utilizing yeasts on starch plates suffers from the same problem (Meaden *et al.*, 1985).

(c) Gene stability and expression

The various recombinant DNA methodologies, such as the use of multi-copy plasmids based upon 2 µm DNA, single copy centromere-containing plasmids or integrative transformation, have been reviewed recently (Knowles and Tubb, 1986; Esser and Kamper, 1988). Most initial transformation experiments with brewing yeasts used multi-copy plasmids such as pJDB207 (Beggs, 1981) containing yeast chromosomal DNA, part of the yeast 2 µm plasmid and DNA from a bacterial plasmid. The presence of bacterial DNA provides the opportunity to replicate the plasmid in *Escherichia coli*, without which transformation experiments would be very tedious. These multi-copy plasmids have the advantage that, due to the number of copies present in the cell, there is usually good expression of the gene. This is, however, sometimes disadvantageous since overproduction of gene product can have deleterious effects on fermentation performance (Perry and Meaden, 1988). A further shortcoming is that such plasmid-based systems are relatively unstable, the desired trait gradually being lost with successive fermentations (Meaden and Tubb, 1985; Cantwell *et al.*,

1985). In contrast, when the required gene is integrated into the yeast chromosome resulting transformants are extremely stable, although the level of expression may be much reduced because of lower gene copy number (Enari *et al.*, 1987). An alternative, intermediate system has been developed which has the expression advantages of a multi-copy plasmid but is less unstable than the typical bacterial–yeast hybrid plasmid (Hinchliffe, Fleming and Vakeria, 1987). All brewing yeasts contain 2 μm DNA which is stably maintained by a highly developed maintenance system. This stability is lost when bacterial sequences are present in the plasmids. By inserting the gene of interest directly into the endogenous 2 μm plasmid, a high copy number transformant can be obtained which is considerably more stable than conventional chimaeric plasmids. Similar effects can be achieved using so-called 2 μm 'disintegration vectors' which contain both yeast and bacterial DNA but which are constructed in such a way that the bacterial sequences are eliminated once they have entered and transformed a yeast cell (Chinery and Hinchliffe, 1989).

Although transformation of brewing strains with native DNA has been reported (Jansen *et al.*, 1979; Barney, Jansen and Helbert, 1980), the use of recombinant methods offers a much higher degree of specificity together with the possibility of amplifying the amount of a gene product made by a yeast strain. Such amplification can be achieved by inserting the gene of interest into a multi-copy plasmid, by integrating the gene at several sites within the yeast chromosomes or by attaching a highly efficient promoter sequence to the gene in order to increase its expression. This last approach is the one that has now been almost universally adopted. Typical highly efficient promoters are those that are involved in controlling the genes for enzymes of the glycolytic pathway, such as alcohol dehydrogenase (*ADH1*; Hitzeman *et al.*, 1981), glyceraldehyde-3-phosphate dehydrogenase (*GAPDH*; Montgomery *et al.*, 1980) and phosphoglycerokinase (*PGK1*; Tuite *et al.*, 1982). As well as using efficient promoters, it is also possible to attach regulated promoters to genes, so that they can be switched on and off in response to external factors. This has potential for controlling accurately the levels of gene expression.

(d) Application of recombinant DNA methods to brewing yeasts

Although a relatively recent technology, recombinant DNA methods have already been used to introduce several new properties into brewing yeasts. These have included the ability to degrade various carbohydrates and proteins as well as modifications to the processes responsible for beer flavour production and flocculation during fermentation. A number of criteria must be met before such strains can be considered for commercial production of beers for sale. The yeasts must be stable under the conditions of fermentation normally employed and must produce sufficient new product in order that the desired phenotypic change is realized. On the other hand, the fermentation behaviour of the yeast and the quality of the product must remain unaffected except, of course, for the desired new attribute. Finally it is desirable that the new strain should contain no DNA other than that necessary for generation of the new property. This will minimize any objections from the regulatory authorities (the topic of regulation is covered in more detail in section 3.5).

(e) Brewing yeasts able to use a wider range of carbohydrates

Glucoamylases

Brewing yeast strains are normally only able to utilize mono-, di- and trihexoses. Larger oligomers (dextrins), which represent about 25% of wort sugars, are not metabolized. Yeasts able to ferment these additional sugars convert raw materials to ethanol much more efficiently than do brewing yeasts. They can be used to produce low carbohydrate diabetic beers or, by dilution to normal ethanol contents, low calorie 'light' beers are produced. Currently such beers are often produced by adding commercial enzymes, either prior to wort boiling or during fermentation. The commercial enzymes used are typically fungal glucoamylases from *Aspergillus niger*.

A yeast capable of producing its own glucoamylase has several advantages. Enzyme addition is much more controllable, mistakes in addition rates are eliminated and deleterious enzyme activities which are often present in commercial enzyme preparations are avoided. Consequently there has been much interest in the generation of brewing yeasts capable of fermenting dextrins. The limited success achieved using rare mating and spheroplast fusion has already been described (sections 3.4.3(a) and 3.4.4). Much greater success has been achieved utilizing recombinant DNA methods by transforming yeast strains with a gene for glucoamylase (α-1,4-glucan glucohydrolase, EC3.2.1.3), an enzyme capable of degrading the normally unfermentable wort dextrins to glucose. A gene for this enzyme (*DEX1* or *STA2*) was cloned from *S. cerevisiae* var. *diastaticus* (Meaden *et al.*, 1985) and inserted into a multi-copy plasmid containing both bacterial and yeast DNA sequences including the *CUP1* selectable marker. The plasmid was used to transform both ale and lager brewing yeast strains (Meaden and Tubb, 1985). The transformants produced extracellular glucoamylase but, because of the lack of α-1,6 debranching activity, superattenuation was limited. As expected for a chimaeric multi-copy plasmid, some degree of plasmid instability was experienced. Additionally, growth and fermentation rates were adversely affected by the production of excess glucoamylase (Perry and Meaden, 1988). The plasmid instability problem was largely overcome by the development of an all-yeast multi-copy plasmid (Vakeria and Hinchliffe, 1989) but the fermentation rate of the transformant was still somewhat slower than that of the parent strain.

More recently another brewing yeast has been transformed using this same plasmid construct. The modified yeast fermented at about the same rate as the parent strain but superattenuated the wort to produce approximately 1% (v/v) more ethanol (Hammond, 1995). The new yeast was judged to be commercially stable in that, although it slowly lost the *STA2* gene, this had no noticeable effect on superattenuation or beer flavour over ten successive fermentations. Under normal circumstances a newly propagated yeast would be introduced before any significant effect on yeast performance would be noticed. Beers have been successfully produced using this yeast at the small commercial (100 hl) scale.

The *STA1* gene, which is closely related to *STA2*, has also been cloned into multi-copy plasmids and used for transformation of brewing yeasts, using both G418 resistance (Sakai *et al.*, 1989) and copper resistance (Park *et al.*, 1990) as the selection systems. Plasmid stability and fermentation performance were comparable to those achieved with *STA2* transformants. Only when the gene was integrated into the yeast chromosome was a stable transformant obtained (Park *et al.*, 1990). A further development has been the transformation of brewing yeasts with a centromeric plasmid containing both *STA2* and the α-amylase gene *AMY* from *Bacillus amyloliquefaciens* (Steyn and Pretorius, 1991). The transformed yeasts were capable of secreting both enzymes and carried out an efficient one-step utilization of starch.

The major problem with all systems based on glucoamylases derived from *S. cerevisiae* var. *diastaticus* is the lack of α-1,6 debranching activity which leaves considerable amounts of wort dextrins undigested. This has been overcome by using genes for glucoamylases from other microbial sources. The *GA* gene from *Aspergillus niger*, which encodes a glucoamylase with both α-1,4 and α-1,6 hydrolytic activities, has been cloned and inserted into an integrating vector (Yocum, 1986). The vector was designed to achieve integration of the *GA* gene into the yeast genome whilst ensuring that any bacterial DNA sequences were eliminated from the transformants (Fig. 3.5). In order to ensure genetic stability three copies of the *GA* gene were integrated into three different copies of the yeast *HO* gene. Good expression was achieved by using a yeast glycolytic enzyme promoter and good secretion was obtained by using a yeast secretion signal sequence. Fermentation trials at scales up to 100 hl have been extremely successful; high levels of secreted extracellular glucoamylase have been achieved and superattenuated beer of good quality has been produced on a regular basis (Gopal and Hammond, 1992).

One disadvantage of the glucoamylase from *A. niger* is that it is relatively heat stable and retains activity in beer after pasteurization. Consequently there has been some interest in the glucoamylase from *Schwanniomyces occidentalis* which, although not currently used commercially, has the advantage of being much less heat stable than the fungal enzyme, as well as possessing debranching activity. The *GAM1* gene for this enzyme has been cloned (Lancashire *et al.*, 1989; Dohmen *et al.*, 1990) and, for transforming brewing yeast strains, placed under the control of the *Saccharomyces ADH1*

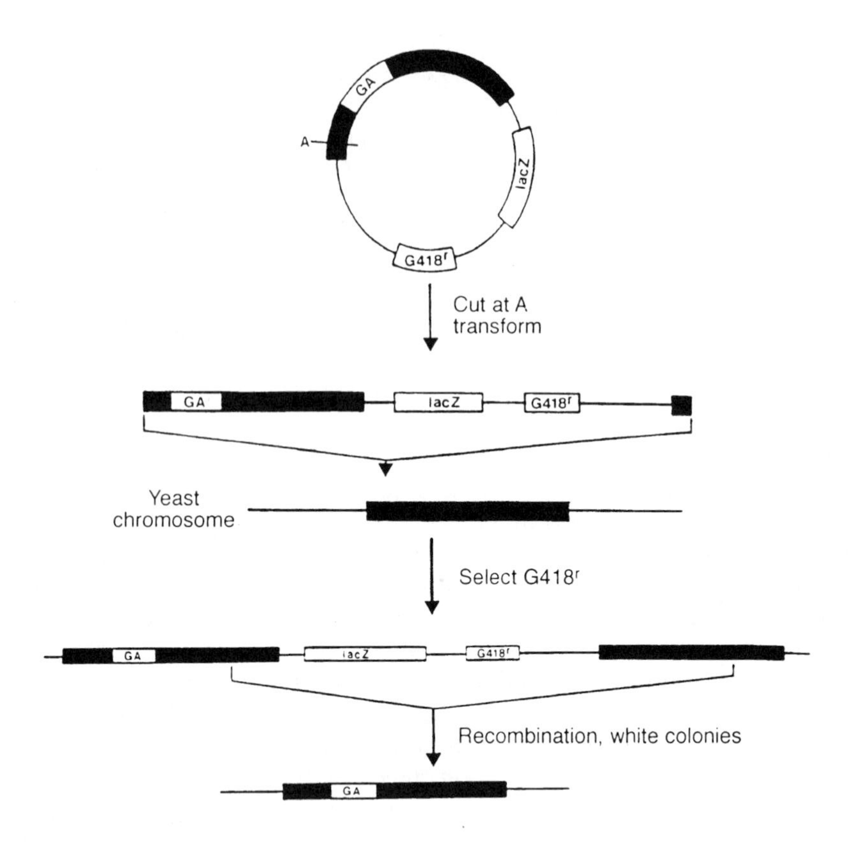

Fig. 3.5 Transformation protocol for achieving stable integration of *Aspergillus niger* glucoamylase gene into brewing yeasts. The *GA* gene is inserted into a copy of the redundant yeast *HO* gene contained on a plasmid which also possesses copies of the bacterial *LacZ* and G418 resistance genes. The lack of a yeast origin of replication ensures transformation by integration. The plasmid is cut at A and used to transform brewing yeasts. Transformants are selected by their ability to grow in the presence of G418. Recombination between the two copies of the *HO* gene leads to either complete loss of the insert or loss of all extraneous bacterial DNA with retention of the *GA* gene. These recombination events can be detected by growing the cells on a medium containing X-gal (5-bromo-4-chloro-3-indolyl-β-D-galactopyranoside). Initial transformants produce blue colonies because of the presence of the *LacZ* gene. Once this has been lost the colonies become white. By screening white colonies for glucoamylase activity the desired transformants can be isolated.

promoter (Lancashire *et al.*, 1989). Vectors for both multi-copy and integrative transformation have been constructed using *SMR1* as the selectable marker. Both types of transformant produced glucoamylase, the fermentations proceeding normally to a lower gravity than usual to produce acceptable beers.

β-Glucanases

Barley β-glucanase is heat labile and easily destroyed during malting. As a result, barley β-glucans are often found in beer, where they present filtration difficulties and can give rise to hazes, sediments and gels. To overcome this problem, microbial β-glucanases are often added to mashes or worts. Consequently there has been considerable interest in transferring genes coding for β-glucanase into brewing yeasts. Genes have been cloned from *Bacillus subtilis* (Cantwell and McConnell, 1983; Hinchliffe and Box, 1984; Lancashire and Wilde, 1987), from *Trichoderma reesei* (Knowles *et al.*, 1985) and from barley (Jackson, Ballance and Thomsen, 1986). Initial attempts to express the *B. subtilis* gene in yeast were disappointing, yielding only low levels of intracellular activity (Hinchliffe and Box, 1985). Expression was increased dramatically by placing the β-glucanase gene under the control of the yeast *ADH1* promoter (Cantwell *et al.*, 1985). Transformed brewing yeasts produced high levels of β-glucanase intracellularly but only very low levels were detected in the beer, probably due to cell lysis. Only when a yeast secretion signal sequence was also introduced (Lancashire and Wilde, 1987) were significant levels of extracellular β-glucanase obtained. For this last work the bacterial gene was placed under the control of the yeast *MFα* promoter and the α mating factor signal sequence. Selection of transformants was based on chloramphenicol resistance. The fermentation performance of the transformed yeasts was identical to that of the parent brewing yeast. However, the β-glucan contents and viscosities of the experimental beers were considerably reduced and their filterabilities much improved.

All reported work involving transformation of brewing yeasts with genes for bacterial glucanases has involved the use of multi-copy plasmids with the inevitable instability problems. However it did seem that instability was much greater in transformed ale yeasts than in lager strains (Cantwell *et al.*, 1985). This was confirmed by Perry and Meaden (1988) using different plasmids. If this is generally true it supports the hypothesis that the two yeast types are fundamentally different (see section 3.4.5(j)).

In order to ensure adequate expression in transformed yeasts, the β-glucanase gene from *T. reesei* was placed in the yeast *PGK* expression cassette which was in turn inserted into both multi-copy and integrating plasmids (Enari *et al.*, 1987). Specific yeast secretion signal sequences were not required since fungal extracellular enzymes, unlike their bacterial equivalents, are efficiently excreted by yeasts. Selection of transformants was based on copper resistance. With both types of transformant, wort β-glucans were effectively degraded and beer filterability was much improved whilst fermentation and beer characteristics were largely unaltered (Enari *et al.*, 1987; Penttila *et al.*, 1987). Although the multi-copy plasmid constructs were more unstable than the integrated transformants, they produced considerably higher β-glucanase activities.

The β-glucanase from barley, which is normally responsible for β-glucan degradation during brewing, shows highest activity at a pH of 4.7, in contrast to the bacterial enzyme which is most active at pH 6.7. It is thus better suited to the conditions found during beer production and is more appropriate for introduction into brewing strains. In order to clone the gene for this enzyme, two separate DNA sequences isolated from barley aleurone, which together encompassed the barley β-glucanase gene, were joined together. The gene was then fused to a mouse α-amylase secretion signal sequence and inserted into an expression vector between yeast *ADH1* promoter and terminator sequences (Jackson, Ballance and Thomsen, 1986). Subsequently an integrating vector was constructed along the same lines as that used for expression of *A. niger* glucoamylase (see Fig. 3.5). Using resistance to G418 as the selection marker, a number of yeast transformants were produced and all produced extracellular β-glucanase (Thomsen, Jackson and Brenner, 1988). A similar integrative transformant of a brewing yeast has been used for pilot-scale beer production. The β-glucan content of the beer was much lower than normal, flavour and foam stability were unaffected but filtration rates were markedly increased (Berghof and Stahl, 1991).

The cloning of the barley gene opens up a host of possibilities outside yeast transformation. For instance, the gene could be changed *in vitro* to produce a more heat-stable enzyme which could then be reintroduced into barley. The new enzyme would be more able to survive the temperatures found during malting and mashing and would probably eliminate the need for β-glucanase-producing yeasts. The techniques for the manipulation of barley in this way have only recently been developed (Mannonen *et al.*, 1993) and, as yet, no barley plants producing recombinant β-glucanases have been reported.

Other carbohydrases

Other carbohydrase genes which have been expressed in *S. cerevisiae* include cellobiohydrolases from *T. reesei* (Penttila *et al.*, 1988), α-amylase from *A. niger* (Knowles and Tubb, 1986), from *Schwanniomyces occidentalis* (Strasser *et al.*, 1989), from wheat (Lancashire, 1986), from rice (Kumagai *et al.*, 1990) and from barley (Sogard and Svensson, 1990), α-glucosidase from *Candida tsukubaensis* (Kinsella and Cantwell, 1991) and melibiase from *S. uvarum* (Post-Beittenmiller *et al.*, 1984; Liljestrom, 1985). The last enzyme, normally found only in lager yeasts, has been expressed in ale yeasts (Tubb *et al.*, 1986), thus enabling the melibiase pasteurization test (Enevoldsen, 1981, 1985) to be applied to ales as well as lagers.

(f) Brewing yeasts with proteolytic activity

So far the least successful application of recombinant DNA methods to yeast has been in the quest for proteolytic yeast strains. There is a need to reduce the level of polypeptides in beer to prevent the formation of hazes on chilling and during storage. Such haze stabilization is conventionally carried out using papain. A gene for an extracellular protease has been cloned and inserted into a centromeric plasmid (Young and Hosford, 1987). Transformation of a lager yeast strain was successfully carried out using G418 resistance as the selection procedure. However, although the fermentation characteristics of the new yeast were acceptable, the flavour of the beer produced was not satisfactory (Sturley and Young, 1988). The introduction of proteolytic activity into yeast would also raise concerns about the hydrolysis of foam-promoting polypeptides in wort and beer.

(g) Modification of flocculation properties of brewing yeasts

The flocculation of brewing yeasts is a complex phenomenon and the genetics involved are far from clear (Calleja, 1987). A number of genes (*FLO1*, *FLO5*, *FLO8* and *tup1*) have been identified as being associated with flocculation but their presence in brewing yeasts is still the subject of research interest. Stratford and Assinder (1991) examined 42 flocculent yeast strains and concluded that brewing yeasts showed two types of flocculation behaviour. The flocculation behaviour of bottom-fermenting lager strains and laboratory strains, containing the known flocculation genes, was clearly different from that of top-fermenting ale strains, suggesting that flocculation of ale strains may not be caused by the so-far identified *FLO* genes. Hybridization studies support the probable involvement of *FLO1* in the flocculation of bottom-fermenting yeasts (Watari *et al.*, 1991a). The *FLO1* gene product has been tentatively identified as a hydrophobic cell wall protein (van der Aar, Straver and Teunissen, 1993).

The *FLO1* gene has been cloned (Watari *et al.*, 1989), inserted into a multi-copy plasmid and used to transform both top- and bottom-fermenting non-flocculent yeasts. The transformed yeasts were more flocculent than their parents (Watari *et al.*, 1991b) but since the character was encoded on a multi-copy plasmid the new property was easily lost. Nevertheless the transformed yeasts fermented satisfactorily (Watari *et al.*, 1991a). The same gene has, more recently, been integrated into the genome of a non-flocculent bottom-brewing yeast (Watari *et al.*, 1994) and the modified yeast stably expressed flocculence much more strongly than did its parent strain.

(h) Yeasts with anti-contaminant properties

The breeding of killer yeasts by cytoduction (section 3.4.3(b)) has been superseded by another technique in which dsRNA is isolated from killer yeasts and inserted directly into protoplasts of *S. cerevisiae* by means of electroinjection (Salek, Schnettler and Zimmermann, 1990). The transformants stably maintain their newly obtained killer activity through many generations.

Killer wine yeasts have also been produced by more normal transformation procedures. A DNA copy of the K1 killer gene was cloned and placed into an *ADH1* expression cassette which was then integrated into the yeast genome. The parent wine yeasts used for these experiments already contained K2 dsRNA and so the resulting transformants expressed both K1 and K2 killer factors (Boone *et al.*, 1990). Such double killer strains cannot normally exist since the K1 and K2 dsRNA molecules appear to be mutually incompatible (Wickner, 1981). The transformants were stable and they fermented grape must in exactly the same manner as the parent yeast. Such transformants have not, as yet, been made from brewing yeasts but the same protocol should be equally applicable.

(i) Elimination of the necessity for flavour maturation of beer

Vicinal diketones, especially diacetyl, are a major off-flavour in finished beers. Diacetyl is formed during fermentation by the chemical oxidation of α-acetolactate, which diffuses from yeast cells into the fermenting wort. Its presence in green beer is the major reason for an expensive, time-consuming flavour maturation period at the end of fermentation. In order to accelerate beer production, various genetic approaches have been used to reduce or eliminate diacetyl formation. Since the conversion of α-acetolactate to diacetyl is the rate limiting step, the best approaches to this problem have involved either preventing α-acetolactate formation or removing it rapidly before it is converted to diacetyl.

The enzyme acetolactate decarboxylase (ALDC), which occurs in a number of bacteria, catalyses the conversion of α-acetolactate to acetoin, thus preventing its oxidation to diacetyl. The gene for this enzyme has been cloned from *Enterobacter aerogenes* (Sone *et al.*, 1987, 1988), from *Klebsiella terrigena* (Blomqvist *et al.*, 1991), from *Lactococcus lactis* (Goelling and Stahl, 1988), from *Acetobacter pasteurianus* (U. Stahl, personal communication) and from *Acetobacter aceti* var. *xylinum* (Yamano, 1994). In initial experiments the *E. aerogenes* gene was expressed off the yeast *ADH1* promoter on a multi-copy plasmid and this was used to transform brewing strains. In fermentation trials a considerable reduction in the level of diacetyl was observed when the modified yeast was used in both batch and continuous fermentations (Shimizu, Sone and Inoue, 1989). There was no effect on other aspects of yeast metabolism since the ALDC made by the yeast was located in the cytoplasm. Production of α-acetolactate occurs in yeast mitochondria and so only that material which escaped into the cytoplasm was converted to acetoin. Any α-acetolactate required for valine synthesis remained in the mitochondria where it was unaffected by the decarboxylase enzyme.

Subsequently the ALDC gene was integrated into the ribosomal DNA (rDNA) of yeast. Because of the tandem repeat nature of rDNA more than 20 copies of the ALDC gene were integrated and all transformants showed a high enzyme activity (Fujii *et al.*, 1990). Because the new genes were integrated, the transformants were extremely stable even under non-selective conditions. Again beers from laboratory-scale fermentation trials exhibited very low final diacetyl concentrations. In an extension of this work, cloned ALDC genes from two strains of *E. aerogenes* and from one strain of *K. terrigena* were placed under the control of both *ADH1* and *PGK* promoters and terminators and were used to transform brewing yeasts by integration into the *rDNA*, *ADH1* and *PGK* loci. All the transformants fermented well and in the case of two of them (both integrated at the *PGK* site and under the control of the *PGK* promoter and terminator) no maturation was necessary since the diacetyl concentration in the beer at the end of primary fermentation was below the taste threshold. Even the beers produced by the other strains had markedly reduced diacetyl levels. The quality of the finished beer was uniformly good and the transformed yeasts were completely stable (Suihko *et al.*, 1989, 1990; Blomqvist *et al.*, 1991). For food approval purposes the ALDC genes from *L. lactis* or *Acetobacter* sp. would be preferable to those from other sources since these organisms are already used for food production (Goelling and Stahl, 1988). Since it occurs as part of the natural flora of Lambic beer, similar status has been claimed for *E. aerogenes* (Blomqvist *et al.*, 1991).

Because of the importance of diacetyl for beer flavour, much research effort has been concentrated on the amino acid biosynthetic pathway leading to the formation of α-acetolactate (Fig. 3.6). The enzymes of the pathway are regulated by quite complex control mechanisms, in particular acetolactate synthase is subject to very strong feedback inhibition by valine (de Robichon-Szulmajster and Magee, 1968). Since the pathway has been studied in considerable detail and the *ILV* genes coding for the various enzymes have been cloned (Petersen *et al.*, 1983), there have been a number of attempts to control the level of diacetyl in beer by manipulating these genes in brewing yeasts. Diacetyl production can be completely eliminated in mutants lacking threonine deaminase and acetolactate synthase but, because of their inability to synthesize a number of amino acids including valine, such yeasts ferment poorly (Ryder and Masschelein, 1983). Partial reduction in acetolactate synthase activity can be achieved by the use of dominant mutation. Sulphometuron methyl inhibits acetolactate synthase activity and dominant mutations of the *ILV2* gene are known which bring about resistance to the herbicide by reducing the binding affinity of the enzyme for the herbicide (Falco and Dumas, 1985). Some of the resistant forms of the enzyme are less active. Spontaneous dominant mutants of this type have been isolated but reductions in diacetyl levels were disappointingly small (Galvan *et al.*, 1987; Gjermansen *et al.*, 1988). More recently, a procedure has been devised whereby recessive mutations in the *ILV2* locus of brewing yeasts can be produced (Kielland-Brandt *et al.*, 1989). This involves mutagenic treatment of spontaneous herbicide-resistant meiotic segregants from brewing yeasts followed by screening for slow growth on media lacking isoleucine and valine. In this way strains have been

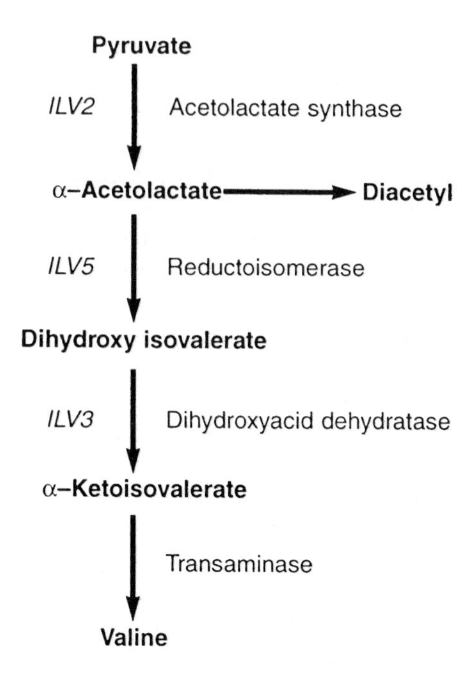

Fig. 3.6 Enzymes and genes of the valine synthetic pathway in *Saccharomyces cerevisiae.*

produced which have very low acetolactate synthase activities. Mating of these strains has yielded hybrids that ferment well, produce little diacetyl and make acceptable beer (P. Sigsgaard, personal communication). Another way of reducing the activity of acetolactate synthase has been described by Vakeria, Box and Hinchliffe (1991). A fragment of the *ILV2* gene was used to transform a brewing yeast in such a way that antisense mRNA was made. This had the effect of reducing the level of normal acetolactate synthase mRNA and hence the amount of enzyme synthesized. Although the amount of diacetyl produced during fermentation was less, the transformed yeast fermented wort much more slowly.

An alternative approach to reducing diacetyl levels is to increase the flux through the pathway leading to valine production. This has been achieved by transforming yeasts with multi-copy plasmids containing copies of *ILV3* and *ILV5* genes. Yeasts transformed with *ILV5* showed an increase in acetolactate reductoisomerase activity and a considerable decrease in diacetyl levels (Dillemans *et al.*, 1987; Villanueba, Goossens and Masschelein, 1990; Goossens *et al.*, 1991, 1993) whereas those transformed with *ILV3* displayed no effect on diacetyl level despite a considerable increase in the corresponding enzyme activity (Goossens *et al.*, 1987). Another way of increasing the copy number of the *ILV5* gene has been suggested by Gjermansen *et al.* (1988). Using a gene replacement technique, functional *ILV5* genes can be stably integrated at the *ILV2* locus with the simultaneous loss of *ILV2*. In this way the level of acetolactate synthase would be reduced at the same time as the level of reductoisomerase is increased.

(j) Molecular biological approaches to yeast differentiation

Traditionally, yeasts strains have been differentiated by a mixture of biochemical, morphological and physiological parameters. With the advent of recombinant DNA techniques, it is now possible to analyse yeast DNA in detail and to distinguish strains on the basis of some form of DNA fingerprint.

Probably the most widely applied technique is that which involves digestion of DNA using specific restriction endonucleases, separation of the fragments by agarose gel electrophoresis and then either direct examination of the banding patterns or examination of the patterns produced by hybridizing labelled 'probes' to the DNA fragments. The probes are themselves short lengths of DNA labelled with either a radioactive marker (typically ^{32}P) or some other marker which can be detected easily. Non-radioactive methods are becoming more popular especially as the technique is now being applied in routine quality control applications (for further details see Meaden (1990)). Direct examination of mitochondrial DNA, without the use of a labelled probe, can be used to differentiate ale yeasts from lager yeasts (Martens, van den Berg and Harteveld, 1985; Lee, Knudsen and Poyton, 1985; Casey, Pringle and Erdmann, 1990) but is unable to distinguish between strains of lager yeasts (Aigle, Erbs and Moll, 1984; Martens, van den Berg and Harteveld, 1985; Casey, Pringle and Erdmann, 1990) except for flocculation variants (Donhauser, Vogeser and Springer, 1989). Direct examination of total DNA after cutting with the restriction enzyme *Hpa*I has also been used to distinguish ale from lager yeasts (Panchal *et al.*, 1987). Normally, however, the probing approach has to be used.

A detailed analysis of a wide range of industrial yeasts has been carried out using labelled probes. With an *RDN1* probe three basic patterns were obtained but almost all of the brewing yeasts examined produced the same pattern (Pedersen, 1983a,b). In contrast, with a *HIS4* probe, ale yeasts could be distinguished from lager strains (Pedersen, 1983a,b). Similar results were obtained using a *LEU2* probe although in this case great heterogeneity was observed amongst the ale strains (Pedersen, 1985a,b, 1986a). In general the results obtained with lager yeasts were so homogeneous as to suggest that they are all closely related, whereas the ale strains showed considerable variability. This general conclusion is further supported by work involving the use of the *Ty1* transposable element as a probe. The *Ty1* element is found distributed about the yeast chromosomal DNA at about 30–35 copies per haploid genome, and would seem to be a good candidate for DNA fingerprinting since changes in the number and location of these elements should lead to alteration in the restriction patterns observed. Great heterogeneity was observed amongst ale strains (Pedersen, 1986a), whilst differentiation of lager yeasts was possible with some *Ty1* probes (Decock and Iserentant, 1985; Martens, van den Berg and Harteveld, 1985; Pedersen, 1985a,b, 1986a) but not with another such probe (Aigle, Erbs and Moll, 1984; Decock and Iserentant, 1985). The heterogeneity of ale strains has been confirmed using a *CUP1* probe and moreover this has been used to distinguish two very closely related lager yeasts (Taylor, Hammond and Meaden, 1990). The *DEX1 (STA2)* gene has also proved extremely useful for analysing brewing yeasts since it cross reacts with other related sequences found in brewing yeasts to produce quite complex patterns which can provide a DNA fingerprint (Taylor, Hammond and Meaden, 1990). Even more complex fingerprints are produced using synthetic polyGT probes, and these have been used to differentiate a number of industrial yeast strains (Walmsley, Wilkinson and Kong, 1989).

Another technique which has been used to investigate yeast genome structure is that of pulsed field electrophoresis of whole chromosomes. This method (karyotyping) has not been widely applied to brewing yeasts but the work carried out so far suggests that it will prove to be a powerful tool, especially for distinguishing lager strains. Initial work in this field (Johnston and Mortimer, 1986; Pedersen, 1986b, 1987; Takata *et al.*, 1989) used orthogonal field-alternation gel electrophoresis (OFAGE; Carle and Olsen, 1984, 1985) which yields curved tracks and so makes interpretation of the results difficult. Nevertheless, distinct karyotypes were distinguished among the brewing yeasts examined. More recently, contour-clamped homogeneous electric field electrophoresis (CHEF; Chu, Vollrath and Davis, 1986) and transverse alternating pulsed field electrophoresis (TAFE; Gardiner, Laces and Patterson, 1986) which yield straight tracks have found widespread acceptance. By careful examination of patterns obtained using these methods, either by densitometry (Casey, Xiao and Rank, 1988b) or by computer image processing (Pedersen, 1989), it has proved possible to distinguish different strains of brewing yeasts. Subsequent work has refined the techniques used, particularly the speed at which the chromosome samples can be prepared (Sheehan *et al.*, 1991) and the resolution of the bands obtained in the electrophoresis gel (Casey, Pringle and Erdmann, 1990). It is clear from a detailed examination of pulsed field electrophoresis gels of brewing yeasts that they contain many more bands than do gels from laboratory strains and that the intensity of the bands varies. This is presumably a reflection of the polyploid nature of the yeasts and their possession of a mixture of different but similar chromosomes (see section 3.4.3(c)). This great number and variety of bands enables karyotyping to be used for distinguishing many closely related industrial yeast strains (Casey, Pringle and Erdmann, 1990; Sheehan *et al.*, 1991).

The techniques of genetic manipulation may also provide the means whereby unique selectable markers can be introduced into individual yeast strains so that the strains can be distinguished by simple laboratory tests. Alternatively small changes can be made to the genome of a brewing yeast which have no effect whatever on its behaviour but which can be detected by means of the DNA fingerprinting techniques described above.

3.5 THE COMMERCIAL USE OF GENETICALLY MODIFIED BREWING YEASTS

3.5.1 Safety considerations

When genetic modification methods were first developed the scientists involved raised a number of concerns about the technology (Berg *et al.*, 1974). These included the colonization of the environment by genetically modified organisms, the unwanted transfer of genes to other organisms in the wild and the possible inadvertent production of new pathogenic organisms. With the passage of time it has generally been agreed that these potential problems were much exaggerated. However, as a result of the expression of these concerns, a diversity of legislation has developed worldwide to address such issues. These regulations are mainly concerned with laboratory-scale work and, although differing in detail, are broadly similar in most countries. They recognize the need for different laboratory facilities for work with different genetically modified organisms, the degree of containment required varying with the perceived risk. All genetic modification work carried out with brewing yeasts has always fallen into the lowest containment category, merely requiring the facilities which are found in any good microbiological laboratory.

Until 1993, regulations applying to genetically modified organisms were concerned solely with human health and safety. Now environmental protection also has to be considered. In the United Kingdom, all work with genetically modified organisms is controlled by two sets of regulations: the Genetically Modified Microorganisms (Contained Use) Regulations, 1992 and the Genetically Modified Organisms (Deliberate Release) Regulations, 1992. These apply to the construction, use and release of genetically modified organisms and involve prior notification of an intention to carry out genetic modification work, together with risk assessments (further details can be found in Hammond and Bamforth, 1994).

3.5.2 Novel foods approval

In the United Kingdom, before beers made with genetically modified yeasts can be produced on a large scale and sold commercially, the approval of several Government committees is required. The primary approval must come from the Advisory Committee on Novel Foods and Processes (ACNFP) which reports to the Ministry of Agriculture, Fisheries and Food (MAFF). Guidelines have been issued to enable manufacturers to decide whether any new product comes within the category of a 'novel food' and, if so, what information will be required by ACNFP for their consideration (Department of Health, 1991). It is clear from this document that beers made from genetically modified yeasts do fall into the category of a 'novel food'. The information requirements include data on the genetic procedures used and the properties of the modified strain including its stability and toxicology.

As part of their assessment, ACNFP may consult with two other committees, the Advisory Committee on Genetic Modification (ACGM) of the Health and Safety Executive who provide advice on all aspects of human health and safety, and the Advisory Committee on Release to the Environment (ACRE) of the Department of Environment who provide advice on waste disposal and incidental release of organisms in fermenter off-gases. Following these consultations, ACNFP makes the decision as to whether the new yeast can be used commercially to make beer.

Novel food approval has recently been obtained for the commercial use of a brewer's yeast genetically modified with the *STA2* gene to produce extracellular glucoamylase. This is the first such permission that has been granted anywhere in the world. Batches of beer have been made with the new yeast and distributed to interested parties worldwide (Hammond, 1994).

At the time of writing (November 1994) the novel food approval system is voluntary although the Food Safety Act (Ministry of Agriculture, Fisheries and Food, 1990) does contain provisions that will enable the scheme to be put on a statutory basis. However, this is unlikely to happen until parallel EC novel food regulations are approved. The current state of affairs in Europe has been described recently (Hammond and Bamforth, 1994), as have the likely effects of European legislation when compared with the situation in the USA (Hammond, 1991).

3.6 CONCLUSIONS

Since the previous edition of this book was written there have been huge advances in the application of modern genetic methods to brewing yeasts. The structure of the genome of brewing yeasts is being slowly unravelled and details of the control of metabolic processes are beginning to be understood. Additionally, the techniques of modern yeast genetics have been applied to the construction of new yeast strains for the more efficient production of beer. Many of these novel yeasts have been assessed at the pilot brewery scale and have been technically successful. One such strain has now been approved for the production of beer for consumption by the general public but has not, as yet, been used commercially. This need for approval is a major barrier to commercial exploitation. The rigorous 'novel food' requirements are clearly a response to perceived consumer concerns about the risk of genetically modified organisms. These same perceptions have also led to proposals that foods produced with genetically modified organisms should be labelled 'Products of Gene Technology' (Food Advisory Committee of MAFF, quoted as an appendix in Department of Health, 1991). Pressure groups and administrators between them are making the application of recombinant DNA technology in the food and drink sector very difficult. From numerous surveys it is clear that the average consumer is singularly ill-informed about the benefits and risks of biotechnology. It is essential that the food and drink sector addresses this problem and educates the public about the factors involved in the use of genetically modified organisms for food and drink production. There is a clear need to convince public opinion that the risks are small, before beer manufactured with genetically modified yeasts will become generally acceptable. All food and drink biotechnologists must devote time and effort to ensuring that any risks are seen in the correct perspective and are not wildly exaggerated by pressure groups. Otherwise the application of novel biotechnology in the food and drink sector will be severely curtailed for many years and the significant available benefits will be lost.

ACKNOWLEDGEMENTS

The author thanks Dr C.W. Bamforth for critical reading of the manuscript and the Director General of BRF International for permission to publish.

REFERENCES

Aigle, M., Erbs, D. and Moll, M. (1984) *Journal of the American Society of Brewing Chemists*, **42,** 1.

Anderson, E. and Martin, P.A. (1975) *Journal of the Institute of Brewing*, **81,** 242.

Barney, M.C., Jansen, G.P. and Helbert, J.R. (1980) *Journal of the American Society of Brewing Chemists*, **38,** 1.

Becker, D.M. and Guarente, L. (1991) In *Methods in Enzymology*, Vol. 194 (eds C. Guthrie and G. Fink), Academic Press, London, p. 182.

Beggs, J.D. (1981) In *Molecular Genetics in Yeast, Alfred Benzon Symposium*, Vol. 16 (eds D. von Wettstein, A. Stenderup, M. Kielland-Brandt and J. Friis), Munksgaard, Copenhagen, p. 383.

Berg, P., Baltimore, D., Boyer, H.W. *et al.* (1974) *Proceedings of the National Academy of Sciences, USA* **71,** 2593.

Berghof, K. and Stahl, U. (1991) *BioEngineering*, **7,** 27.

Bilinski, C.A., Russell, I. and Stewart, G.G. (1986) *Journal of the Institute of Brewing*, **92,** 594.

Bilinski, C.A., Russell, I. and Stewart, G.G. (1987a) *Journal of the Institute of Brewing*, **93,** 216.

Bilinski, C.A., Russell, I. and Stewart, G.G. (1987b) *Proceedings of the 21st Congress of the European Brewery Convention, Madrid*, IRL Press, Oxford, p. 497.

Blomqvist, K., Suihko, M.-L., Knowles, J. and Penttila, M. (1991) *Applied and Environmental Microbiology*, **57**, 2796.

Boone, C., Sdicu, A.-M., Wagner, J. *et al.* (1990) *American Journal of Enology and Viticulture*, **41**, 37.

Broach, J.R. (1981) In *The Molecular Biology of the Yeast Saccharomyces, Life Cycle and Inheritance* (eds J.N. Strathern, E.W. Jones and J.R. Broach), Cold Spring Harbor Laboratory, Cold Spring Harbor, NY, p. 445.

Brown, S.W. and Oliver, S.G. (1982) *European Journal of Applied Microbiology and Biotechnology*, **16**, 119.

Calleja, G.B. (1987) In *The Yeasts* (eds A.H. Rose and J.S. Harrison), Vol. II, Academic Press, London, p. 165.

Cantwell, B.A. and McConnell, D.J. (1983) *Gene*, **23**, 211.

Cantwell, B., Brazil, G., Hurley, J. and McConnell, D. (1985) *Proceedings of the 20th Congress of the European Brewery Convention, Helsinki*, IRL Press, Oxford p. 259.

Carle, G.F. and Olsen, M.V. (1984) *Nucleic Acids Research*, **12**, 5647.

Carle, G.F. and Olsen, M.V. (1985) *Proceedings of the National Academy of Sciences, USA*, **82**, 3756.

Casey, G.P. (1986a) *Carlsberg Research Communications*, **51**, 327.

Casey, G.P. (1986b) *Carlsberg Research Communications*, **51**, 343.

Casey, G.P. and Pedersen, M.B. (1988) *Carlsberg Research Communications*, **53**, 209.

Casey, G.P., Pringle, A.T. Erdmann, P.A. (1990) *Journal of the American Society of Brewing Chemists*, **48**, 100.

Casey, G.P., Xiao, W. and Rank, G.H. (1988a) *Journal of the Institute of Brewing*, **94**, 93.

Casey, G.P., Xiao, W. and Rank, G.H. (1988b) *Journal of the Institute of Brewing*, **94**, 239.

Casey, G.P., Xiao, W. and Rank, G.H. (1988c) *Journal of the American Society of Brewing Chemists*, **46**, 67.

Chu, G., Vollrath, D. and Davis, R.W. (1986) *Science*, **234**, 1582.

Chinery, S.H. and Hinchliffe, E. (1989) *Current Genetics*, **16**, 21.

Clarke, L. and Carbon, J. (1979) In *Methods in Enzymology*, Vol. 68 (ed. R. Wu), Academic Press, New York, p. 396.

Conde, J. and Mascort, J.L. (1981) *Proceedings of the 18th Congress of the European Brewery Convention, Copenhagen*, IRL Press, Oxford, p. 177.

Crumplen, R.M., D'Amore, T., Russell, I. and Stewart, G.G. (1990) *Journal of the American Society of Brewing Chemists*, **48**, 58.

Debourg, A., Goossens, E., Villanueba, K.D. *et al.* (1991) *Proceedings of the 23rd Congress of the European Brewery Convention, Lisbon*, IRL Press, Oxford, p. 265.

Decock, J.P. and Iserentant, D. (1985) *Proceedings of the 20th Congress of the European Brewery Convention, Helsinki*, IRL Press, Oxford, p. 195.

de Figueroa, L.I.C. and de van Broock, M.R.G. (1985) *Applied Microbiology and Biotechnology*, **21**, 206.

Delgado, M.A. and Conde, J. (1983) *Proceedings of the 19th Congress of the European Brewery Convention, London*, p. 465.

Department of Health (1991) *Guidelines on the Assessment of Novel Foods and Processes, Report on Health and Social Subjects*, No. 38, HMSO, London.

de Robichon-Szulmajster, H. and Magee, P.T. (1968) *European Journal of Biochemistry*, **3**, 497.

de van Broock, M.R., Sierra, M. and de Figueroa, L. (1981) In *Current Developments in Yeast Research* (eds G.G. Stewart and I. Russell), Pergamon Press, Toronto, p. 171.

Dillemans, M., Goossens, E., Goffin, O. and Masschelein, C.A. (1987) *Journal of the American Society of Brewing Chemists*, **45**, 81.

Dohmen, R.J., Strasser, A.W.M., Dahlems, U.M. and Hollenberg, C.P. (1990) *Gene*, **95,** 111.
Donhauser, S., Vogeser, G. and Springer, R. (1989) *Monatsschrift fur Brauwissenschaft*, **42,** 4.
Dujon, B. (1981) In *The Molecular Biology of the Yeast Saccharomyces, Life Cycle and Inheritance* (eds J.N. Strathern, E.W. Jones and J.R. Broach), Cold Spring Harbor Laboratory, Cold Spring Harbor, NY, p. 505.
Dutcher, S.K. (1981) *Molecular and Cellular Biology*, **1,** 245.
Enari, T.M., Knowles, J., Lehtinen, U. *et al.* (1987) *Proceedings of the 21st Congress of the European Brewery Convention, Madrid*, IRL Press, Oxford, p. 529.
Enevoldsen, B.S. (1981) *Carlsberg Research Communications*, **46,** 37.
Enevoldsen, B.S. (1985) *Journal of the American Society of Brewing Chemists*, **43,** 183.
Esposito, R.E. and Klapholz, S. (1981) In *The Molecular Biology of the Yeast Saccharomyces, Life Cycle and Inheritance* (eds J.N. Strathern, E.W. Jones and J.R. Broach), Cold Spring Harbor Laboratory, Cold Spring Harbor, NY, p. 211.
Esser, K. and Kamper, J. (1988) *Process Biochemistry*, April, 36.
Evans, I.H. and Wilkie, D. (1976) *Genetic Research*, **27,** 89.
Falco, S.C. and Dumas, K.S. (1985) *Genetics*, **109,** 21.
Fangman, W.L. and Zakian, V.A. (1981) In *The Molecular Biology of the Yeast Saccharomyces, Life Cycle and Inheritance* (eds J.N. Strathern, E.W. Jones and J.R. Broach), Cold Spring Harbor Laboratory, Cold Spring Harbor, NY, p. 27.
Fink, G.R. and Styles, C.A. (1973) *Proceedings of the National Academy of Sciences, USA*, **69,** 2846.
Freeman, R.F. (1981) *Proceedings of the 18th Congress of the European Brewery Convention, Copenhagen*, IRL Press, Oxford, p. 497.
Fujii, T., Kondo, K., Shimizu, F. *et al.* (1990) *Applied and Environmental Microbiology*, **56,** 997.
Galvan, L., Perez, A., Delgado, M. and Conde, J. (1987) *Proceedings of the 21st Congress of the European Brewery Convention, Madrid*, IRL Press, Oxford, p. 385.
Gardiner, K., Laces, W. and Patterson, D. (1986) *Somatic Cell and Molecular Genetics*, **12,** 185.
Gjermansen, C. (1983) *Carlsberg Research Communications*, **48,** 557.
Gjermansen, C. and Sigsgaard, P. (1981) *Carlsberg Research Communications*, **46,** 1.
Gjermansen, C., Nilsson-Tillgren, T., Petersen, J.G.L. *et al.* (1988) *Journal of Basic Microbiology*, **28,** 175.
Goelling, D. and Stahl, U. (1988) *Applied and Environmental Microbiology*, **54,** 1889.
Goodey, A.R. and Tubb, R.S. (1982) *Journal of General Microbiology*, **128,** 2615.
Goodey, A.R., Brown, A.J.P. and Tubb, R.S. (1981) *Journal of the Institute of Brewing*, **87,** 239.
Goossens, E., Dillemans, M., Debourg, A. and Masschelein, C.A. (1987) *Proceedings of the 21st Congress of the European Brewery Convention, Madrid*, IRL Press, Oxford, p. 553.
Goossens, E., Debourg, A., Villanueba, K.D. and Masschelein, C.A. (1991) *Proceedings of the 23rd Congress of the European Brewery Convention, Lisbon*, IRL Press, Oxford, p. 289.
Goossens, E., Debourg, A., Villanueba, K.D. and Masschelein, C.A. (1993) *Proceedings of the 24th Congress of the European Brewery Convention, Oslo*, IRL Press, Oxford, p. 251.
Gopal, C.V. and Hammond, J.R.M. (1992) *Proceedings of the 5th International Brewing Technology Conference, Harrogate*, Brewing Technology Services Ltd, London, p. 297.
Guinova-Stoyanova, T.A. and Lahtchev, K.L. (1991) *Proceedings of the 23rd Congress of the European Brewery Convention, Lisbon*, IRL Press, Oxford, p. 313.
Gunge, N. and Nakatomi, Y. (1972) *Genetics*, **70,** 41.

Guthrie, C. and Fink, G.R. (eds) (1991) *Methods in Enzymology*, Vol. 194, Academic Press, London.
Gyllang, H. and Martinson, E. (1971), *Proceedings of the 13th Congress of the European Brewery Convention, Estoril*, IRL Press, Oxford, p. 265.
Hadfield, C., Cashmore, A.M. and Meacock, P.A. (1986) *Gene*, **45,** 149.
Halfmann, H.J., Rocken, W., Emeis, C.C. and Zimmermann, U. (1982) *Current Genetics*, **6,** 25.
Hammond, J.R.M. (1991) *Proceedings of the 23rd Congress of the European Brewery Convention, Lisbon*, IRL Press, Oxford, p. 393.
Hammond, J.R.M. (1995) In *Proceedings of the 4th Aviemore Conference on Malting, Brewing and Distilling* (eds F.G. Priest and I. Campbell), Institute of Brewing, London, in press.
Hammond, J. and Bamforth, C. (1994) *The Brewer*, **Feb.**, 65.
Hammond, J.R.M. and Eckersley, K.W. (1984) *Journal of the Institute of Brewing*, **90,** 167.
Hammond, J.R.M. and Wenn, R.V. (1985) *Proceedings of the 20th Congress of the European Brewery Convention, Helsinki*, IRL Press, Oxford, p. 315.
Hansen, M., Rocken, W. and Emeis, C.C. (1990) *Journal of the Institute of Brewing*, **96,** 125.
Hartley, J.L., and Donelson, J.E. (1980) *Nature*, **286,** 860.
Henderson, R.C.A., Cox, B.S. and Tubb, R.S. (1985) *Current Genetics*, **9,** 133.
Herskowitz, I. and Oshima, Y. (1981) In *The Molecular Biology of the Yeast Saccharomyces, Life Cycle and Inheritance* (eds J.N. Strathern, E.W. Jones and J.R. Broach), Cold Spring Harbor Laboratory, Cold Spring Harbor, NY, p. 181.
Hinchliffe, E. and Box, W.G. (1984) *Current Genetics*, **8,** 471.
Hinchliffe, E. and Box, W.G. (1985) *Proceedings of the 20th Congress of the European Brewery Convention, Helsinki*, IRL Press, Oxford, p. 267.
Hinchliffe, E. and Daubney, C.J. (1986) *Journal of the American Society of Brewing Chemists*, **44,** 98.
Hinchliffe, E., Fleming C.J. and Vakeria, D. (1987) *Proceedings of the 21st Congress of the European Brewery Convention, Madrid*, IRL Press, Oxford, p. 505.
Hinchliffe, E., Kenny, E. and Leaker, A. (1987) *European Brewery Convention Monograph*, **XIII,** 139.
Hinnen, A., Hicks, J.B. and Fink, G.R. (1978) *Proceedings of the National Academy of Sciences, USA*, **75,** 1929.
Hitzeman, R.A., Hagie, F.E., Levine, H.L. *et al.* (1981) *Nature*, **293,** 717.
Hockney, R.C. and Freeman, R.F. (1980) *Journal of General Microbiology*, **121,** 479.
Holmberg, S. (1982) *Carlsberg Research Communications*, **47,** 233.
Ito, H., Fukuda, Y., Murata, K. and Kimura, A. (1983) *Journal of Bacteriology*, **153,** 163.
Jackson, E.A., Ballance, G.M. and Thomsen, K.K. (1986) *Carlsberg Research Communications*, **51,** 445.
Jansen, G.P., Barney, M.C., Helbert, J.R. and Esposito, M.S. (1979) *Journal of Applied Biochemistry*, **1,** 369.
Jimenez, A. and Davies, J. (1980) *Nature*, **287,** 869.
Johnston, J.R. and Mortimer, R.K. (1986) *International Journal of Systematic Bacteriology*, **36,** 569.
Johnston, J.R. and Oberman, H. (1979) *Progress in Industrial Microbiology*, **15,** 151.
Jones, R.M., Russell, I. and Stewart, G.G. (1986) *Journal of the American Society of Brewing Chemists*, **44,** 161.
Khan, N.C. and Sen, S.P. (1974) *Journal of General Microbiology*, **83,** 237.
Kielland-Brandt, M.C., Peterson, J.G.L. and Mikkelsen, J.D. (1979) *Carlsberg Research Communications*, **44,** 27.
Kielland-Brandt, M.C., Gjermansen, C., Nilsson-Tillgren, T. and Holmberg, S. (1989)

Proceedings of the 22nd Congress of the European Brewery Convention, Zurich, IRL Press, Oxford, p. 37.
Kinsella, B.T. and Cantwell, B.A. (1991) *Yeast*, **7,** 445.
Knowles, J.K.C. and Tubb, R.S. (1986) *European Brewery Convention Monograph*, **XII,** 169.
Knowles, J.K.C., Penttila, M., Teeri, T.T. *et al.* (1985) *Proceedings of the 20th Congress of the European Brewery Convention, Helsinki*, IRL Press, Oxford, p. 251.
Korhola, M. (1983) In *Gene Expression in Yeast, Proceedings of the Alko Yeast Symposium* (eds M. Korhola and E. Vaisanen), Foundation for Biotechnical and Industrial Fermentation Research, Helsinki, p. 231.
Kreil, H., Kleber, W. and Teuber, M. (1975) *Proceedings of the 15th Congress of the European Brewery Convention, Nice*, IRL Press, Oxford, p. 323.
Kumagai, M.H., Shah, M., Terashima, M. *et al.* (1990) *Gene*, **94,** 209.
Lampen, J.O. (1968) *Antonie van Leeuwenhoek*, **34,** 1.
Lancashire, W.E. (1986) *The Brewer*, **72,** 345.
Lancashire, W.E. and Wilde, R.J. (1987) *Proceedings of the 21st Congress of the European Brewery Convention, Madrid*, IRL Press, Oxford, p. 513.
Lancashire, W.E., Carter, A.T., Howard, J.J. and Wilde, R.J. (1989) *Proceedings of the 22nd Congress of the European Brewery Convention, Zurich*, IRL Press, Oxford, p. 491.
Lee, S.Y., Knudsen, F.B. and Poyton, R.O. (1985) *Journal of the Institute of Brewing*, **91,** 169.
Lewis, C.W., Johnston, J.R. and Martin, P.A. (1976) *Journal of the Institute of Brewing*, **82,** 158.
Liljestrom, P.L. (1985) *Nucleic Acids Research*, **13,** 7257.
Lindegren, C.C. and Lindegren, G. (1943) *Proceedings of the National Academy of Sciences, USA*, **29,** 306.
Mahler, H.R. and Wilkie, D. (1978) *Plasmid*, **1,** 125.
Mannonen, L., Kurten, V., Ritala, A. *et al.* (1993) *Proceedings of the 24th Congress of the European Brewery Convention, Oslo*, IRL Press, Oxford, p. 85.
Martens, F.B., van den Berg, R. and Harteveld, P.A. (1985) *Proceedings of the 20th Congress of the European Brewery Convention, Helsinki*, IRL Press, Oxford, p. 211.
Meaden, P. (1990) *Journal of the Institute of Brewing*, **96,** 195.
Meaden P.G, and Taylor N.R. (1991) *Journal of the Institute of Brewing*, **97,** 353.
Meaden, P.G. and Tubb, R.S. (1985) *Proceedings of the 20th Congress of the European Brewery Convention, Helsinki*, IRL Press, Oxford, p. 219
Meaden, P., Ogden, K., Bussey, H. and Tubb, R.S. (1985) *Gene*, **34,** 325.
Ministry of Agriculture, Fisheries and Food (1990) Food Safety Act, HMSO, London.
Molzahn, S.W. (1977) *Journal of the American Society of Brewing Chemists*, **35,** 54.
Montgomery, D.L., Leung, D.W., Smith, M., *et al.*, (1980) *Proceedings of the National Academy of Sciences, USA*, **77,** 541.
Morrison, K.B. and Suggett, A. (1983) *Proceedings of the 19th Congress of the European Brewery Convention, London*, IRL Press, Oxford, p. 489.
Mortimer, R.K., Schild, D., Contopoulou, C.R. and Kans, J.A. (1989) *Yeast*, **5,** 321.
Mowshowitz, D.B. (1979) *Journal of Bacteriology*, **137,** 1200.
Navas, L., Esteban, M. and Delgado, M.A. (1991) *Journal of the Institute of Brewing*, **97,** 115.
Newlon, C.S. and Fangman, W.L. (1975) *Cell*, **5,** 423.
Nilsson-Tillgren, T., Petersen, J.G.L., Holmberg, S. and Kielland-Brandt, M.C. (1980) *Carlsberg Research Communications*, **45,** 113.
Nilsson-Tillgren, T., Gjermansen, C., Kielland-Brandt, M.C., *et al.* (1981) *Carlsberg Research Communications*, **46,** 65.
Nilsson-Tillgren, T., Gjermansen, C., Holmberg, S., *et al.* (1986) *Carlsberg Research Communications*, **51,** 309.

Oppenoorth, W.F.F. (1959) *Proceedings of the 7th Congress of the European Brewery Convention, Rome*, IRL Press, Oxford, p. 29.
Panchal, C.J., Peacock, L. and Stewart, G.G. (1982) *Biotechnology Letters*, **4**, 639.
Panchal, C.J., Bast, L., Dowhanick, T. and Stewart, G.G. (1987) *Journal of the Institute of Brewing*, **93**, 325.
Park, C.S., Park, Y.J., Lee, Y.H., *et al.* (1990) *Technical Quarterly of the Master Brewers Association of the Americas*, **27**, 112.
Pedersen, M.B. (1983a) *Proceedings of the 19th Congress of the European Brewery Convention, London*, IRL Press, Oxford, p. 457.
Pedersen, M.B. (1983b) *Carlsberg Research Communications*, **48**, 485.
Pedersen, M.B. (1985a) *Proceedings of the 20th Congress of the European Brewery Convention, Helsinki*, IRL Press, Oxford, p. 203.
Pedersen, M.B. (1985b) *Carlsberg Research Communications*, **50**, 263.
Pedersen, M.B. (1986a) *Carlsberg Research Communications*, **51**, 163.
Pedersen, M.B. (1986b) *Carlsberg Research Communications*, **51**, 185.
Pedersen, M.B. (1987) *Proceedings of the 21st Congress of the European Brewery Convention, Madrid*, IRL Press, Oxford, p. 489.
Pedersen, M.B. (1989) *Proceedings of the 22nd Congress of the European Brewery Convention, Zurich*, IRL Press, Oxford, p. 521.
Penttila, M.E., Suihko, M.L., Lehtinen, U., *et al.* (1987) *Current Genetics*, **12**, 413.
Penttila, M.E., Andre, L., Lehtovaara, P., *et al.* (1988) *Gene*, **63**, 103.
Perez, C., Vallin, C. and Benitez, J. (1984) *Current Genetics*, **8**, 575.
Perry, C. and Meaden, P. (1988) *Journal of the Institute of Brewing*, **94**, 64.
Petersen, J.G.L., Holmberg S., Nilsson-Tillgren, T., *et al.* (1983) *European Brewery Convention Monograph*, **IX**, 30.
Petersen, J.G.L., Nilsson-Tillgren, T., Kielland-Brandt, M.C., *et al.* (1987) *Current Genetics*, **12**, 167.
Petes, T.D. (1980) *Annual Review of Biochemistry*, **49**, 845.
Philippsen, P., Eibel, H., Gafner, J. and Stotz, A. (1983) In *Gene Expression in Yeast, Proceedings of the Alko Yeast Symposium* (eds M. Korhola and E. Vaisanen), Foundation for Biotechnical and Industrial Fermentation Research, Helsinki, p. 189.
Post-Beittenmiller, M.A., Hamilton, R.W. and Hopper, J.E. (1984) *Molecular and Cellular Biology*, **4**, 1238.
Rocken, W. (1984) *Monatschrift fur Brauwissenschaft*, **37**, 384.
Russell, I., Hancock, I.F. and Stewart, G.G. (1983) *Journal of the American Society of Brewing Chemists*, **41**, 45.
Russell, I., Jones, R. and Stewart, G.G. (1985) *Proceedings of the 20th Congress of the European Brewery Convention, Helsinki*, IRL Press, Oxford, p. 235.
Russell, I., Stewart, G.G., Reader, H.P., *et al.* (1980) *Journal of the Institute of Brewing*, **86**, 120.
Ryder, D.S. and Masschelein, C.A. (1983) *European Brewery Convention, Monograph*, **IX**, 2.
Sakai, K., Fukui, S., Yabuuchi, S., *et al.* (1989) *Journal of the American Society of Brewing Chemists*, **47**, 87.
Salek, A., Schnettler, R. and Zimmermann, U. (1990) *FEMS Microbiological Letters*, **70**, 67.
Sasaki, T., Watari, J., Kohgo, M., *et al.* (1984) *Journal of the American Society of Brewing Chemists*, **42**, 164.
Sheehan, C.A., Weiss, A.S., Newsom, I.A., *et al.* (1991) *Journal of the Institute of Brewing*, **97**, 163.
Sherman, F. and Wakem, P. (1991) In *Methods in Enzymology*, Vol. 194 (eds C. Guthrie and G. Fink), Academic Press, London, p. 38.
Shimizu, F., Sone, H. and Inoue, T. (1989) *Technical Quarterly of the Master Brewers Association of the Americas*, **26**, 47.

Sogaard, M. and Svensson, B. (1990) *Gene*, **94**, 173.
Sone, H., Kondo, K., Fujii, T., *et al.* (1987) *Proceedings of the 21st Congress of the European Brewery Convention, Madrid*, IRL Press. Oxford, p. 545.
Sone, H., Fujii, T., Kondo, K., Shimizu, F., Tanaka, J. and Inoue, T. (1988) *Applied and Environmental Microbiology*, **54**, 38.
Stewart, G.G. and Russell, I. (1986) *European Brewery Convention Mongraph*, **XII**, 53.
Stewart, G.G., Goring, T.E. and Russell, I. (1977) *Journal of the American Society of Brewing Chemists*, **35**, 168.
Stewart, G.G., Russell, I. and Panchal, C. (1981) In *Current Developments in Yeast Research* (eds G.G. Stewart and I. Russell), Pergamon Press, Toronto, p. 17.
Steyn, A.J.C. and Pretorius, I.S. (1991) *Gene*, **100**, 85.
Strasser, W.M., Selk, R., Dohmen, R.J., *et al.* (1989) *European Journal of Biochemistry*, **184**, 699.
Stratford, M. and Assinder, S. (1991) *Yeast*, **7**, 559.
Sturley, S.L. and Young, T.W. (1988) *Journal of the Institute of Brewing*, **94**, 133.
Suihko, M.-L., Penttila, M., Sone, H., *et al.* (1989) *Proceedings of the 22nd Congress of the European Brewery Convention, Zurich*, IRL Press, Oxford, p. 483.
Suihko, M.-L., Blomqvist, K., Penttila, M., *et al.* (1990) *Journal of Biotechnology*, **14**, 285.
Takata, Y., Watari, J., Nishikawa, N. and Kamada, K. (1989) *Journal of the American Society of Brewing Chemists*, **47**, 109.
Taylor, N.R., Hammond, J.R.M. and Meaden, P.G. (1990) In *Proceedings of the 3rd Aviemore Conference on Malting, Brewing and Distilling* (ed. I. Campbell), Institute of Brewing, London, p. 403.
Thomsen, K.K., Jackson, E.A. and Brenner, K. (1988) *Journal of the American Society of Brewing Chemists*, **46**, 31.
Thorne, R.S.W. (1970) *Process Biochemistry*, **4**, 15.
Thorner, J. (1981) In *The Molecular Biology of the Yeast Saccharomyces, Life Cycle and Inheritance* (eds J.N. Strathern, E.W. Jones and J.R. Broach), Cold Spring Harbor Laboratory, Cold Spring Harbor, NY, p. 143.
Tubb, R.S. (1980) *Journal of the Institute of Brewing*, **86**, 76.
Tubb, R.S., Searle, B.A., Goodey, A.R. and Brown A.J.P. (1981) *Proceedings of the 18th Congress of the European Brewery Convention, Copenhagen*, IRL Press, Oxford, p. 487.
Tubb, R.S., Liljestrom, P.L., Torkkeli, T. and Korhola, M. (1986) In *Proceedings of the 2nd Aviemore Conference on Malting, Brewing and Distilling* (eds F.G. Priest and I. Campbell), Institute of Brewing, London, p. 298.
Tuite, M.F., Dobson, M.J., Roberts, N.A., *et al.* (1982) *EMBO Journal*, **1**, 603.
Urano, N., Nishikawa, N. and Kamimura, M. (1990) *Proceedings of the Institute of Brewing (Australia and New Zealand Section) Convention, Auckland*, Australian Industrial Publishers, Adelaide, p. 154.
Vakeria, D. and Hinchliffe, E. (1989) *Proceedings of the 22nd Congress of the European Brewery Convention, Zurich*, IRL Press, Oxford, p. 475.
Vakeria, D., Box, W.G. and Hinchliffe, E. (1991) *Proceedings of the 23rd Congress of the European Brewery Convention, Lisbon*, IRL Press, Oxford, p. 305.
van der Aar, P.C., Straver, M.H. and Teunissen, W.R.H. (1993) *Proceedings of the 24th Congress of the European Brewery Convention, Oslo*, IRL Press, Oxford, p. 259.
van Solingen, P. and van der Plaat, J.B. (1977) *Journal of Bacteriology*, **130**, 946.
Villanueba, K.D., Goossens, E. and Masschelein, C.A. (1990) *Journal of the American Society of Brewing Chemists*, **48**, 111.
von Wettstein, D. (1983) *Proceedings of the 19th Congress of the European Brewery Convention, London*, IRL Press, Oxford, p. 97.
Walmsley, R.M., Wilkinson, B.M. and Kong, T.H. (1989) *Biotechnology*, **7**, 1168.
Watari, J., Takata, Y., Ogawa, M., *et al.* (1989) *Agricultural and Biological Chemistry*, **53**, 901.

Watari, J., Takata, Y., Murakami, J. and Koshino, S. (1991a) *Proceedings of the 23rd Congress of the European Brewery Convention, Lisbon*, IRL Press, Oxford, p. 297.
Watari, J., Takata, Y., Ogawa, M., *et al.* (1991b) *Agricultural and Biological Chemistry*, **55,** 1547.
Watari, J., Nomura, M., Sahara, H., *et al.* (1994) *Journal of the Institute of Brewing*, **100,** 73.
Wickner, R.B. (1981) In *The Molecular Biology of the Yeast Saccharomyces, Life Cycle and Inheritance* (eds J.N. Strathern, E.W. Jones and J.R. Broach), Cold Spring Harbor Laboratory, Cold Spring Harbor, NY, p. 415.
Wickner, R.B. (1983) *Archives of Biochemistry and Biophysics*, **222,** 1.
Xiao, W. and Rank, G.H. (1989) *Gene*, **76,** 99.
Yamano, S., Tanaka, J. and Inoue, T. (1994) *Journal of Biotechnology*, **32,** 165.
Yocum, R.R. (1986) *Proceedings of BioExpo86*, 171.
Young, T.W. (1981) *Journal of the Institute of Brewing*, **87,** 292.
Young, T.W. (1983) *Journal of the American Society of Brewing Chemists*, **41,** 1.
Young, T.W. and Hosford, E.A. (1987) *Proceedings of the 21st Congress of the European Brewery Convention, Madrid*, IRL Press, Oxford, p. 521.
Young, T.W. and Yagiu, M. (1978) *Antonie van Leeuwenhoek*, **44,** 59.

CHAPTER 4

The microflora of barley and malt

B. Flannigan

4.1 THE MICROFLORA OF BARLEY

4.1.1 Introduction

The nature and the magnitude of the microflora of barley depend on both the field conditions under which the crop was grown and the post-harvest history of the grain. The microflora includes bacteria, actinomycetes, yeasts and filamentous fungi which contaminate and colonize the grain in the field, as well as others, particularly filamentous fungi, which are associated with storage. The filamentous fungi, or moulds, in the first category are usually referred to as field fungi, and include species of *Alternaria, Cladosporium, Epicoccum, Fusarium, Cochliobolus, Drechslera, Pyrenophora* (the latter three formerly known as *Helminthosporium*) and several other genera. The moulds in the second category are known as storage fungi, and are mainly species of *Aspergillus, Eurotium* and *Penicillium.* Although Pepper and Kiesling (1963) listed a wide range of microorganisms which had been isolated from barley kernels (and subsequent publications have added to this range), the microfloras of different barleys are remarkably similar to each other, and to other cereals, generally being dominated by the same limited number of species. It should be said at this juncture, that since Pepper and Kiesling (1963) compiled their list, and even since more recent work cited below has been published, there have been considerable changes in the nomenclature of fungi. Where nomenclatural changes have taken place, the organisms will be from here on referred to by their current names, with the older names used by the authors of cited papers following in parentheses. Notice is also taken that the International Commission on Botanical Nomenclature (ICBN) has decreed that, where fungi have both asexual (anamorph) and

Brewing Microbiology, 2nd edn. Edited by F. G. Priest and I. Campbell.
Published in 1996 by Chapman & Hall, London. ISBN 0 412 59150 2

sexual (teleomorph) phases, the teleomorphic name has priority. In this chapter, therefore, teleomorphic names will be used for those members of the anamorphic genera *Aspergillus*, *Fusarium*, *Helminthosporium* and *Penicillium* which have sexual stages, with the anamorphic name appearing in parentheses after the abbreviation 'ana.', e.g. *Cochliobolus sativus* (ana. *Helminthosporium sativum*) or *Pyrenophora teres* (ana. *H. teres*).

4.1.2 The field microflora

(a) Contamination and colonization

As in wheat (Flannigan and Campbell, 1977), barley kernels are first contaminated by airborne bacteria, yeast and moulds soon after the ears emerge from the enveloping leaf sheaths (Hill and Lacey, 1983a). At first, the developing caryopsis in the opened floret is fully exposed to contamination from the air, as is the inside surface of the lemma and palea, which later mature to form the husk. However, as the caryopsis swells, filling the space enclosed by these bracteoles, further direct aerial contamination is prevented, although the outer surface of the lemma and palea remains exposed. Climatic conditions appear to play an important part in the deposition of microorganisms on the surface of the barley kernel. Warnock (1973a) has found that the number of *Cladosporium* spores deposited on the outside of the lemma and palea is related to the numbers in the atmosphere, and both increase markedly after rain. The number of spores germinating on the outside of the bracteoles also increases after rain, but there appears to be little if any invasion of underlying tissues from the outside surface of these organs.

The environment between the caryopsis and the bracteoles, however, is favourable for microbial growth, which may then extend into adjacent tissues. For example, whilst some spores deposited when the inside surface of the bracteoles was exposed remain dormant, the majority of *Cladosporium* spores appear to germinate. Initially, the dark-coloured hyphae developing from these spores spread to form a superficial mycelium, which later invades the parenchymal layer of the lemma and palea (Warnock, 1973a). Mycelium of other field fungi, such as *Alternaria* and *Cochliobolus sativus* (ana. *Helminthosporium sativum*), has also been noted on the inner surface of, and within, the bracteoles (Mead, 1943; Warnock, 1973b). The pericarp layer of the caryopsis itself may also be extensively invaded by mycelium (Mead, 1943; Warnock and Preece, 1971), but perhaps less so than the lemma and palea (Tuite and Christensen, 1955). Although most reports appear to suggest that colonization arises from spores germinating on the inner surface of the bracteoles and caryopsis, there is some evidence that mycelium of *Cladosporium* invading the bracteoles and the outer layer of the caryopsis may originate in trapped anthers which have previously been contaminated and colonized (Warnock, 1973a).

In addition to affecting spore deposition and germination on the outer surface of the kernel, climatic conditions have a profound effect on the colonization of the kernel by microorganisms. The prevalence of seedborne phytopathogens, including various species of *Cochliobolus* and *Pyrenophora* (ana. *Helminthosporium*), in barley in eastern Canada has been attributed to a moister climate than in other parts of the country (Machacek *et al.*, 1951). The presence of larger numbers of microorganisms, including ciliated protozoa and the slime-mould *Physarum polycephalum*, in stained and weathered barley (Kotheimer and Christensen, 1961) has been ascribed to high relative humidity (RH) or rainfall after heading (Follstad and Christensen, 1962). Lodging of crops frequently leads to profuse mould growth; a high RH around the fallen ears may exist long enough to allow development of the dark-coloured sporing structures of, particularly, *Cladosporium* on the surface of the kernel. However, it has been the experience in this laboratory that the viable microbial counts for stained or weathered barleys are not invariably higher than those for bright barleys.

The determination of the degree to which kernels have been invaded by microorganisms requires painstaking microscopical examination of serial sections and excised organs. It is therefore not surprising there has been only one investigation reported in which attempts have been made to obtain quantitative estimates of the amount of fungal mycelium present (Warnock, 1971; Warnock and Preece, 1971).

Warnock (1971) examined 20 kernels in detail and estimated that the total length of hyphae present in individual kernels ranged from 19 to 177 cm, the mean values for two ten-kernel samples being 60 and 74 cm. It was suggested that these means represented a mycelial dry mass of 1.5–1.9 μg. Whilst the spores of a number of field fungi, including *Cladosporium* (Warnock, 1973a), are distinctive enough to be identified and enumerated, the mycelium of one species can seldom be readily distinguished from that of another. The location of the mycelium of a particular species requires the use of fluorescent antibody technique in conjunction with exacting microscopical examination (Warnock, 1971, 1973b). In addition to the time-consuming nature of such work, another disadvantage is that these methods cannot distinguish between living and dead mycelium or spores. Furthermore, although clusters of yeast cells have been observed on the inner surface of the lemma and palea (Tuite and Christensen, 1955), it is not possible to identify them. Consequently, since many workers have been concerned with the presence of viable microorganisms which are phytopathogens or spoilage agents, most of what is known about the components of the microflora derives from studies using methods to enumerate such viable microorganisms, i.e. direct and dilution plating (section 4.6).

(b) Bacteria

Viable counts obtained by dilution plating show that even during the early stages of development of the kernel in the field the microflora is numerically dominated by bacteria (Kotheimer and Christensen, 1961; Follstad and Christensen, 1962; Hill and Lacey, 1983a), and as many as 10^8 bacteria per gram of barley have been isolated by Kotheimer and Christensen (1961) at the late dough stage, or growth stage 87–89 (Zadoks, Chang and Konzak, 1974). No detailed investigation of the bacterial flora of barley has been published, but *Erwinia herbicola* and *Xanthomonas campestris* are often mentioned as being prevalent in pre-harvest barley (Clarke and Hill, 1981; Flannigan *et al.*, 1982). It is likely, however, that other bacteria, such as pseudomonads, micrococci and *Bacillus* spp., which have been noted in dry-stored barley for malting (Haikara, Makinen and Hakulinen, 1977), will also be present before harvest. Generally, the counts of actinomycetes in barley in the field are low (Clarke and Hill, 1981; Hill and Lacey, 1983a). Most appear to be *Streptomyces* spp. (Hill and Lacey, 1983a), *Str. griseus* probably being the commonest, but *Str. albus* and *Str. thermoviolaceus* also being notable. *Thermoactinomyces vulgaris*, one of the causative agents of the respiratory disease farmer's lung, may also be evident in ripening barley (Hill and Lacey, 1983a) and barley at harvest (Clarke and Hill, 1981), as well as in recently harvested grain (Flannigan, 1970).

(c) Yeasts

Yeasts are usually the next most abundant components after bacteria in viable counts prepared from pre-harvest barley, although their numbers may be exceeded by filamentous fungi during the later stage of ripening. By harvest, 50–85% of kernels may be colonized by yeasts (Hill and Lacey, 1983a). The pink yeasts, *Sporobolomyces* and *Rhodotorula* are frequently predominant (Lund, 1956; Clarke, Hill and Niles, 1966; Flannigan, 1974; Clarke and Hill, 1981; Hill and Lacey, 1983a), but *Hansenula, Torulopsis* and *Candida* have been isolated from Danish barley before harvest (Lund, 1956). Although Kotheimer and Christensen (1961) reported the presence of *Saccharomyces* in pre-harvest barley in Minnesota, it is generally regarded that members of this genus are rare in barley (Pepper and Kiesling, 1963). Considering the strong resemblances between other components of the microfloras of the two cereals, it is probable, however, that species of *Cryptococcus* and *Trichosporon* noted in pre-harvest wheat (Flannigan and Campbell, 1977) are also present in barley.

(d) Filamentous fungi

As might be expected from the microscopical studies mentioned previously, viable field fungi appear to be associated mainly with the lemma and palea, judged by their distribution in naked barley before harvest (Flannigan, 1974). Here, approximately two-thirds of the inoculum is associated with the bracteoles and one-third with the caryopsis. After harvest, the high percentages of surface-disinfected kernels which give rise to colonies on plating on agar media confirm that field fungi are present within the outer layers of the kernel as viable mycelium or spores (Machacek *et al.*, 1951; Follstad and Christensen, 1962; Flannigan, 1969, 1970; Haikara, Makinen and Hakulinen, 1977). The field fungus most frequently present in kernels is *Alternaria alternata* (*Alt. tenuis*), both in Europe (Clarke, Hill and Niles, 1966; Jorgensen, 1969; Flannigan, 1969, 1970; Haikara, Makinen and Hakulinen, 1977; Clarke and Hill, 1981) and North America (Machacek *et al.*, 1951; Follstad and Christensen, 1962; Christensen and Kaufmann, 1969), although *Cladosporium* spp. (mainly *Clad. cladosporioides* and *Clad. herbarum*) may also be present in a large number of kernels. For example, Hill and Lacey (1983a) have found that 90% of kernels are contaminated by *Cladosporium* in some pre-harvest crops, and 100% by *Alt. alternata*. Other saprophytic fungi which are common, but usually less abundant than *Alt. alternata* and *Cladosporium* spp., are *Epicoccum nigrum* (*E. purpurascens*) and *Aureobasidium* (*Pullularia*) *pullulans* (Machacek *et al.*, 1951; Flannigan, 1969, 1970). *Acremonium* (*Cephalosporium*) spp. (Flannigan, 1970; Haikara, Makinen and Hakulinen, 1977; Ylimaki *et al.*, 1979) may be prominent, and *Verticillium lecanii* may be present on every kernel of some crops in the field (Hill and Lacey, 1983a).

Among seedborne phytopathogens, *Cochliobolus sativus* (ana. *Helminthosporium sativum*) is rarer in the British Isles than in other parts of Europe and North America, but *Pyrenophora graminea* (ana. *H. gramineum*) and *P. teres* (ana. *H. teres*) are found (Flannigan, 1970). *Fusarium* spp. may be frequent in wetter regions of North America (Machacek *et al.*, 1951; Follstad and Christensen, 1962), and in northern Europe, where the harvest is late (Uoti and Ylimaki, 1974; Haikara, Makinen and Hakulinen, 1977; Ylimaki *et al.*, 1979; Flannigan and Healy, 1983). The most frequent fusaria in European barleys appear to be the root-rot species, *F. avenaceum* (Ylimaki *et al.*, 1979) and *F. culmorum* (Haikara, Makinen and Hakulinen, 1977), but in some studies the snow mould, *Monographella nivalis* (ana. *F. nivale*) has been noteworthy, (e.g. Flannigan, 1969, 1970; Jorgensen, 1969). *F. sporotrichioides* (*F. tricinctum*) has also been reported as prominent in Finnish barley (Ylimaki *et al.*, 1979). In a Canadian survey, *F. poae*, *F. equiseti* and *F. acuminatum* (*F. scirpi* var. *acuminatum*) were the commonest species (Machacek *et al.*, 1951), and in Japan, *F. graminearum* (Ichinoe, Hagiwara and Kurata, 1984).

If the total amount of inoculum is assessed by viable count methods, it usually appears that the mould flora immediately before harvest is dominated in Europe by *Cladosporium* spp. (Flannigan, 1974; Hill and Lacey, 1983a), and in the United States by *Alt. alternata* (Follstad and Christensen, 1962). Studies on the development of the microflora of wheat (Flannigan and

Campbell, 1977) suggest that in Europe *Aureobasidium pullulans* and *Cladosporium* spp. are likely to be the earliest contaminants of barley and to increase rapidly in numbers after anthesis, followed by *Alt. alternata* and often *Verticillium lecanii*, and finally by *Epicoccum nigrum*. At this stage, a single kernel may carry approximately 1.6×10^4 colony forming units (CFU) of bacteria, 8×10^3 yeast CFU and 2.5×10^3 field fungi (Flannigan, 1974).

4.1.3 The storage microflora

(a) Bacteria in stored barley

Unless there has been extensive contamination by the excretory products of rodents, insects or mites, the bacteria found in stored barley are likely to be the hardier remnants of the field bacterial flora. Petters, Flannigan and Austin (1988) reported that the flora of stored English malting barley comprised representatives of 13 taxa, mostly Gram-positive pigmented organisms. The Gram-positive bacteria included *Aureobacterium flavescens*, *Bacillus* spp., *Brevibacterium linens*, *Corynebacterium* spp., *Clavibacterium iranicum*, *Microbacterium imperiale* and *Oerskovia xanthineolytica*, whilst among the Gram-negative bacteria were *Erwinia herbicola*, *Pseudomonas fluorescens* and *Chromobacterium* sp. The highest populations of individual types were of *Corynebacterium* sp. (18% of total isolates) and *Ps. fluorescens* (16%).

(b) The source of storage fungi

Storage fungi are not found exclusively in stored grain; they are to be found as contaminants of grain in the field. In studies where kernels have been surface disinfected, after the manner of seed pathologists, it has been concluded that these fungi are virtually absent, e.g. by Clarke, Hill and Niles (1966) and Clarke and Hill (1981). When kernels have been plated directly onto agar media without surface disinfection, however, small numbers have yielded penicillia and aspergilli, with the latter being reported to be most frequently in the *Aspergillus glaucus* group (Tuite and Christensen, 1957; Flannigan, 1978; Hill and Lacey, 1983a). This very important *Asp. glaucus* group of species possess sexual stages and are now referred to by the teleomorphic name *Eurotium*. Very occasionally, as many as 20% of kernels may be contaminated by *Eurotium* spp. in the field (Flannigan, 1978), but seldom more than 5% bear these storage fungi (Christensen and Kaufmann, 1969). During harvesting there is apparently little, if any, increase in the level of contamination by *Eurotium*, but it appears there can be massive contamination by *Penicillium* spp. The number of kernels yielding penicillia may increase from < 1% on the ear to approximately 60% in the combine harvester (Flannigan, 1978). The source of this inoculum seems to be soil thrown up by the combine harvester, culms and leaves which are threshed in the drum along with the grain and residues of previous crops in the auger and storage tank of the combine. If newly harvested grain is surface disinfected, it is likely that < 0.5% of kernels will yield storage fungi (Christensen and Kaufmann, 1969), confirming that at this stage contamination is almost entirely superficial, and that few kernels will have been actively colonized by *Eurotium* spp. and *Penicillium* spp.

Although they have received most attention, the *Eurotium* species *E. amstelodami, E. chevalieri, E. repens* and *E. rubrum* (ana. *Asp. amstelodami, Asp. chevalieri, Asp. repens* and *Asp. ruber*) are not the only storage aspergilli which contaminate barley in the field. *Asp. candidus, Asp. flavus* (Flannigan, 1978), *Asp. fumigatus, Asp. nidulans, Asp. niger, Asp. ochraceus* and *Asp. versicolor* (Hill and Lacey, 1983a) may also be present. *Asp. fumigatus*, a thermotolerant species associated with the spoilage of moist-stored feed barley (Lacey, 1971; Clarke and Hill, 1981; Hill and Lacey, 1983b), has been noted as being surprisingly widespread in pre-harvest barley. Other such spoilage species which may also be present as contaminants in the field (Hill and Lacey, 1983a) include mesophilic *Penicillium piceum* and *P. roqueforti* and thermophilic *Rhizomucor* (*Mucor*) *pusillus, Talaromyces thermophilus* (ana. *P. dupontii*) and *Thermomyces lanuginosus* (*Humicola lanuginosa*).

(c) Post-harvest colonization

The moisture content (MC) of barley at harvest may range from <10% (Lutey and Christensen, 1963) in drier areas of North America and Australia to 25–30% in northern Europe (Hellberg and Kolk, 1972). Grain with a high MC must be dried to <14% if it is to be stored for any period of time, and to <12.5% to exclude the possibility of any mould growth in storage. Whilst the need to dry grain is clearly understood, it is often the case that the rate at which it can be dried cannot keep pace with the rate at which it is harvested. The resulting delay in drying the harvested grain can allow microorganisms to develop and the grain to heat. One method of reducing microbial growth and cooling the grain in such an event is to blow air at ambient temperature through the bulk, i.e. low-volume ventilation (LV). LV is frequently employed in pre-drying silos until the grain can be dried to a safe level using a hot-air drier, but it can only be regarded as a temporary holding operation. Since many farmers do not have the capacity to satisfactorily dry barley for malting, grain merchants and maltsters often prefer to dry the grain themselves. Although, by having specifications which state that, for example, 'Grain must be sound, sweet and free from heat, (visible) mould and infestation' (Dolan, 1979), the maltster appears to safeguard his interests, it would be possible for a farmer to present for purchase barley which had heated and then been cooled using LV. The maltster is well advised 'to be wary of barley held on the farm for any length of time before drying' (Flannigan and Healy, 1983).

Depending on the moisture content of the harvested grain, there may be some growth of field fungi, e.g. *Fusarium* spp., if drying is delayed. However, such delays are usually characterized by the development of storage fungi. It has been found that, compared with barley delivered for drying to a maltster directly after harvesting, the viable mould count of barley held on the farm for a few days can rise from 3×10^3 to 1.3×10^4 CFU per kernel, with species of *Eurotium, Penicillium* and especially *Absidia, Mucor* and *Rhizopus* being prominent in the farm-held grain. Numbers of yeasts can increase from 5×10^3 to 2.2×10^4 per kernel (B. Murphy, unpublished results, cited in Flannigan and Healy (1983)). The development of these storage fungi usually appears to be at the expense of the field species. Mills and Wallace (1979) noted that the numbers of kernels bearing the predominant field fungi *Alt. alternata* and *Cladosporium* spp. apparently declined as those contaminated by *Eurotium* spp. (*Asp. glaucus*) and *P. aurantiogriseum* (*P. verrucosum* var. *cyclopium*) increased in piles of undried grain after a wet harvest in Manitoba. In the UK, Flannigan and Healy (1983) have observed that undried barleys received in a 'hot' or 'hot and smelly' state are characterized by lower levels of field fungi, particularly *Cladosporium* spp., and higher levels of storage fungi, especially *Eurotium* (*Asp. glaucus*), than are present in barleys acceptable to the maltster for purchase and drying.

Even when there is no delay in drying, the number of kernels contaminated by storage fungi increases by the time the grain enters storage. Much of the increase results from cross-contamination as the grain is moved during the post-harvest operations leading up to, and including, drying and cleaning, but part may be the result of proliferation during the early part of drying. Mills and Wallace (1979) have found that the drying process can reduce the numbers of kernels bearing viable myxomycetes, and lead to increases in the levels of *Absidia, Eurotium* and *Penicillium*. Where, because of a low MC at harvest, the grain does not require drying, it has been demonstrated (Tuite and Christensen, 1957) that there will still be a large increase between the combine harvester and elevator in the number of kernels contaminated by *Eurotium* spp. Christensen and Kaufmann (1969) have drawn attention to the large numbers of spores of storage fungi which are habitually present in dust and the air of elevators and act as the inoculum for incoming grain. There is no practical method of reducing the number of these spores, far less eliminating them, so that control of storage fungi, which can initiate spoilage of dry-stored grain, lies in ensuring that storage conditions prevent their growth.

(d) Storage fungi as agents of spoilage

Whilst barley entering storage is contaminated or colonized by a wide range of microorganisms, the storage conditions determine which, if any, of these organisms can grow on the grain. The moisture content of the grain is clearly important, but other interacting factors, including temperature, aeration, inclusion of chaff and foreign material such as weed seeds and infestation by insects and mites, play a part in selecting which microorganisms proliferate.

Although maltsters and grain merchants think in terms of the MC of grain as far as safe storage is concerned, it is perhaps more appropriate from the microbiological standpoint to consider the water activity (a_w) of the grain. In this context, it is defined as the ratio of the vapour pressure of water in the grain to that of pure water at the same temperature, and is equivalent to equilibrium relative humidity (ERH). If seeds are stored, for example, in an atmosphere at 80% RH they will eventually come to equilibrium with that atmosphere, their a_w being 0.80, but their MC depending on the chemical nature of the seed. Oily seeds will have a lower MC than starchy grain, e.g. in groundnuts the MC will be approximately 9.2%, and in barley 16.6%. The relationship between MC, a_w and ERH for barley over a range relevant to mould growth is shown in Table 4.1, and between MC and a_w for barley and finished malt in Fig. 4.1. At a_w 1.0, the grain is fully hydrated; at a_w 0.6, even the moulds least demanding in their water requirements cannot grow.

The interacting parameters of a_w and temperature define whether the spores of different microorganisms germinate or not, and the rate at which the organisms grow. In Table 4.2, the temperature range and minimum a_w of some of the commoner species of actinomycetes and moulds which have been isolated from stored barley are listed, the physiological classification of Lacey, Hill and Edwards (1980) being adopted. Of the moulds, *Asp. restrictus* is the most extreme xerophile: Hill and Lacey (1983b) have recorded that it grows slowly in barley at a_w 0.65–0.70. *Eurotium* spp. are less extreme than *Asp. restrictus*, although nevertheless strongly xerophilic. Most other storage aspergilli are regarded as being moderately xerophilic, but the thermotolerant species *Asp. fumigatus* is only slightly xerophilic, like the storage penicillia. As stated previously, drying barley to an MC of 12% (a_w 0.6) prevents any microbial growth. Drying to 14% (a_w 0.7) should be sufficient to prevent growth of all but the most xerophilic of moulds, and clearly precludes growth of field fungi and yeasts, generally with a

Table 4.1 The relationship between equilibrium relative humidity (ERH), water activity (a_W) and moisture content (MC) of barley

ERH (%)	a_w	*Approx. MC (%)*
60	0.6	12.3
70	0.7	14.0
80	0.8	16.6
90	0.9	20.3
95	0.95	24.8

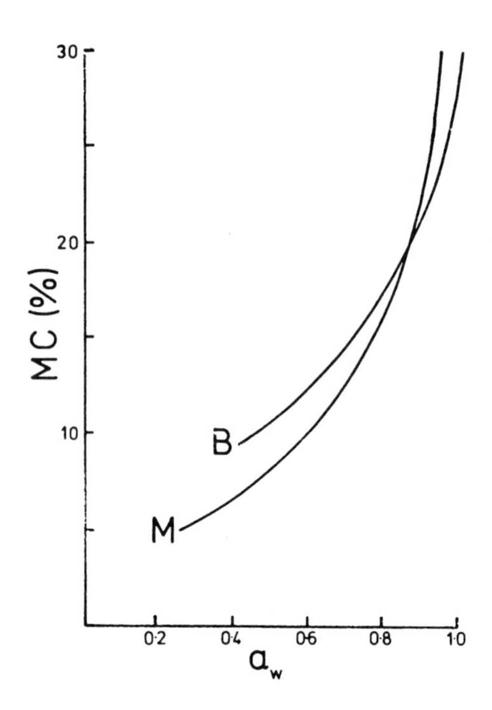

Fig. 4.1 The relationship at 25°C between moisture content (MC) and water activity (a_w) of barley (B) (after Pixton and Warburton, 1971) and finished malt (M) (after Pixton and Henderson, 1981).

minimum $a_w > 0.90$, and bacteria, with a minimum a_w usually > 0.93. Some field fungi, including *Fusarium* (Welling, 1969) and *Cladosporium*, are able to grow in grain with high a_w which has been stored at low temperatures. In such situations, psychrotolerant penicillia (with a temperature range −10°C to 35°C) seem also to be prominent, e.g. *P. brevicompactum*, *P. chrysogenum*, *P. aurantiogriseum* (*P. verrucosum* var. *cyclopium*) and *P. verrucosum* (*P. verrucosum* var. *verrucosum*).

By their metabolic activity, microorganisms generate water and heat, so they have the capacity to raise both the a_w and temperature of stored grain, a poor conductor of heat. As has been mentioned above, growth of extreme xerophiles at a_w 0.65–0.70 is slow and the small amount of heat that is generated is dissipated, so that the temperature of the grain does not rise. However, by generating water these organisms gradually raise the moisture level in the grain, and in so doing make possible the growth of other less xerophilic species, as well as allowing acceleration of their own growth. As further species grow, the build-up of heat is such that thermophilic organisms are able to develop rapidly and the temperature of the grain may reach 65°C or more.

Table 4.2 Physiological classification of storage moulds and actinomycetes according to their temperature limits and minimum a_W for growth (after Lacey, Hill and Edwards, 1980)

Category	*Lower mesophilic (2–37°C)*	*Upper mesophilic (5–50°C)*	*Thermotolerant/ thermophilic (10–57°C)*	*Extremely thermophilic (25–70°C)*
Extremely xerophilic (min. a_W <0.75)	–	*Aspergillus restrictus* *Eurotium* spp.*	–	–
Moderately	*Asp. versicolor*	*Asp. candidus* *Asp. flavus* *Asp. niger* *Asp. ochraceus* *Asp. terreus*	–	–
Slightly xerophilic (min. a_W 0.80 –0.89	*Penicillium frequentans* *P. rugulosum*	*P. capsulatum* *P. citrinum* *P. funiculosum* *P. piceum*	*Asp. fumigatus*	–
Hydrophilic (min. a_W >0.90)	*Fusarium* spp. *Mucor hiemalis* *M. racemosus* *R. stolonifer* *S. aureofaciens* *S. olivaceus*	*Absidia corymbifera* *Rhizopus arrhizus* *Streptomyces griseus*	*Rhizomucor pusillus* *Thermoascus crustaceus* *S. albus*	*Thermomyces lanuginosus* *Talaromyces thermophilus* *Faenia rectivirgula* *Thermoactinomyces vulgaris*

* 2–50°C.

In storage experiments, Hill and Lacey (1983b) observed that after 9–12 months the slow growth of *Asp. restrictus* in barley at a_w 0.65–0.70 gave way to *Eurotium* spp. (*Asp. glaucus* group) at a_w 0.75 (MC > 14.4%), and deterioration of the grain (judged by percentage germination) accelerated. This group continued to increase in numbers as the a_w rose and reached a maximum at a_w 0.90–0.93 (MC 20.0–22.3%), although they were then no longer the predominant organisms. Above a_w 0.85 (MC 17.2%), spontaneous heating occurred and *P. aurantiogriseum* (*P. verrucosum* var. *cyclopium*) and *Asp. candidus* became dominant. At a_w 0.90–0.95 (MC 20.0–25.3%) the temperature reached a maximum of 50°C, and *P. funiculosum*, *P. capsulatum*, *Asp. flavus*, *Asp. nidulans*, streptomycetes and bacteria predominated. Together with the respiratory pathogen *Asp. fumigatus*, *Absidia corymbifera* increased in numbers, but only reached a maximum above a_w 0.95, when the temperature rose to a maximum of 65°C and thermophilic organisms were present in large numbers. These included the actinomycetes *Thermoactinomyces vulgaris* and *Micropolyspora faeni* (causative agents of farmer's lung) and fungi such as *Rhizomucor pusillus* and *Thermomyces lanuginosus*.

Although the development of a microbial succession culminating in massive proliferation of thermophilic organisms is usually associated with the faulty storage of undried feed barley (Clarke, Niles and Hill, 1967; Clarke *et al.*, 1969; Lacey, 1971; Clarke and Hill, 1981) high temperatures may be reached in dry-stored grain when 'hot spots' develop. Temperatures as high as 53°C have been recorded in hot spots within grain bulks during the Canadian winter (Wallace and Sinha, 1962), and in one hot spot a maximum of 64°C in May (Sinha and Wallace, 1965). The initiation of heating in this last case was attributed to the growth of *P. aurantiogriseum* (*P. verrucosum* var. *cyclopium*) and *P. funiculosum* at –5°C to 8°C during the winter after harvest. *Asp. flavus*, *Asp. versicolor*, *Absidia* spp. and *Streptomyces* spp. succeeded the penicillia as the temperature rose. In addition to *Asp. flavus* and *Asp. versicolor*, *Asp. fumigatus* was prominent in other hot spots (Wallace and Sinha, 1962). Since Gilman and Barron (1930) observed in laboratory experiments that in less than 1 week *Asp. flavus* and *Asp. fumigatus* can raise the temperature of barley, oats or wheat by as much as 26°C, it appears that after the initial stages these aspergilli play an important part in the accelerating development of such hot spots.

Sinha and Wallace (1965) found that, after the maximum had been reached, the temperature of the hot spot fell to that of the surrounding grain in about 3 weeks, and there was no later resurgence of growth. Although the time scale may vary, this can be regarded as being typical of the course of events in a hot spot. The fall in temperature presumably results from the cessation of microbial growth due to the combined effects of the elevated temperature and associated drying of the grain. Sinha and Wallace (1965) observed that the MC at the focus of the hot spot fell in 4 weeks from 20.0–23.5% at the peak temperature to 16.0–18.9%, and that the size of the central area with an MC > 16.0% was halved between the second and the fourth weeks. Growth may cease within the heated pocket of grain, but many viable spores remain and can present a health hazard (Lacey, 1975). Whilst storage fungi survive as spores (Sinha and Wallace, 1965), field fungi do not, despite being able to exist (in decreasing numbers) for a period of several years in dry-store grain (Machacek and Wallace, 1952).

Development of a large hot spot can start from mould growth in a very small patch of moister grain within a bulk. Moisture migrates from this small focus of activity, as it does throughout the development of the hot spot, so that the zone of fungal growth extends and more grain is affected. There are various reasons for some grain initially being moister than the remainder of the bulk, and consequently capable of supporting mould growth. In the first instance, it may not have been dried to a safe level. MC measurements are average values, but grain lots are not homogeneous and there can be marked differences between the true moisture content of one kernel and the next. It may be that in a batch dried on the basis of the MC of grain harvested in the middle of the day, there is barley which was harvested at either end of the day and consequently had a higher initial

MC. It can be that some grain is not cooled sufficiently after drying, so that when it cools within a bin or silo, condensation forms on the surface of the kernels. Sufficient water for mould growth can also become available at the surface of the kernel as a result of the diurnal evaporation/condensation cycle on the insolated side of a storage bin. There is also the possibility that structural faults may allow water to leak into a grain bulk.

There is often a close association between spoilage fungi and the presence of mites and insects (Howe, 1973; Sinha, 1973). If insects are present in large numbers they can increase the moisture of grain and initiate the development of hot spots. Mites and insects, including the grain weevil *Sitophilus granarius*, may feed on moulds, and they act as vectors for the spread of moulds by carrying spores on the surface of their bodies as they migrate. In a typical hot spot developing as a result of insect infestation (Jacobson and Thomas, 1981) the temperature at the origin may reach 39°C and the adult insects migrate upwards. As moisture also migrates upwards, barley at the top of the bulk may become moist enough to sprout and support profuse mould growth.

Whichever is the case, there is a rapid drop in the germinative capacity of kernels as storage fungi develop and grain heats (Sinha and Wallace, 1965; Lund, Pedersen and Sigsgaard, 1971; Hill and Lacey, 1983b). Where the MC of barley is 15–18% or greater, it has been found (Tuite and Christensen, 1955; Armolik, Dickson and Dickson, 1956; Follstad and Christensen, 1962) that even when there is no self-heating in storage there is reduced germination and an associated build-up of the more xerophilic moulds, particularly the *Eurotium* spp. (*Asp. glaucus* group). From numerous experiments (reviewed in Christensen and Kaufmann (1969, 1974) and Christensen (1973)) in which grain was heavily inoculated with *Eurotium* (*Asp. glaucus*), Christensen and his colleagues concluded that the storage fungi are directly responsible for this reduction in germination. It cannot be disputed that in such experiments, and in many other cases where there is a low germinative capacity, *Eurotium* spp. will grow out from the embryo and outer layers if the kernels are incubated in a moist chamber. However, Harrison and Perry (1976) noted various deteriorative changes of endogenous origin, developing in advance of any substantial invasion by fungi, and concluded that, although *Eurotium rubrum* (*Asp. repens*), *P. aurantiogriseum* (*P. cyclopium*) and, at higher moisture levels, *Fusarium culmorum* increase in numbers, they are not the primary causes of deterioration.

Flannigan and Bana (1980) have drawn attention to characteristics of the embryo which may make it a favoured site for colonization by fungi. Lipid- and ethanol-soluble sugars and oligosaccharides, including sucrose, raffinose and fructosans, are concentrated in the embryo and aleurone. As experimental work showed, *Eurotium* spp. (*Asp. glaucus* group) show lipolytic activity and are able to utilize sucrose and raffinose as sole sources of carbon, although they show little or no capacity to degrade starch or structural polysaccharides in grain. In addition to providing a readily assimilable source of carbon for these moulds, the concentration of sugars in the embryo means that the embryo will more readily absorb moisture than the starchy endosperm, e.g. from a humid atmosphere. In correctly dried grain, the MC of the embryo should be similar to that for the whole kernel; in badly stored grain, the MC of the embryo will be higher than that of the whole. This raising of the MC of the embryo will in theory make it more susceptible to invasion by storage fungi, but it may be that it must be debilitated to some degree before it is invaded.

4.2 THE MICROFLORA OF MALT

4.2.1 The microflora during malt production

During steeping of grain, dormant microorganisms are activated: mould spores germinate, mycelium grows and yeasts and bacteria multiply. In laboratory studies, increases in bacterial numbers range from ×5 (Follstad and Christensen, 1962; Flannigan *et al.*, 1982) to ×36 (Kotheimer and Christensen, 1961), and in yeasts from ×1.1 in stained and weathered barley and ×5 in bright barley (Kotheimer and Christensen, 1961) to ×200 in Finnish barley (Haikara, Makinen and Hakulinen, 1977). In commercial malting, Douglas (1984) found that where steeping is interrupted by an air-rest, there may be increases of ×4 to ×25 in bacterial numbers in the grain by the end of the first steep and ×35 to ×150 by the time it is couched. However, during a different year in the same maltings, the number of viable heterotrophic bacteria on the grain actually fell during the first steep (Table 4.3), but had increased to nearly ×4 by the end of the second (Petters, Flannigan and Austin, 1988). The range of bacteria on the barley at the first steep was narrower than in dried barley, consisting of *Aureobacterium flavescens*, *Alcaligenes* sp., *Clavibacterium iranicum* and *Clav. michiganense*, *Flavobacterium esteroaromaticum*, *Microbacterium imperiale*, *Erwinia herbicola*, *Pseudomonas fluorescens* and *Ps. putida*. By the end of the second steep, the

Table 4.3 Viable counts of microorganisms associated with kernels during commercial production of a sulphured malt (after Petters, Flannigan and Austin, 1988)

Stage	*Aerobic heterotrophic bacteria*	*Lactobacilli*	*Moulds*	*Yeasts*
Stored barley	1.8×10^6	2.0×10^2	2.0×10^2	4.7×10^3
First steep	6.7×10^5	4.2×10^3	8.0×10^2	4.6×10^5
Second steep	6.6×10^6	7.8×10^4	1.7×10^3	1.1×10^6
Green malt (5 days)	5.7×10^7	8.7×10^6	1.5×10^2	3.9×10^6
Kilned malt	5.6×10^6	1.6×10^5	2.0×10^2	3.2×10^4
Screened malt	5.5×10^6	5.7×10^4	8.3×10^2	1.8×10^4

range had widened, with Gram-positive rods being particularly prominent. Lactobacilli appeared in substantial numbers (Table 4.3), the increase being proportionally much greater than for total viable bacteria (Petters, Flannigan and Austin, 1988). This is in line with the earlier observation of Sheneman and Hollenbeck (1960) that mesophilic lactic acid bacteria rose during steeping from $0–10^2$ to $>5 \times 10^5$ CFU g^{-1}, while total bacterial numbers increased from around 10^6 g^{-1} to 7×10^6 g^{-1}.

Just as bacterial counts generally increase with steeping, so do yeast numbers. For example, Douglas (1984) reported that viable yeast numbers at the end of the first and second steeps could be respectively ×3 to ×20 and ×5 to ×1300 those of dry barley. Petters, Flannigan and Austin (1988) observed increases of nearly ×100 and more than ×230 from a level of 4.7×10^3 CFU per dry barley kernel at the first and second steep, respectively (Table 4.3). In comparison, for filamentous fungi the corresponding increases from the original 2×10^2 CFU per dry kernel were only 4- and 8-fold.

Examination of steep liquor shows that, in addition to activating dormant organisms, steeping washes superficial contaminants off kernels (Healy, 1985; Petters, Flannigan and Austin, 1988), e.g. *Aureobasidium*, *Cladosporium*, *Mucor* and *Penicillium* (Healy, 1985). A proportion of these contaminants is deposited on other kernels. In micro-malting, Haikara, Makinen and Hakulinen (1977) reported an increase from 15% to 90% in kernels bearing *Fusarium* spp., although Follstad and Christensen (1962) earlier noted a fall with steeping from 47% to 3% in kernels contaminated by *Cladosporium* spp. Increases may, of course, result in part from the growth of mycelium from one kernel to another, but it seems unlikely that this would account for all the cross-contamination observed by Haikara, Makinen and Hakulinen (1977). Douglas (1984) observed that, as well as increases in the numbers of kernels contaminated by *Fusarium* spp. by the end of the first steep, in some commercial malting runs there could be increases in viable counts, indicating proliferation during steeping. Healy (1985) noted increased contamination by *F. avenaceum* and *F. culmorum* during steeping, and detected *F. poae* and other fusaria which had not been evident in the original dry barley, although presumably they had been present in a small number of kernels. Increases have also been noted in both viable counts of, and percentage frequency of kernels bearing, *Aureobasidium pullulans* and *Cladosporium* spp. (Douglas and Flannigan, 1988). Together with *Trichosporon beigelii* (*T. cutaneum*), *Geotrichum candidum*, a yeast-like species rarely found in dry barley (Pepper and Kiesling, 1963), was reported by Healy (1985) to be present on appreciable numbers of steep-out kernels. Although Haikara, Makinen and Hakulinen (1977) observed an increase in the percentage frequency for what is usually the commonest field fungus, *Alt. alternata*, during laboratory steeping, Gyllang and Martinson (1976b) and Healy (1985) recorded falls during steeping in the course of production of a commercial malt. It was found by Douglas (1984) that in some cases there was little change and in others a fall in the percentage of kernels yielding *Alt. alternata*, even when there was an apparent increase in the total amount of inoculum (viable count) of the organism. Haikara, Makinen and Hakulinen (1977) recorded increases in the percentage frequency of kernels yielding the storage aspergilli and penicillia, and Gyllang and Martinson (1976b) reported an increase for the latter, but Douglas (1984) found that these two types can increase or decrease independently of each other.

Whether in laboratory or commercial malting, viable counts of bacteria and yeasts reach a maximum during germination of the barley. Although Follstad and Christensen (1962) recorded smaller increases, Haikara, Makinen and Hakulinen (1977) reported bacterial counts for germinated barley ×400 those for the original dry barley, and yeasts ×100, and Kotheimer and Christensen (1961) reported corresponding increases of ×4300 and ×79, respectively, in laboratory trials. Sheneman and Hollenbeck (1960) noted a peak in bacterial numbers in green malt after 5 days, when the mean total count was 7×10^8 g^{-1} (×700 the count for barley). Numbers of mesophilic lactic acid bacteria increased in parallel with the total numbers, but accounted for < 2% of the total counts. Douglas (1984) noted ×85 to ×440 more bacteria on green malt in box maltings than on the dry barley, and ×40 to ×1800 more yeasts. Corresponding rises (Table 4.3) were observed by Petters, Flannigan and Austin (1988), with the range of Gram-negative bacteria isolated being greater than earlier. The dominant species was *Flav. esteroaromaticum* (26% of isolates), and *Erw. herbicola*, *Ps. fluorescens*, *Ps. putida* and *Serratia plymuthica* were also prominent. Lactobacilli increased 110-fold from the end of the second steep, *L. alimentarius* and *L. plantarum* being among those noted. It should be noted here that there can be large differences between the relative increases in different malting systems (Flannigan *et al.*, 1982). The increases in bacterial counts between casting and the end of the germination period are proportionally greater in floor malting than in saladin box malting, and in germination vessels there is apparently little or no change. The corresponding increases in yeast numbers are, however, apparently greater in box malting than in floor malting, but in germination vessels the increases are much smaller.

During laboratory malting, Kotheimer and Christensen (1961) found that there was no change in the viable mould count as bright barley germinated, although, compared with the dry grain, there was a 4-fold increase in the case of stained and weathered barley. Follstad and Christensen (1962) reported that in their laboratory studies counts in some cases increased by roughly ×6 relative to the dry barley, but in others there were either no or considerably smaller changes. Flannigan *et al.* (1982) also recorded a ×6 increase during germination, but noted in floor malting a corresponding 13-fold increase, and in some saladin boxes and germination vessels apparent falls of 20–70% and 60–90%, respectively. On some occasions, Douglas (1984) recorded lower counts in green malt produced in saladin boxes than in the original barley, but also found an increase of ×11 during one malting run, largely due to *Fusarium* spp.

When the literature is examined, it is difficult to discern any consistent pattern of development of moulds from the evidence on viable counts, and this is compounded by the results for direct plating. Haikara, Makinen and Hakulinen (1977) found that during germination, *Fusarium* spp. already being present in 90% of kernels after steeping, the largest increase in the number of kernels bearing any particular category of mould during germination was for those bearing *Penicillium* spp. (from 30% to 77%). Smaller proportional increases were observed in those contaminated by, in descending order, *Mucor*, *Eurotium* (*Asp. glaucus*), *Alt. alternata* and *Rhizopus*. In commercial malting, Gyllang and Martinson (1976b) again found that the greatest increase, although smaller, was in the percentage frequency of *Penicillium* spp. in kernels, followed by *Pyrenophora teres* (*H. teres*) and *Alt. alternata*. However, Douglas (1984) reported that during germination the greatest increases in frequency were for *Geotrichum candidum*, *Mucor* spp. and, once, *Alt. alternata*. Direct and dilution platings (Douglas, 1984; Douglas and Flannigan, 1988) indicated that the level of contamination by xerophilic storage aspergilli (*Eurotium* spp.) and penicillia may fall during the production of green malt in Saladin boxes. This decrease in xerophilic fungi was also observed by Healy (1985) in other maltings. Although the levels of field fungi such as *Cladosporium* spp., *Epicoccum nigrum* (*E. purpurascens*) and sometimes *Alt. alternata* may decrease during production of green malt in Saladin boxes (Flannigan *et al.*, 1984), increases in *Cladosporium* and *Alternaria*, as well as *Fusarium*, have also been observed in commercial maltings (Healy, 1985). However, as Haikara, Makinen and Hakulinen (1977) have indicated, the initial level of contamination in the dry barley is important in determining how successful a mould is in colonizing other kernels during the production of green malt.

Kilning has a marked effect on bacterial numbers. Follstad and Christensen (1962) found that the very high numbers on green malt were reduced to between 63% of the count for the barley before steeping and ×8 the barley count. Haikara, Makinen and Hakulinen (1977) recorded a drop of 95% in numbers, but they were still ×20 those in the barley. In contrast, Haikara, Makinen and Hakulinen (1977) did not observe any reduction in yeast numbers, which remained ×100 those on barley, but Follstad and Christensen (1962) found that there was a fall to ×6–27 the original count for barley.

Having noted 15 yeast species in eight genera during malting in Saladin boxes, the commonest being species of *Candida*, *Cryptococcus*, *Debaryomyces*, *Pichia* and *Rhodotorula*, Healy (1985) observed a reduction to < 10% of former levels as a result of kilning. In another system, in which the same vessel was used for grain drying, steeping, germination and kilning, she recorded 22 yeast species, adding *Kluyveromyces* to those genera observed in the Saladin boxes. The total viable yeast count in this case was ×2.5 that of the original barley, and nearly one-half of the species survived kilning, to varying extents. Petters, Flannigan and Austin (1988) only recorded eight species, with the predominant species on green malt being *Candida catenulata* and *Debaryomyces hansenii*, and on screened malt the former and *Rhodotorula mucilaginosa*.

The combination of sulphuring (to produce a pale malt or reduce nitrosamine formation) and kilning results in greater reductions in numbers of bacteria (Graff, 1972; Flannigan, 1983). Sheneman and Hollenbeck (1960) found that bacterial counts for finished malt were similar to those for the original barley, and Douglas (1984) observed that they were ×0.5 to ×2 those in barley. Later, a fall greater than one order of magnitude in both total viable aerobic heterotrophic bacteria and lactobacilli was observed by Petters, Flannigan and Austin (1988). The dominant types on the kilned malt were Gram-positive, viz. *Clav. iranicum*, *Flav. esteroaromaticum*, *Arthrobacter globiformis* and, among the lactobacilli, *L. alimentarius*. Flannigan (1983) found in the laboratory that yeast numbers were also affected by sulphuring, and Douglas (1984) observed in a maltings that yeasts were reduced to < 1–20% of the number in barley, although Petters, Flannigan and Austin (1988) later reported a 6-fold increase relative to the barley (Table 4.3).

Follstad and Christensen (1962) found that, in general, kilning brought about reductions in the numbers of moulds on germinated grain with the final counts being 63–170% of those for the original barley. However, in a malt prepared from weathered barley the count rose from ×5 the original to ×7 with kilning. As far as the percentage frequency of kernels contaminated by different moulds is concerned, Haikara, Makinen and Hakulinen (1977) noted large increases with kilning for *Mucor* spp. and *Rhizopus* spp., and smaller increases for *Penicillium* spp., *Eurotium* spp. (*Asp. glaucus* group) and, from a very low level, *Cladosporium* spp. However, there was no change for *Alt. alternata* and *Fusarium* spp. Gyllang and Martinson (1976b) also noted increases in the number of kernels bearing *Rhizopus* spp. and *Cladosporium* spp. after kilning in a maltings for 24 h, together with an increase in the frequency of the respiratory pathogen *Asp. fumigatus*. Kilning with sulphur appears to have a profound effect in reducing both viable counts and the percentage of kernels bearing moulds. Douglas (1984) found that most field fungi and storage penicillia were frequently reduced to a level below the limits of detection in dilution plating, although they might still be detected in small numbers of kernels by direct plating. However, *Eurotium* spp. appear to be much less affected, and the number of kernels yielding *Mucor* spp. may increase. An increase in *Mucor* spp. with kilning can be responsible for the total mould count being greater than that for the original barley, although the counts usually appear to be lower (Douglas, 1984). Despite the proliferation of *Geotrichum candidum* resulting in contamination of nearly all kernels of green malt, it has been reported that viable inoculum of this species may be virtually eliminated by kilning with sulphur (Douglas, 1984; Douglas and Flannigan, 1988), although other work (Healy, 1985) indicates that this species may survive on malt kernels.

It is worth noting that the design of the kiln may influence whether moulds survive. Heaton, Callow and Butler (1992) reported that a *Phoma* and two species of *Fusarium* with a maximum temperature for growth of 40–45°C were able to survive kilning at an air temperature of 70°C and caused disfigurement of the painted wall in a concrete malting tower. It appeared that the temperature of the poorly insulated kiln wall was below 40°C for most of the kiln cycle during the winter months and only reached the kiln air temperature of 70°C for short periods during the summer. Although agar plate cultures of the organism totally lost viability when placed in the centre of the kiln, it is not known whether malt kernels adjacent to the wall failed to attain high kilning temperatures or whether the fungi survived in these kernels.

4.2.2 The microflora of finished malt

Reports indicate that the total numbers of bacteria in distiller's malts (Boruff, Claasen and Sotier, 1938) are likely to be in the range $0.7–7.7 \times 10^6\ g^{-1}$ for barley malt and $1.6–4.6 \times 10^6\ g^{-1}$ for rye malt, although Flannigan *et al.* (1982) have recorded a maximum of $9.8 \times 10^7\ g^{-1}$ for barley malt. Sheneman and Hollenbeck (1960) obtained totals of $1.4–7.0 \times 10^6\ g^{-1}$ for brewer's malts, with the maximum count for lactic acid bacteria being $1.4 \times 10^4\ g^{-1}$. They reported that *Lactobacillus leishmanii* was the predominant mesophilic lactic acid species, and that *Pediococcus acidilactici* and *L. delbrueckii* were the main thermophilic species, although they were present in very small numbers. The laboratory studies of Haikara, Makinen and Hakulinen (1977) indicated that the principal Gram-negative bacteria present are likely to be *Erwinia herbicola* and *Pseudomonas* spp., with micrococci and *Bacillus* spp. being the main Gram-positive types.

Yeast numbers in finished malt have been reported to lie in the range 7.2×10^2 to $1.1 \times 10^4\ g^{-1}$ (Flannigan *et al.*, 1982), with the pink yeasts in the genera *Sporobolomyces* and *Rhodotorula* often being abundant. Other yeasts usually present include those found in grain (section 4.1.2(c)).

Counts for fungi, including yeast-like species such as *Aureobasidium pullulans* and *Geotrichum candidum*, and filamentous yeasts in the genus *Trichosporon*, may be $10^3–10^4\ g^{-1}$ (Flannigan *et al.*, 1982). In a survey of finished malts, the moulds isolated most frequently from kernels by Gyllang and Martinson (1976b) were *Eurotium* (*Aspergillus*) *amstelodami*, *Asp. fumigatus* and *Rhizopus* spp. These authors also found that the percentages of kernels bearing the three categories quoted were much greater in malt from one Swedish malting plant than from another, where the most frequently isolated types were *Penicillium* spp. and *Alt. alternata*. Relatively high frequencies have been noted for *Alt. alternata* and *Fusarium* spp. in some Scandinavian malts (Gyllang *et al.*, 1981), and *Cochliobolus sativus*

(*H. sativum*) in a distiller's malt (Flannigan *et al.*, 1982), but in sulphured malts the moulds most often on kernels appear to be mucoraceous types and *Eurotium* spp., with *Penicillium* spp. usually being less common and field fungi present on small numbers of kernels (Douglas, 1984). It is to be expected that kilning temperature and procedure will strongly influence the final microflora: more microorganisms are likely to survive the lower kilning temperatures used in the production of distiller's or high-diastatic malts.

The microbiological status of malt reaching a brewery or distillery will depend very much on its handling after production. Handling operations result in cross-contamination between kernels, and also fresh aerial contamination. In addition, malt in transit usually picks up moisture. It was observed that malt exported to Nigeria in sacks which became damaged showed greater levels of storage fungi than in the reference sample retained in the UK (Flannigan *et al.*, 1984). Badly stored malt is also likely to show increasing colonization by storage fungi. At RH > 79%, malt becomes progressively more colonized by storage aspergilli, including *Asp. candidus* and *Asp. versicolor* (Flannigan *et al.*, 1982), and can support growth of the toxigenic species *Asp. flavus* (Flannigan, Healy and Apta, 1986).

4.3 EFFECTS OF MICROORGANISMS ON MALTING

4.3.1 Water sensitivity and dormancy

One explanation put forward to account for water sensitivity in barley is the presence in the lemma, palea and pericarp testa of large populations of field microorganisms which compete for available oxygen during and immediately after harvest, when the numbers of viable field fungi and bacteria on and in the kernel are also greatest, and water sensitivity disappears with time, just as with time the viability of the field microflora declines (Lutey and Christensen, 1963). When water-sensitive barley is placed in an excess of water, the limitation of entry of oxygen to the embryo results in the kernel either failing to germinate or germinating only slowly. Under these circumstances, some microorganisms are able to invade the embryo, and the kernel loses its viability.

In a paper where they reviewed earlier work and discussed various methods of dealing with water sensitivity, Gaber and Roberts (1969) reported that combinations of antibacterial and antifungal antibiotics were effective in overcoming the inhibition of germination. They concluded that it was necessary to suppress the activity of both bacteria and fungi. However, Matthews and Collins (1975) found that antimicrobial compounds did not eliminate differences in oxygen uptake between water sensitive and control barleys, although they did reduce water sensitivity to some extent, and considered that it was characteristics of the kernel, rather than the presence of microorganisms, which determined whether it was water sensitive or not. Nevertheless, in experiments on field germination with *Gliocladium roseum*, a saprotrophic mould which has been isolated from barley (Pepper and Kiesling, 1963), Lynch and Prynn (1977) noted that the effects of low oxygen availability on germination and viability were exacerbated by the presence of fungi. Harper and Lynch (1981) have since observed that this organism grows at the embryo end of the kernel, i.e. at the point of entry of oxygen. From measurements of oxygen uptake of kernels and isolated moulds they have concluded that, if in imbibed barley 100 μg of mycelium of *G. roseum* are present below the husk, the requirement of the fungus for oxygen will be nearly twice that of the embryo; with the rapidly growing *Mucor hiemalis* it will be approximately × 4; and with a slower growing *Penicillium* sp. about two-thirds of the requirement of the embryo. Earlier, Harper and Lynch (1979) had demonstrated that the nitrogen-fixing bacterium *Azotobacter chroococcum* could inhibit germination by competing for oxygen, its effects being greater at low oxygen concentrations. It therefore appears that both bacteria and fungi could play a part in causing water sensitivity, but the evidence that they are the sole cause is far from conclusive.

In a recent series of papers, Briggs and his colleagues (Kelly and Briggs, 1992a, b, 1993; Doran and Briggs, 1993; Briggs and McGuiness, 1993) have re-examined the role of microorganisms, noting that during malting the germination of undried dormant barley appears to be the result of excessive microbial growth occasioned by aeration and mixing. They have demonstrated that controlling microbial growth by antimicrobials leads to increased germinability and vigour after germination, as well as improving α-amylase levels. Experiments with de-embryonated grain indicate that the surface microflora is one factor influencing responsiveness of the aleurone to gibberellic acid. The work has confirmed that steeping with antibiotics results in improved germination, especially in the case of dormant barley, and, as microorganisms at the surface of the grain have a substantial uptake of oxygen, it has been concluded that they are a major cause of dormancy.

4.3.2 Seedling blight fungi

In the field, seedling blight and root rot caused by *Fusarium* spp. and *Cochliobolus sativus* are well known. Jorgensen (1983) found that in Denmark, over a 15-year period, *C. sativus* (*H. sativum*), *F. avenaceum*, *F. culmorum*, *F. graminearum* and *Monographella nivalis* (*F. nivale*) were, together with *Pyrenophora graminea* (*H. gramineum*), the most important seedborne pathogens of barley. He suggested that if >15% of kernels were contaminated by *C. sativus* seed barley required treatment with a seed dressing. Much earlier, Christensen and Stakman (1935) found that as many as 91% of kernels in some barleys from Iowa and southern Minnesota were contaminated by *C. sativus* (*H. sativum*), and up to 54% by *Fusarium* spp. They observed that there was a negative correlation between percentage contamination and percentage germination. In special germination tests to assess possible effects on malting, badly infected kernels were set to germinate on moist blotting paper at 16°C. Only a little over two-thirds germinated, and of these more than 25% showed a stunted, rotted or aborted primary rootlet and a contorted or very short acrospire. *C. sativus* was the more destructive of these two types of field fungi. *Fusarium* spp. were nevertheless recognized as lowering germinative capacity and giving uneven growth on malting floors (Mason, 1923). When fusaria were applied in steep by Sloey and Prentice (1962), although most isolates did not, single isolates of *F. oxysporum* f.sp. *lycopersici* (*F. bulbigenum* var. *lycopersici*), *F. graminearum*, *Monographella nivalis* (*F. nivale*) and an unidentified species caused significant reductions in rootlet growth. The presence of *Fusarium* spp. and *C. sativus* (*H. sativum*), and of *Alt. alternata* and *Cladosporium* spp., in greater numbers of kernels than in bright barley may also have accounted for the lower percentage germination and rootlet dry weight in stained and weathered barley observed by Kotheimer and Christensen (1961).

By inoculating barley in the field with several strains of *F. avenaceum* and *F. culmorum*, Haikara (1983) has been able to demonstrate that subsequent decreases in the germinative capacity and germinative energy of the contaminated grain are more dependent on the weather between anthesis and harvest than on the strain. For example, with one strain of *F. culmorum* the germinative capacity was 99.5% after a drier growing season and 48% after the following wetter season. The effects of this species on germination were, in general, greater than those of *F. avenaceum*. In some instances, there was a marked disparity between the germinative capacity and the germinative energy, indicating that the fungi had weakened the kernels, so that they were able to germinate only under optimum conditions (Haikara, 1983).

A possible explanation for these effects on germination might be the production of toxic metabolites, or mycotoxins, by the fungi. *C. sativus* produces toxic substances to which barley is more susceptible than wheat or oats (Ludwig, 1957). Nummi, Niku-Paavola and Enari (1975) and Flannigan, Morton and Naylor (1985) have found that the trichothecene, T-2 toxin, which is produced by strains of *F. sporotrichioides* (*F. tricinctum*), *F. avenaceum*, *F. culmorum* and other fusaria (Smith *et al.*, 1984), retards acrospire and rootlet development, but Haikara (1983) did not detect trichothecenes in her barleys or malt. However, when barley was stored undried before malting, there was a 10-fold increase in the level of zearalenone (F-2) and a reduction in germinative capacity. Low levels of this toxin inhibit germination in maize (Brodnik, 1975), so that it may have had a direct effect in this case.

As indicated in section 4.2.2, and by the fact that Andersen, Gjertsen and Trolle (1967) noted an increase in the number of superficial spores on kernels, conditions during production of green malt may be ideal for the growth of *Fusarium* spp. They form the 'red mould' which can grow on malting floors, and in some cases in the past grew to the extent that production of satisfactory malt was impossible and maltings were temporarily closed for disinfection. Although growth of fusaria can be accompanied by the discoloration of rootlets, noted in germination tests on blotters by Jorgensen (1983), there is no published evidence that they produce mycotoxins during malting.

4.3.3 Stimulation of germination

Although some of the *Fusarium* strains examined by Sloey and Prentice (1962) had a deleterious effect on rootlet growth during malting, the only isolate of *F. culmorum* tested significantly enhanced the production of rootlets. In her experiments with barley inoculated in the field with *F. avenaceum* and *F. culmorum*, Haikara (1983) found that relative to uninoculated controls there was enhanced rootlet growth during malting of grain from a wetter growing season, although there had been a reduction in the malt produced after the previous drier season. This discrepancy was probably related to the level of inoculum on the grain: wetter conditions favour the growth of fusaria. Another species, *F. moniliforme* (teleomorph = *Gibberella fujikuroi*) produces a number of growth-regulating compounds, the gibberellins (Bu'lock, 1984), which are used to accelerate malting. It may be that if this species is present in abundance there can be some stimulation of germination, since Gjertsen, Trolle and Andersen (1965) found that one *F. moniliforme* isolate added at steeping apparently stimulated rootlet growth. However, isolates of other species not recognized as being gibberellin producers gave the same effect. In any event, the effects of gibberellins could be counteracted by other factors. For example, *F. moniliforme* is known to produce a number of toxins (Drysdale, 1984), including fusaric acid, which is injurious to cell membranes.

In contrast to its phytotoxic properties (Ludwig, 1957), *C. sativus* (*H. sativum*) was shown to produce two metabolites which stimulate sugar release in de-embryonated barley (Briggs, 1966). It has also been reported that both living mycelium and extracts of *Alt. alternata* stimulate elongation of wheat coleoptiles (Ponchet, 1966). *Clad. herbarum* produces in culture the auxins indole-3-acetic acid and indole-3-acetonitrile (Valadon and Lodge, 1970), and apparently has a stimulatory effect on the growth of maize seedlings (Pidoplichko, Moskovels and Zholanova, 1960). It appears possible that these last two species, which are often the predominant field fungi on barley, could have similar effects in malting, although the presence of other, antagonistic, species could have a modifying effect.

4.3.4 Malt analysis

Weathered barley not only shows poorer germination (section 4.3.1), it also produces malts with undesirable characteristics, the most widely reported being high protein modification, and consequently high wort nitrogen

(Malt Research Institute, 1955; Kneen, 1963). In reviewing the effects of the microflora of barley on malt and beer properties, Etchevers, Banasik and Watson (1977) reported that, compared with malts prepared directly from high-moisture barley, they had found pronounced changes in the characteristics of those prepared after storage of the barley for 1 month. The extract and wort nitrogen levels were higher, but α-amylase and diastatic power (DP) were much lower.

When they have been used to inoculate barley in the field (Haikara, 1983) or in the steep (Prentice and Sloey, 1960; Sloey and Prentice, 1962; Kneen, 1963; Gjertsen, Trolle and Andersen, 1965), *Fusarium* spp. have been found to affect malt characteristics. Sloey and Prentice (1962) noted a general tendency towards increased steep and respiration losses, with a concomitant reduction in malt recovery. The greatest reduction was seen with an isolate of *F. culmorum*, where the steep and respiration losses were accompanied by greater rootlet production. Earlier, Prentice and Sloey (1960) had found that isolates of *F. graminearum* and the gibberellin producer, *F. moniliforme* increased both α-amylase and DP. In their second paper they reported that malt prepared from barley inoculated with further isolates of these species, and *F. culmorum*, *F. equiseti*, *F. poae* and unidentified species, also showed higher α-amylase and extract values. Using some of these strains and others from Danish barley, Gjertsen, Trolle and Andersen (1965) confirmed that there were increases in α-amylase in most cases, and in DP in a few cases. Kneen (1963) noted that an unidentified *Fusarium* sp. gave a distiller's malt with a greater DP. Moreover, Haikara (1983) found that malts produced from barley inoculated with *F. avenaceum* and *F. culmorum* showed a higher degree of modification, although in some cases the α-amylase activity and DP were lower than in controls. This was attributed to fungal enzyme activity undetected by the analytical methods employed. In the investigations mentioned above, increased protein modification and wort nitrogen have also been observed, and *F. graminearum* and *F. moniliforme* have been noted in particular (Prentice and Sloey, 1960; Sloey and Prentice, 1962).

Few of the wide range of other microorganisms have as yet been implicated in alterations in malt quality. Prentice and Sloey (1960) examined nearly 100 bacteria, yeasts and moulds, and found that, apart from fusaria, only a relatively small number affected wort nitrogen and none raised α-amylase or DP. Among those which increased wort nitrogen were several bacteria, including *Pseudomonas* spp., and unidentified species of *Aspergillus* (two) and *Mucor*. Kneen (1963) reported that *Asp. niger*, *Rhizopus arrhizus* and unidentified species of *Absidia*, *Trichothecium* (*Cephalothecium*), *Coniothyrium*, *Helminthosporium* and *Rhizopus* also had this effect. *Eurotium* (*Aspergillus*) *amstelodami* and *Asp. fumigatus*, which were reported to grow during malting (Gyllang and Martinson, 1976a), were additionally found to raise nitrogen levels and extract yields (Gyllang, Satmark and Martinson, 1977).

The extent to which mycotoxins are involved in the effects discussed above is uncertain. Both Nummi, Niku-Paavola and Enari (1975) and Flannigan, Morton and Naylor (1985) found in laboratory experiments that the trichothecene toxin T-2, well known as a potent inhibitor of protein synthesis, depressed synthesis of α-amylase during malting. Flannigan, Morton and Naylor (1985) also noted reductions in α-amino nitrogen in the wort, although not apparently sufficient to alter the gravity of the wort. In other laboratory experiments, diacetoxyscirpenol (DAS) (Ferguson, Flannigan and Schapira, 1984) and, to a lesser extent, deoxynivalenol (DON) (Schapira, Whitehead and Flannigan, 1989) have also been shown to affect germination and enzyme development during malting of barley. As with T-2, growth of the coleoptile and rootlets growth is inhibited and α-amylase and α-amino nitrogen levels reduced as a result of the presence of these toxins in barley. It can be argued that, although individual mycotoxins apparently affect malting, the levels of toxin used to spike the grain in order to demonstrate this are higher than would be encountered in cereals for malting. However, grain carries mixtures of mycotoxins, not only because a variety of different moulds are present but also because particular species produce different ranges of toxins. For example, Haikara (1983) found that the strain of *F. culmorum* which had the greatest effect on the germinative capacity of barley and the characteristics of malt also produced the greatest amounts of the oestrogenic mycotoxin zearalenone during malting. However, she did not test for production of deoxynivalenol (DON), also known to be produced by this species. These two toxins, and other toxins in natural assemblages, have entirely different modes of action and target different metabolic functions. It is therefore not surprising that experimental animal studies have presented evidence that natural mixtures of mycotoxins are much more potent than the sum of the effects of the individual toxins might suggest (Flannigan, 1991). Such synergy, resulting from combinations of low levels of different toxins, might therefore be of significance to malting, but would make tracing the real effects of particular species or mycotoxins extremely complex.

4.4 EFFECTS OF THE MICROFLORA ON BEER AND DISTILLED SPIRIT

4.4.1 Beer

The best known effect of the microflora of barley and malt is that of reduced gas stability, or gushing. It is particularly associated with late-harvest areas, e.g. Scandinavia, and wetter growing seasons. The earlier literature on the phenomenon was discussed by Gjertsen, Trolle and Andersen (1963), who investigated its causes in Danish beer and concluded that the problem of primary gushing arose from the use of weathered or badly stored barley for malt production. Prentice and Sloey (1960) investigated the problem by preparing malts inoculated at steep with a range of organisms, and found that it was mainly fusaria which created the problem. Further work using 30 *Fusarium* isolates established that inoculation with many of these resulted in increased wort nitrogen and formol nitrogen in wort and beer, and that *F. graminearum* and *F. moniliforme* were particularly notable in causing gushing (Sloey and Prentice, 1962). Using some strains from these workers, Gjertsen, Trolle and Andersen (1965) established that gushing was not dependent on when the inoculations were made during steeping, and also that it was the result of interaction between the mould and the germinating barley. It has since been demonstrated that malt made with field-inoculated barley can give rise to gushing (Haikara, 1983). Haikara found that gushing caused by *F. avenaceum* and *F. culmorum* was both weather and strain dependent and that there appeared to be a relationship between gushing and the ability of the strains to produce the mycotoxin zearalenone. It was noted that gushing induced by *F. graminearum*-contaminated malt was unpredictable; it could occur shortly after bottling or several weeks later (Donhauser *et al.*, 1989). Niessen *et al.* (1992) confirmed that strains of *F. graminearum* and *F. culmorum* in German grain could initiate gushing. It was subsequently noted by Niessen *et al.* (1993) that gushing was associated with DON in malted barley and wheat. Although some gushing beers did not contain DON and some DON-positive beers were non-gushing, taken overall, the concentration of the toxin in gushing beers was found to be significantly higher than in non-gushing beers.

Other field fungi have also been found to be capable of causing gushing. Culture filtrates of *Alternaria* sp., *Nigrospora* sp. and *Stemphylium* sp. caused gushing when added to normal beer (Amaha *et al.*, 1973; Yoshida *et al.*, 1975). Various storage fungi have also been found to have the same potential, including *Eurotium* spp. (*A. glaucus* group), *P. chrysogenum* (including strains previously called *P. notatum*), *P. griseoroseum* (*P. cyaneofulvum*) and *Rhizopus* sp. (Amaha *et al.*, 1973; Yoshida *et al.*, 1975; Fukushima, Kitabatake and Amaha, 1976). *E. amstelodami* and *Asp. fumigatus* added in the steep were also found to cause gushing (Gyllang and Martinson, 1976a), and were considered to be responsible for periodic episodes of this in a Swedish brewery, where they appeared to be the predominant species on malt.

In investigating the cause of gushing, Amaha *et al.* (1973) found that a *Fusarium* isolate produced at least two gushing inducers, one of which was a peptide-containing substance of low molecular weight. A polypeptide (MW about 15 000) which induced gushing was isolated from culture filtrates and grain contaminated with *Nigrospora* sp., which was originally identified wrongly as being *Rhizopus* sp. (Amaha *et al.*, 1973, 1974). The purified substance induced vigorous gushing in normal beer at concentrations as low as 0.05 ppm. The isolate was morphologically very similar to type strains of *N. oryzae*, one of which was also found to produce gushing inducers (Kitabatake and Amaha, 1974). A gushing-inducing peptidoglycan causing gushing at a level of 4 ppm was isolated from *Stemphylium* sp. (Amaha *et al.*, 1973; Kitabatake and Amaha, 1976).

Elevated nitrogen levels have been mentioned above in relation to *Fusarium* spp., but Kneen (1963) found that *Asp. niger* and *Rhizopus arrhizus* also increased nitrogen in beer. This last species and, particularly, *Eurotium rubrum* (*Asp. ruber*) and *Asp. ochraceus* caused reduced haze stability, although unidentified species of *Alternaria*, *Penicillium* and *Geotrichum* increased the stability. In addition, *Asp. ochraceus* inhibited the action of the beer stabilizer papain. *Asp. fumigatus* and *R. oryzae* have also been found to increase α-amino nitrogen and soluble nitrogen in beer (Gyllang, Satmark and Martinson, 1977).

The effect of moulds on beer flavour was studied by Kneen (1963), who observed that *Asp. niger*, *Asp. ochraceus*, *R. arrhizus* and unidentified species of *Cladosporium*, *Coniothyrium* and *Fusarium* were responsible for strong off-flavours in beer. These ranged from 'molasses, burned' in the first case to 'unclean, winey, harsh' in the last. Isolates of *Absidia*, *Trichothecium* (*Cephalothecium*), *Cladosporium* (*Hormodendron*) and *Rhizopus* gave slight off-flavours. Beer brewed with malt contaminated with *R. oryzae* was reported by Gyllang, Satmark and Martinson (1977) to be distinctive, without having any special off-flavour, but *A. fumigatus* gave a pronounced roughness and a stale flavour.

Colour is also affected by the presence of moulds. Kneen (1963) noted that malt prepared from weathered barley produced beer with a high colour, as was the case when malt prepared after inoculation with *Asp. niger*, *Trichothecium* (*Cephalothecium*), *Rhizopus* and, especially, *Fusarium* sp. or *R. arrhizus* was used. Increased colour has subsequently been observed when *Asp. fumigatus*, *R. oryzae* (Gyllang, Satmark and Martinson, 1977), *F. avenaceum* and *F. culmorum* (Haikara, 1983) have been present.

Having mentioned earlier the apparent association between two *Fusarium* mycotoxins and gushing, it is appropriate to consider here the possibility that, as various mycotoxins are known to be toxic to a wide range of organisms, including yeasts, there may be some effect on fermentation. The fact that maltsters in Western countries use high quality cereals might be considered to preclude the possibility of appreciable levels of mycotoxins passing into commercial fermentations. However, grain that is used as a cheap starchy adjunct, e.g. maize, may be of much lower quality and may therefore contribute mycotoxins to the wort. It has been shown that mycotoxins such as zearalenone and trichothecenes can have a concentration-dependent effect on yeast growth (Schappert and Khachatourians, 1983, 1984; Flannigan, Morton and Naylor, 1985). The effects of different trichothecene toxins on log phase growth of *Saccharomyces cerevisiae* (Flannigan, Morton and Naylor, 1985) reflect their potency against animal systems. At comparable concentrations, T-2 strongly inhibited growth, DAS (Flannigan *et al.*, 1986) also retarded growth, but to a lesser extent, and DON had very little effect (Whitehead and Flannigan,

1989). With another yeast, *Kluyveromyces marxianus*, the inhibition produced by T-2, i.e. the reduction in dry mass relative to toxin-free controls over a period of 6 h at 25°C, was approximately 13 times that by DAS (Flannigan and Barnes, 1990). The difference in potency between members of the trichothecene family is also seen when yeast growth is allowed to continue into the stationary phase. For example, at the extremely high DON concentration of 50 μg DON ml^{-1} culture, the dry mass of the yeast was reduced by only 10% compared with toxin-free controls, i.e. not much more than one-half of the reduction caused by 2.5 μg T-2 ml^{-1}. The degree to which yeast viability is affected depends on both the toxin and its concentration, with DON having only slight effects (Whitehead and Flannigan, 1989; Flannigan, 1989).

Although Lafont, Romand and Lafont (1981) reported that T-2 and DAS had greater effects than aflatoxin B_1 and patulin on carbon dioxide evolution by *S. carlsbergensis*, their results for T-2 and DAS actually show that the velocity of gas production began to recover towards the end of the 4 h experiments, even when toxin was present at 50 μg ml^{-1}. In laboratory experiments with *S. cerevisiae* and *K. marxianus*, T-2 toxin was found to inhibit fermentation initially, causing a concentration-dependent lag in the attenuation of glucose and production of ethanol in Wickerham medium (Flannigan *et al.*, 1986; Flannigan and Barnes, 1990); at the high level of 10 μg T-2 ml^{-1} medium a 42 h lag in controls lengthened by around 120 h (Schapira, 1985). Despite this extended lag, however, the rate of attenuation recovered to roughly the same rate as controls (Flannigan *et al.*, 1986), this recovery being associated with a time-dependent increase in viable cell number (Schapira, 1985). Koshinsky, Cosby and Khachatourians (1992) have confirmed that ethanol production *per se* is not affected by T-2, and that any apparent inhibition of production results from inhibition of yeast growth.

As in laboratory malting investigations, DAS had a similar but less potent effect on fermentation than T-2 (Schapira, 1985; Flannigan, 1989), but concentrations of DON as high as 20 μg ml^{-1} had little effect on attenuation and ethanol production (Whitehead and Flannigan, 1989). Also as in malting investigations, the concentrations of these toxins producing the effects mentioned above are considerably higher than those encountered in even very heavily contaminated grain. However, as Flannigan (1989) has suggested, combinations of lower concentrations of different toxins acting in concert might contribute to slow starting or 'sticking' of fermentations. Other experiments (Schapira, 1985) indicate that increasing the pitching rate would probably overcome the effects of these combinations.

4.4.2 Distilled spirit

Using a ratio of 90 parts cooked maize to 10 parts lightly kilned distiller's malt in the mash, Kneen (1963) found that moulds appeared to have little effect on alcohol yield. He did note, however, that even 10% of heavily mould-contaminated malt could give off-flavours in the distillate. Isolates of *Cladosporium, Fusarium, Coniothyrium* and the yeast *Rhodotorula* produced marked off-flavours, with *Asp. niger* and an unidentified species of *Alternaria* giving less pronounced effects. There was not necessarily a correlation between these results and those for beer, e.g. *Rhizopus arrhizus* had an effect on the flavour of beer, but not of distillate.

Although Harrison and Graham (1970) stated that it requires a very high level of contamination in wort – about 10^8 bacteria ml^{-1} – before there is substantial competition with the yeast during fermentation, Dolan (1979) has pointed out the need to have the millroom and grist cases separate from the mash house and tun room in order to reduce contamination by microorganisms present in airborne malt dust.

4.5 HEALTH HAZARDS

4.5.1 Respiratory hazards

The hazards to the health of people in the grain industry arising from inhalation of high concentrations of grain dust over long periods have been reviewed by Dennis (1973). Some idea of the degree of hazard can be gauged from the fact that, in Saskatchewan, Williams, Skoulas and Merriman (1964) found that 54% of some 500 elevator agents had a history of a persistent cough, wheezing, breathlessness, dermatitis or 'grain fever'. In a follow-up investigation (Skoulas, Williams and Merriman, 1964), it was found that there was less impairment of lung function in those without these symptoms than in agents exhibiting them. Chronic bronchitis, pulmonary fibrosis and emphysema regularly develop over a period of years in such elevator agents.

Grain dust consists of more than particles of grain and chaff: soil fragments, bacteria, yeasts, spores and mycelium of moulds and actinomycetes, mites and insects, and their fragments and products are all present. It is now well understood that workers handling grain can become sensitized to microorganisms (Lacey, 1975) and mites (Dennis, 1973; Wraith, Cunnington and Seymour, 1979) in grain and develop allergic conditions. Although skin allergies may develop, allergic reactions most frequently occur in the respiratory system. Where the allergic reaction is provoked in the lungs, repeated exposure to the antigen, or allergen, which causes the allergy leads to degenerative changes in the lung tissue and serious respiratory impairment.

The spores of certain fungi and actinomycetes are potent sources of allergens, which are frequently associated with the spore wall. After inhalation, spores which are larger than 10 μm in diameter are largely deposited on the mucous membranes of the nasopharynx. Any allergic reaction they cause will be of a rhinitis or hay fever type. Spores in the range 4–10 μm are mainly deposited in the bronchi and larger bronchioles and provoke an asthmatic type of response, in which there is difficulty in breathing (dyspnoea). In both these cases the allergic responses are Type I reactions, i.e. the response is evident more or less immediately after exposure to the allergen and the effects last only a few hours. The Type I reaction is characteristic of the 10% of the population who are atopic. Such individuals are usually allergic to a wide range of allergens and react to spore concentrations which do not normally affect the remainder of the population, $<10^4–10^6\ m^{-3}$ air.

The greatest deposition of spores which are less than 4 μm in diameter occurs on the respiratory surface of the terminal bronchioles and air sacs, the alveoli; if the spores provoke an allergic response the condition is described as extrinsic allergic alveolitis. This is a Type III reaction and usually occurs when the non-atopic individual is exposed to greater concentrations ($>10^6\ m^{-3}$) of a particular type of spore containing a single specific allergen. Unlike the Type I reaction, it does not develop for about 4 h and may continue for 48 h or more. The Type III reaction is also unlike Type I because it is mediated by precipitating antibodies, and there is tissue damage. In the lung, there is a perivascular inflammatory response and granulomata are formed. The overt symptoms are fever, malaise, dry cough and breathlessness. With repeated attacks, progressive fibrosis of the lung occurs: it is permanently damaged, its efficiency is considerably reduced and there is increasing dyspnoea. Extreme respiratory distress is experienced if both Type I and Type III responses are provoked in the same individual, although this is a relatively rare event.

The types of spore associated with barley which are most likely to provoke Type I reactions include the common field fungi *Alt. alternata* and *Clad. herbarum*, with *Epicoccum nigrum* (*E. purpurascens*) causing rhinitis only. Type III reactions can be caused by the spores of *Penicillium* spp. and the thermophilic actinomycetes *Micropolyspora faeni* and *Thermoactinomyces vulgaris*, both of which are associated with self-heated barley and fodders (Lacey, 1975) and are the causal agents of farmer's lung. Although its spores are in the size range associated with Type III reactions, the storage fungus *Asp. fumigatus* may provoke both types of reaction. The clumping together of spores, with consequent deposition in the upper part of the respiratory system, may account for its involvement in a Type I reaction.

4.5.2 Maltworker's lung

Although Bridge (1932) noted the incidence and severity of chronic respiratory diseases, such as bronchitis and emphysema, in maltworkers, it was not until 1968 that the condition known as maltworker's lung was described (Riddle *et al.*, 1968). This extrinsic allergic alveolitis is now recognized as an occupational disease, like farmer's lung. It is caused by exposure to high concentrations of the spores of *Asp. clavatus*. This organism appears to be relatively rare in dry grain. It has, however, been found in barley in Canada (Wallace, 1973), Egypt (Abdel-Kader *et al.*, 1979) and Romania (Stankushev, 1969).

After first reports of the occurrence of this type of allergic alveolitis among workers in floor maltings (Riddle *et al.*, 1968; Channell *et al.*, 1969), 56 Scottish maltings were surveyed by Grant *et al.* (1976). Grant and his colleagues only detected *Asp. clavatus* in 12 of these. Of 711 maltworkers in the 56 maltings, nearly as many in box maltings (6.2%) as in floor maltings (6.8%) were affected by the disease. Overall, 5.2% of the workers surveyed had symptoms of maltworker's lung, but the incidence was lowest (1.1%) among those where modern mechanical systems were employed, e.g. drum maltings.

Where *Asp. clavatus* has become established in floor maltings, it can be seen sporing profusely on the surface of kernels. In consequence, clouds of spores are released into the atmosphere during turning of green malt (Riddle *et al.*, 1968). Other operations which further increase the spore load in the atmosphere are loading and stripping the kiln, and screening off the culms and rootlets. It is to be expected that enclosure of the malting barley, e.g. in drums, will result in a reduction of the number of spores in the air. The spore walls are particularly rich in allergenic substances which induce alveolitis (Blyth, 1978). Although it appears that maltworker's lung results from a Type III reaction, it has been suggested that the allergens in the spores also provoke a Type I reaction, which may be a component of the disease (Grant *et al.*, 1976).

As *Asp. clavatus* is rarely found in dry barley, the primary source of contamination of malting premises remains uncertain. Riddle *et al.* (1968) considered that pigeons which had free access to the grain store might have been implicated in one maltings, and it was suggested that the fungus might have been introduced with a particular consignment of grain in another (Channell *et al.*, 1969). Whatever the case, it is certain that once *Asp. clavatus* is introduced into a maltings it meets with favourable conditions for growth and sporulation. Although it only grows during germination of the grain, and perhaps also in the early stages of kilning, its spores are easily spread throughout the premises. In addition to isolating the organism from all parts of a distillery maltings, Channell *et al.* (1969) detected the organism in the mash house. They also isolated it from the sputum of employees who did not work as maltmen.

The growth of the mould becomes a particular problem when grain is germinated at elevated temperatures, e.g. when the temperature is deliberately raised to speed up malting (Riddle *et al.*, 1968), or when, during warm weather, there is no means of preventing it from rising above the normal 16°C (Flannigan *et al.*, 1984). It has been found in experimental malts that there may be a 10^4-fold increase in the viable counts of this organism if the malting temperature is raised from 16°C to 25°C (Flannigan *et al.*, 1984), the optimum temperature for growth of the fungus. Consequently, it is not surprising that the organism is one of the most frequently isolated toxigenic fungi (see section 4.5.3) in sorghum malt in South Africa (Rabie, Thiel and Marasas, 1983; Rabie and Lubben, 1984), where the temperature of malting may be 28°C. The presence of cracked corns (Riddle *et al.*, 1968), pre-germinated grain or accidentally introduced finished malt (Flannigan *et al.*, 1984), which provide easily colonized loci from which contamination can spread, may also lead to greater growth on the germinating barley. When the malting industry is taken as a whole, it would appear that the incidence of this allergenic fungus is low. However, when it does appear, the fact that its spores are spread so readily makes control of this organism extremely difficult. Even dry grain stored on the premises becomes contaminated by spores prior to malting, and addition of hypochlorite to the steep water in the concentrations often employed is not effective in killing these spores (Flannigan *et al.*, 1984). Riddle *et al.* (1968) have suggested that dilute hypochlorite solutions may directly stimulate growth of *Asp. clavatus*, but it appears more likely that hypochlorite promotes growth by reducing the numbers of other microorganisms competing with it. The higher concentrations used in cleaning floors or plant generally give only temporary relief from the problem, since the spores are still present on inaccessible surfaces. Caustic disinfectants are probably more effective than hypochlorite, but scrupulous attention to dust control and hygiene is necessary if the organism is to be controlled. Measures taken to control *Asp. clavatus* would also reduce the prevalence of other airborne fungi which were commonly isolated from the sputum of Scottish maltworkers and acted as sensitizing agents (Blyth *et al.*, 1977).

4.5.3 Mycotoxins

In the preceding sections, the effects of recognized mycotoxins and other mould metabolites on germination and malt characteristics were mentioned, and it is therefore appropriate to consider mycotoxins associated with brewing materials as they relate to health. Viewed on a worldwide scale, the most important and best researched of mycotoxins found in cereals (Flannigan, 1991) are the aflatoxins produced by *Asp. flavus* and *Asp. parasiticus*. They are not only toxic but also potent carcinogens. Other important mycotoxins are produced by various species of *Fusarium* associated with grain, including the oestrogenic toxin zearalenone, various trichothecenes, which are potent inhibitors of protein synthesis and are also immunosuppressive, and a group of toxins first reported in 1988, the fumonisins, which if not directly carcinogenic are cancer promoters. Two nephrotoxins, citrinin and ochratoxin A, produced by *P. verrucosum* and some other penicillia are not uncommonly found in European barley.

In section 4.3 it was noted that *Asp. clavatus* is toxigenic as well as allergenic, and it is therefore important to recognize that inhalation of the spores which cause the allergic disease maltworker's lung may also be a means by which its toxins enter the human system. Indeed, the author has been told that some victims of maltworker's lung also exhibited neurological symptoms, including tremor. Among the known mycotoxins which the fungus can synthesize are patulin, cytochalasin E and two tremorgenic compounds, tryptoquivaline and tryptoquivalone (Glinsukon *et al.*, 1974; Clardy *et al.*, 1975). Recently, it has been shown in a laboratory investigation of patulin and cytochalasin E production that, although there are strain differences, the fungus can produce these toxins during malting of both barley and wheat (T.M. Lopez, unpublished results). It is not known what the effects of natural mixtures of these various toxins are on humans, but mycotoxicosis and tumour development have been observed in experimental mice inoculated nasally with spores (Blyth and Hardy, 1982). This finding gives further reason for the use of efficient facemasks, respirators or air-flow helmets in maltings.

There may also be hazards to stock fed on malting by-products contaminated with this organism. Blyth *et al.* (1977) were able to isolate the mould from screened-off culms and rootlets, and in a recent investigation of a mycotoxicosis in the UK (Gilmour *et al.*, 1989) culms were found to be contaminated by the organisms at a level of 15×10^6 CFU g^{-1}. In this case, a supplementary feed containing culms from a distillery maltings as the principal ingredient was the cause of death among cattle, but not among sheep consuming the feed. However, in another very recent case there was 96% mortality in sheep fed on sprouted barley grains from an Israeli factory producing malt extract (Shlosberg *et al.*, 1991). In addition to *Asp. clavatus* growing and contaminating coleoptiles and rootlets during malting, it is also possible that it could adventitiously contaminate culms subsequent to kilning and later grow under damp storage conditions. It has been reported that the organism is able to grow on culms stored at RH 92.5% or higher, i.e. culms with an a_w of 0.925 or above will support growth and toxin production (Flannigan and Pearce, 1994).

It is not just culms or sprouted grains which have been involved in mycotoxicoses, however. Wet residues from sorghum beer production which had been spread out to dry and were found to be contaminated with *Asp. clavatus* were the cause of fatal tremorgenic disease among cattle (Kellerman *et al.*, 1976). In this case, the residues had been spread out to dry, but it is not known whether the mould was present in the residues beforehand or whether it contaminated the material during drying. Whichever, the practice of spreading the residues out to dry was considered to favour growth of the mould (Kellerman *et al.*, 1976). In laboratory experiments, Flannigan and Pearce (1994) found that *Asp. clavatus* could also grow on brewer's grains, but treatment of this material with propionic acid, which is applied to moist-stored barley for animal feed, could prevent growth of the fungus.

As far as beer is concerned, Flannigan (1989) has pointed out that the effects of some natural toxins in beer were recognized a long time ago, although their existence was not. For example, the use of the grass *Lolium temulentum* (temulentia = drunkenness) to 'fortify' or enhance the effects of beer was prohibited in France in the thirteenth century during the reign of Louis IX. In 1669, it was written that the presence of this 'weed or grayne' in barley crops in the west of Scotland had 'such a mischievous effect that one gill of ale or bere wherin such grayn hath been will fuddle a man more than a gallon of other drink' (Stones, 1984). Although it has not yet been demonstrated, there is a distinct possibility that the toxins in *L. temulentum* are alkaloid mycotoxins produced by endophytic fungi similar to those responsible for outbreaks of mycotoxicosis in stock feeding on infected pasture grasses. Somewhat more recently, circumstantial evidence indicated that a case of acute human illness in the USA was attributable to mycotoxins in beer (Cole *et al.*, 1983). An individual who drank a single can of beer developed neurological and other symptoms of toxicosis, which persisted for more than a day. A pellicle-like growth of *P. crustosum* was subsequently found in the empty can. When grown in a medium containing beer, the fungus produced four alkaloid mycotoxins which are active against the central nervous system, viz. festuclavine and three roquefortines. Since the can was reportedly normal at opening, being fully filled and carbonated, the circumstances under which the organisms entered, grew and produced toxin remain a mystery.

It is clear from various surveys which will be mentioned later that a variety of mycotoxins do reach the final product, beer, but generally only in limited concentrations. In experiments in which toxin was added at various stages during the malting and brewing processes, Gjertsen *et al.* (1973), Krogh *et al.* (1974), Chu *et al.* (1975) and Nip *et al.* (1975) found that little aflatoxin B_1, citrinin or ochratoxin A could be detected in beer. It appeared that about 40% of ochratoxin A in malt remained in the spent grains, with a further third being degraded and up to 20% being recoverable from the yeast after fermentation (Nip *et al.*, 1975). It was also noted (Gjertsen *et al.*, 1973; Krogh *et al.*, 1974) that, where high levels of ochratoxin A or citrinin were naturally present, the grain would have been unacceptable for malting because of its appearance or failure to germinate. Flannigan, Morton and Naylor (1985) deduced from work with toxin-spiked barley that most zearalenone and trichothecene T-2 toxin would be lost or degraded during malting, and El-Banna (1987) reported that 77% of the trichothecene DON added to barley was destroyed during germination. Mannio and Enari (1973) observed that no zearalenone or T-2 appeared to pass into beer from malt prepared from barley heavily contaminated with *F. culmorum*.

A group of mycotoxins which are currently the object of considerable research effort are the fumonisins produced by *F. moniliforme*, which were first discovered in South Africa in 1988. These toxins, which can cause fatal mycotoxicoses in pigs and horses, are of particular concern because they are cancer promoters and are widespread in maize in South Africa, Europe and the USA. Studies carried out at the USA Department of Agriculture fermentation laboratory in Peoria by Bothast *et al.* (1992) on α-amylase-treated maize with heavy natural fumonisin B_1 contamination showed that little of the toxin was degraded during fermentation to produce industrial ethanol. Some 85% of the toxin was recoverable, with most of this being in the distiller's dried grains, and much smaller quantities in the thin stillage and distiller's solubles. These results were in line with earlier investigations at that laboratory on utilization of aflatoxin- and zearalenone-contaminated maize, in that there was accumulation of toxin in the spent grains and none in the distilled ethanol. Taking into account this and other evidence, it is clear that mycotoxins are most unlikely ever to appear in distilled spirit.

However, zearalenone, DON and another trichothecene, nivalenol, have been found together in Korean malt (Lee *et al.*, 1985, 1986), and Lovelace and Nyathi (1977) detected up to 4.6 mg zearalenone l^{-1} in some home-brewed opaque maize beers in Zambia. In a survey of pito beer prepared from millet and/or red guinea corn in Nigeria, zearalenone was found in the products of 28 out of 46 breweries, although at 12.5–200 $\mu g\ l^{-1}$ the amounts were much lower than the extreme values in Zambia (Okoye, 1986). Subsequently, it was discovered that 51% of zearalenone in malt carries over into the final product (Okoye, 1987).

The survey of Okoye (1986) also confirmed earlier work which showed that contamination by aflatoxins was extensive (Okoye and Ekpenjong, 1984). In this first study, all 23 beers examined contained aflatoxin B_1. The concentrations ranged from 1.7 to 137 $\mu g\ l^{-1}$ and compared with the 92–262 μg aflatoxin l^{-1} reported by Alozie, Rotimi and Oyibo (1980) for traditional Nigerian beers. These concentrations contrast with those in a 5 year survey of commercial sorghum beer brewing in South Africa (Trinder, 1988). On average, the aflatoxin B_1 content of sorghum malt in this much larger survey was 2.18 $\mu g\ kg^{-1}$, with 68% of malts containing $\leq 1\ \mu g\ kg^{-1}$, 20% having 1–3 $\mu g\ kg^{-1}$ and only 4.5% yielding more than 10 $\mu g\ kg^{-1}$. While 39% of samples of strainings, normally separated from the wort before fermentation and bottling and then dried, contained the toxin (0.3–6.0 $\mu g\ kg^{-1}$), it was only detected in 5% of sorghum beers (0.05–0.13 $\mu g\ kg^{-1}$). On the basis of these results, it was concluded that consumption of sorghum beer was unlikely to constitute a health hazard in South Africa. In Europe, no aflatoxins were detected in 86 beers from various countries and 88 samples of grain, malt, hop pellets and other brewing materials (Woller and Majerus, 1982). However, using an immunochemical method with a detection limit of 0.2–1.0 $\mu g\ kg^{-1}$, Fukal, Prosek and Rakosova (1990) did detect aflatoxins in one-third of 37 malting barleys and one-fifth of 42 malts, although none was found in any of 34 beers examined.

In contrast to the apparent widespread incidence of zearalenone in regional African beers, only one of four surveys of beers in Europe has detected the toxin, and even then only in one sample (Payen *et al.*, 1983). A possible reason for this is that zearalenone in wort is apparently rapidly reduced to zearalenol during fermentation (Scott *et al.*, 1992). In their study of 49 European light lagers, Payen *et al.*. (1983) did not find other *Fusarium* toxins in more than a few samples either, i.e. the trichothecenes DON, DAS and T-2 were present in only one, two and three samples, respectively. When Cerrutti *et al.* (1987) carried out a thin layer chromatography examination of 24 beers imported into Italy, as in an earlier investigation of Italian beers, they failed to detect any of these *Fusarium* toxins or aflatoxins, sterigmatocystin, citrinin and ochratoxin A. However, the last-named toxin, most probably originating from grainborne *P. verrucosum* or other penicillia, but possibly from *Asp. ochraceus*, was found in five lagers tested by Payen *et al.* (1983). Tressl, Hommel and Helak (1989) were also able to detect ochratoxin in beer by high performance liquid chromatography, as well as in barley and malt. The detection limit in all three was 5 $\mu g\ kg^{-1}$.

The European evidence on zearalenone is backed up by a recent Canadian survey of 50 samples of beer brewed in or imported into Canada (Scott, Kanhere and Weber, 1993). No zearalenone was detected, but the investigation did reveal DON in 29 samples, at levels ranging from 0.33 to 50.3 ng ml^{-1}, and nivalenol in three, at 0.10–0.84 ng ml^{-1}. The incidence of DON in Canadian cereals has been of concern in recent years, and Scott *et al.* (1992) showed that in Germany gushing in beer was associated with the abundance of this toxin in cereals in 1987 (Lepschy-von Gleissenthal *et al.*, 1989; Müller and Schwadorf, 1993) and 1991 (Niessen *et al.*, 1993). Other *Fusarium* toxins found in the German crops included 3-acetyldeoxynivalenol (3-AcDON), DAS, nivalenol, T-2 and zearalenone (Lepschy-von Gleissenthal *et al.*, 1989; Müller and Schwadorf, 1993). In a survey of 196 samples of beers from several breweries, Niessen *et al.* (1993) found that three-quarters of the beers brewed from wheat malt (Weiß Bier) and around one-quarter brewed from barley malt contained DON. As the breweries were mostly in southern Germany and around 20% of the beers had been selected on suspicion of gushing, the amounts of toxin detected give rather a skewed picture. Nevertheless, the levels in the wheat beers (mean 0.25 ng ml^{-1}; maximum 0.57 ng ml^{-1}) were significantly higher than in beers derived from barley malt (0.15 ng ml^{-1}; 0.48 ng ml^{-1}).

It is impossible to assess the real importance of mycotoxins in beer to human health, but it is worth noting again that yeasts may bring about conversion of mycotoxins to both less and more toxic compounds, which may not be assayed. The oestrogenic and tumour-promoting toxin zearalenone can be taken as an example. In suggesting that there was a risk from mycotoxins in beer Schoental (1984) commented on the fact that the toxin is relatively common in barley and may be produced during malting (Haikara, 1983), but yet is apparently uncommon in beer. She suggested that a reason for this is that during fermentation it is reduced to α-zearalenol, a compound which apparently is ten times more oestrogenic than the zearalenone from which it is derived and therefore presents a greater health risk. As mentioned earlier in this chapter, zearalenone does appear to be reduced rapidly to zearalenol during fermentation. However, the principle derivative was β-zearalenol, not its isomer, α-zearalenol (Scott *et al.*, 1992), which is much less oestogenic.

At present, the evidence suggests that there is a greater risk to animal health than to human health from mycotoxins associated with beer production. Aflatoxins, fumonisins and zearalenone (Bothast *et al.*, 1992) have been shown to accumulate, and trichothecenes (Mannio and Enari, 1973) and ochratoxin A (Nip *et al.*, 1975) to be present in significant amounts in spent grains. As Bothast *et al.* (1992) has pointed out, it might be necessary to detoxify spent grains containing large amounts of mycotoxins, e.g. by formaldehyde or ammoniation, before they are used to feed farm animals.

4.6 ASSESSMENT OF MOULD CONTAMINATION

Since it is the evaluation of the fungal contamination of barley and malt which is likely to present the greatest problem to the brewing microbiologist, it is worth noting that methods for assessment of the percentage contamination of kernels by different moulds have been set out in the European Brewery Convention *Analytica Microbiologica*, Part II (Moll, 1981). In addition to direct plating without surface disinfection, Flannigan (1977) has described dilution plating methods for viable mould counts. A general text of use for the identification of moulds is *Smith's Introduction to Industrial Mycology* (Onions, Allsopp and Eggins, 1981), which includes references to monographs for the identification of species in large or difficult genera such as *Aspergillus, Fusarium* and *Penicillium.* Another extremely useful aid to identification is *Introduction to Food-Borne Fungi* by Samson *et al.* (1995). In addition to having extremely well-illustrated keys, this book has chapters on detection and quantification of moulds and on mycotoxins and their detection, as well as appended mycological media.

Although for some time there will continue to be reliance on traditional methods for detection, quantification and identification of moulds, rapid methods are being developed and adopted. For example, an enzyme-linked immunosorbent assay (ELISA) has been developed and used at Carlsberg Breweries for detection of *Fusarium* in barley and malt (Vaag, 1991; Vaag and Pedersen, 1993), with very high levels of *Fusarium* antigen indicating a high risk of gushing. Another immunological method has been employed by Schwabe *et al.* (1993, 1994), who have used a latex agglutination test employing an antibody to extracellular polysaccharide to assess the prevalence of *Fusarium* in barley, and have detected a quantitative relationship between test results and DON/AcDON levels. Whilst these immunological methods are genus specific, there is scope for developing species-specific methods. Highly specific monoclonal antibody-based ELISA and dip-stick methods for a single grainborne storage fungus, *Penicillium islandicum*, have already been developed (Dewey *et al.*, 1990). In addition to such immunological methods, research in various laboratories in Europe and North America is being directed to using molecular methods in plant pathology and food microbiology. One type of approach is to develop a rapid amplified polymorphic DNA method (RAPD) for typing particular organisms. For example, Ouellet and Seifert (1993) have employed RAPD to distinguish the head blight fungus *F. graminearum* from other species of *Fusarium* and to track the organism in the field.

REFERENCES

Abdel-Kader, M.I.A., Moubasher, A.H. and Abdel-Hafez, S.I.I. (1979) *Mycopathologia*, **68**, 143.

Alozie, T.C., Rotimi, C.N. and Oyibo, B.B. (1980) *Mycopathologia*, **70**, 125.

Amaha, M., Kitabatake, K., Nakagawa, A. *et al.* (1973) *Proceedings of the 14th Congress of the European Brewery Convention, Salzburg*, IRL Press, Oxford p. 381.

Amaha, M., Kitabatake, K., Nakagawa, A., Yoshido, J. and Harada, T. (1974) *Bulletin of Brewing Science*, **20**, 35.

Andersen, K., Gjertsen, P. and Trolle, B. (1967) *Brewers' Digest*, **42**, 76.

Armolik, N., Dickson, J.G. and Dickson, A.D. (1956) *Phytopathology*, **46**, 457.

Blyth, W. (1978) *Clinical and Experimental Immunology*, **32**, 272.

Blyth, W. and Hardy, J.C. (1982) *British Journal of Cancer*, **45**, 105.

Blyth, W., Grant, I. W.B., Blackadder, E.S. and Greenberg, M. (1977) *Clinical Allergy*, **7**, 549.

Boruff, C.S., Claassen, R.I. and Sotier, A.L. (1938) *Cereal Chemistry*, **15**, 451.

Bothast, R.J., Bennett, G.A., Vancauwenberge, J.E. and Richard, J.L. (1992) *Applied and Environmental Microbiology*, **58**, 233.

Bridge, J.C. (1932) *Annual Report of the Chief Inspector of Factories*, HMSO, London.

Briggs, D.E. (1966) *Nature*, **210**, 418.

Briggs, D.E. and McGuiness, G. (1993) *Journal of the Institute of Brewing*, **99**, 249.

Brodnik, T. (1975) *Seed Science and Technology*, **3**, 691.

Bu'lock, J.D. (1984) In *The Applied Mycology of Fusarium* (eds M.O. Moss and J.E. Smith), Cambridge University Press, Cambridge, p. 215.

Cerrutti, G., Vecchio, A., Finoli, C. and Trezzi, A. (1987) *Monatsschrift für Brauwissenschaft*, **40**, 455.

Channell S., Blyth, W., Lloyd, M. *et al.* (1969) *Quarterly Journal of Medicine*, **38**, 351.

Christensen, C.M. (1973) *Seed Science and Technology*, **1**, 547.

Christensen, C.M. and Kaufmann, H.H. (1969) *Grain Storage, The Role of Fungi in Quality Loss*, University of Minnesota Press, Minneapolis.

Christensen, C.M. and Kaufmann, H.H. (1974) In *Storage of Cereal Grain and its Products* (ed. C.M. Christensen), American Association of Cereal Chemists, St Paul, p. 158.

Christensen, J.J. and Stakman, E.C. (1935) *Phytopathology*, **25**, 309.

Chu, F.S., Chang C.C., Ashoor, S.H. and Prentice, N. (1975) *Applied Microbiology*, **29**, 313.

Clardy, J., Springer, J.P., Buechi, G. *et al.* (1975) *Journal of the American Chemical Society*, **97**, 663.

Clarke, J.H. and Hill, S.T. (1981) *Transactions of the British Mycological Society*, **77**, 557.

Clarke, J.H., Hill, S.T. and Niles, E.V. (1966) *Pest Infestation Research, 1965*, 13.

Clarke, J.H., Niles, E.V. and Hill, S.T. (1967) *Pest Infestation Research, 1966*, 14.
Clarke, J.H., Hill, S.T., Niles, E.V. and Howard, M.A.R. (1969) *Pest Infestation Research, 1968*, 17.
Cole, R.J., Dorner, J.W., Cox, R.H. and Raymond, L.W. (1983) *Journal of Agricultural and Food Chemistry*, **31**, 655.
Dennis, C.A.R. (1973) In *Grain Storage: Part of a System* (eds R.N. Sinha and W.E. Muir), Avi, Westport, CT, p. 367.
Dewey, F.M., MacDonald, M.M., Phillips, S.I. and Priestley, R.A. (1990) *Journal of General Microbiology*, **136**, 753.
Dolan, T.C.S. (1979) *The Brewer*, **65**, 60.
Donhauser, S., Weideneder, A., Winnewisser, W. and Geiger, E. (1989) *Brauwelt*, **129**, 1658.
Doran, P.J. and Briggs, D.E. (1993) *Journal of the Institute of Brewing*, **99**, 165.
Douglas, P.E. (1984) A microbiological investigation of industrial malting of barley, MSc Thesis, Heriot-Watt University.
Douglas, P.E. and Flannigan, B. (1988) *Journal of the Institute of Brewing*, **94**, 85–88.
Drysdale, R.B. (1984) In *The Applied Mycology of Fusarium* (eds M.O. Moss and J.E. Smith), Cambridge University Press, Cambridge, p. 95.
El-Banna, A.A. (1987) *Mycotoxin Research*, **3**, 37.
Etchevers, G.C., Banasik, O.J. and Watson, C.A. (1977) *Brewers' Digest*, **52**, 46.
Ferguson, P.E., Flannigan, B. and Schapira, S.F.D. (1984) *SIM News* **34 (4)**, 45–46.
Flannigan, B. (1969) *Transactions of the British Mycological Society*, **53**, 371.
Flannigan, B. (1970) *Transactions of the British Mycological Society*, **55**, 267.
Flannigan, B. (1974) *Transactions of the British Mycological Society*, **62**, 51.
Flannigan, B. (1977) In *Biodeterioration Investigation Techniques* (ed. A.H. Walters), Applied Science, London, p. 185.
Flannigan, B. (1978) *Transactions of the British Mycological Society*, **71**, 37.
Flannigan, B. (1983) *Journal of the Institute of Brewing*, **89**, 364.
Flannigan, B. (1989) In *Biodeterioration Research II* (eds G.C. Llewellyn and C.E. O'Rear), Plenum, New York, p. 191.
Flannigan, B. (1991) In *Toxic Substances in Crop Plants* (eds J.P.F. D'Mello, C.M. Duffus and J.H. Duffus), Royal Society of Chemistry, London, p. 226.
Flannigan, B. and Bana, M.S.O. (1980) In *Biodeterioration, Proceedings of the Fourth International Symposium, Berlin* (eds T.A. Oxley, D. Allsopp and G. Becker), Pitman, London, p. 229.
Flannigan, B. and Barnes, S.D. (1990) *Proceedings of the Japanese Association of Mycotoxicology*, **31**, 25.
Flannigan, B. and Campbell, 1. (1977) *Transactions of the British Mycological Society*, **69**, 485.
Flannigan, B. and Healy, R.E. (1983) *Journal of the Institute of Brewing*, **89**, 341.
Flannigan, B. and Pearce, A.R. (1994) In *Biology of Aspergillus* (eds K.A. Powell, J. Peberdy and E. Renwick), Plenum, New York, p. 115.
Flannigan, B., Healy, R.E. and Apta, R. (1986) In *Biodeterioration 6* (eds S. Barry, D.R. Houghton, G.C. Llewellyn and C.E. O'Rear), Commonwealth Agricultural Bureaux International, Farnham Royal, p. 300.
Flannigan, B., Morton, J.G. and Naylor, R.J. (1985) In *Trichothecenes and Other Mycotoxins* (ed. J. Lacey), Wiley, New York, p. 171.
Flannigan, B., Okagbue, R.N., Khalid, R. and Teoh, C.K. (1982) *Brewing and Distilling International*, **12**, 31.
Flannigan, B., Day, S.W., Douglas, P.E. and McFarlane, G.B. (1984) In *Toxigenic Fungi – Their Toxins and Health Hazard* (eds K. Kurata and Y. Ueno), Kodansha/ Elsevier, Tokyo, p. 52.
Flannigan, B., Naylor, R.J., Prescott, G.M. and Schapira, S.F.D. (1986) In *Biodeterioration 6* (eds S. Barry, D.R. Houghton, G.C. Llewellyn and

C.E. O'Rear), Commonwealth Agricultural Bureaux International, Farnham Royal, p. 238.
Follstad, M.N. and Christensen, C.M. (1962) *Applied Microbiology*, **10**, 331.
Fukal, L., Prosek, J. and Rakosova, A. (1990) *Monatsschrift für Brauwissenschaft*, **43**, 212.
Fukushima, S., Kitabatake, K. and Amaha, M. (1976) *Bulletin of Brewing Science*, **22**, 37.
Gaber, S.D. and Roberts, E.H. (1969) *Journal of the Institute of Brewing*, **75**, 303.
Gilman, J.C. and Barron, D.H. (1930) *Plant Physiology*, **5**, 565.
Gilmour, J.S., Inglis, D.M., Robb, J. and Maclean, M. (1989) *Veterinary Record*, **124**, 133.
Gjertsen, P., Trolle, B. and Andersen, K. (1963) *Proceedings of the 9th Congress of the European Brewery Convention, Stockholm*, IRL Press, Oxford, p. 320.
Gjertsen, P., Trolle, B. and Andersen, K. (1965) *Proceedings of the 10th Congress of the European Brewery Convention, Brussels*, IRL Press, Oxford, p. 428.
Gjertsen, P., Myken, F., Krogh, P. and Hald, B. (1973) *Proceedings of the 14th Congress of the European Brewery Convention, Salzburg*, IRL Press, Oxford, p. 373.
Glinsukon, T., Yuan, S.S., Wightman, R., Kitaura, Y., Buechi, G., Shank, R.C., Wogan, G.N. and Christensen, C.M. (1974) *Plant Foods for Man*, **1**, 113
Graff, A.R. (1972) *MBAA Technical Quarterly*, **9**, 18.
Grant, I.W.B., Blackadder, E.S., Greenberg, M. and Blyth, W. (1976) *British Medical Journal* 1, 490.
Gyllang, H. and Martinson, E. (1976a) *Journal of the Institute of Brewing*, **82**, 182.
Gyllang, H. and Martinson, E. (1976b) *Journal of the Institute of Brewing*, **82**, 350.
Gyllang, H., Satmark, L. and Martinson, E. (1977) *Proceedings of the 16th Congress of the European Brewery Convention, Amsterdam*, IRL Press, Oxford, p. 245.
Gyllang, H., Kjellen, K., Haikara, A. and Sigsgaard, P. (1981) *Journal of the Institute of Brewing*, **87**, 248.
Haikara, A. (1983) *Proceedings of the 20th Congress of the European Brewery Convention, London*, IRL Press, Oxford, p. 401.
Haikara, A., Makinen, V. and Hakulinen, R. (1977) *Proceedings of the 16th Congress of the European Brewery Convention, Amsterdam*, IRL Press, Oxford, p. 35.
Harper, S.H.T. and Lynch, J.M. (1979) *Journal of General Microbiology*, **112**, 45.
Harper, S.H.T. and Lynch, J.M. (1981) *Journal of General Microbiology*, **122**, 55.
Harrison, J.S. and Graham, J.C.J. (1970) In *The Yeasts*, Vol. 3, *Yeast Technology* (eds A.H. Rose and J.S. Harrison), Academic Press, London, p. 283.
Harrison, J.G. and Perry, D.A. (1976) *Annals of Applied Biology*, **84**, 57.
Healy, R.E. (1985) Effects of the microflora of barley on malt characteristics, PhD Thesis, Heriot-Watt University.
Heaton, P.E., Callow, M.E. and Butler, G.M. (1992) *International Biodeterioration and Biodegradation*, **29**, 135.
Hellberg, A. and Kolk, H. (1972) *Acta Agriculturae Scandinavica*, **22**, 137.
Hill, R.A. and Lacey, J. (1983a) *Annals of Applied Biology*, **102**, 455.
Hill, R.A. and Lacey, J. (1983b) *Annals of Applied Biology*, **103**, 467.
Howe, R.V. (1973) *Seed Science and Technology*, **1**, 563.
Ichinoe, M., Hagiwara, H. and Kurata, H. (1984) In *Toxigenic Fungi – Their Toxins and Health Hazard* (eds H. Kurata and Y. Ueno), Kodansha / Elsevier, Tokyo, p. 190.
Jacobson, R.J. and Thomas, K.P. (1981) *Plant Pathology*, **30**, 54.
Jorgensen, J. (1969) *Friesia*, **9**, 97.
Jorgensen, J. (1983) *Seed Science and Technology*, **11**, 615.
Kellerman, T.S., Pienaar, J.G., van der Westhuizen, G.C.A., Anderson, L.A.P. and Naude, T.W. (1976) *Onderstepoort Journal of Veterinary Research*, **43**, 147.
Kelly, L. and Briggs, D.E. (1992a) *Journal of the Institute of Brewing*, **98**, 329.
Kelly, L. and Briggs, D.E. (1992b) *Journal of the Institute of Brewing*, **98**, 395.
Kelly, L. and Briggs, D.E. (1993) *Journal of the Institute of Brewing*, **99**, 57.

Kitabatake, K. and Amaha, M. (1974) *Bulletin of Brewing Science*, **20**, 1.
Kitabatake, K. and Amaha, M. (1976) *Bulletin of Brewing Science*, **22**, 37.
Kneen, E. (1963) *Proceedings of the Irish Maltsters Technical Conference, 1963*, Irish Maltsters Association, Dublin, p. 51.
Koshinsky, H.A., Cosby, R.H. and Khachatourians, G.G. (1992) *Biotechnology and Applied Biochemistry*, **16**, 275.
Kotheimer, J.B. and Christensen, C.M. (1961) *Wallerstein Laboratory Communications*, **24**, 21.
Krogh, P., Hald, B., Gjertsen, P. and Myken, F. (1974) *Applied Microbiology*, **28**, 31.
Lacey, J. (1971) *Annals of Applied Biology*, **69**, 187.
Lacey, J. (1975) *Transactions of the British Mycological Society*, **65**, 171.
Lacey, J., Hill, S.T. and Edwards, M.A. (1980) *Tropical Stored Products Information*, **39**, 19.
Lafont, J., Romand, A. and Lafont, P. (1981) *Mycopathologia*, **74**, 119.
Lee, U.-S., Jang, H.-S., Tanaka, T. *et al.* (1985) *Food Additives and Contaminants*, **2**, 185.
Lee, U.-S., Jang, H.-S., Tanaka, T. *et al.* (1985) *Food Additives and Contaminants*, **3**, 253.
Lepschy-von Gleissenthal, J., Dietrich, R., Martlbauer, E. *et al.*(1989) *Zeitschrift für Lebensmittel Untersuchung und Forschung*, **188**, 521.
Lovelace, C.E.A. and Nyathi, C.B. (1977) *Journal of the Science of Food and Agriculture*, **28**, 288.
Ludwig, R.A. (1957) *Canadian Journal of Botany*, **35**, 291.
Lund, A. (1956) *Friesia*, **5**, 297.
Lund, A., Pedersen, H. and Sigsgaard, P. (1971) *Journal of the Science of Food and Agriculture*, **22**, 458.
Lutey, R.W. and Christensen, C.M. (1963) *Phytopathology*, **53**, 713.
Lynch, J.M. and Prynn, S.J. (1977) *Journal of General Microbiology*, **103**, 193.
Machacek, J.E. and Wallace, H.A.H. (1952) *Canadian Journal of Botany*, **30**, 164.
Machacek, J.E., Cherewick, W.J., Mead, H.W. and Broadfoot, W.C. (1951) *Scientific Agriculture*, **31**, 193.
Malt Research Institute (1955) *Publication of the Malt Research Institute*, **12**, 2.
Mannio, M. and Enari, T. (1973) *Brauwissenschaft*, **26**, 134.
Mason, F.A. (1923) *Bulletin of the Bureau of Biotechnology*, **11/12**, 78.
Matthews, S. and Collins, M.T. (1975) *Seed Science and Technology*, **3**, 863.
Mead, H.W. (1943) *Wallerstein Laboratory Communications*, **6**, 26.
Mills, J.T. and Wallace, H.A.H. (1979) *Canadian Journal of Plant Science*, **59**, 645.
Moll, M. (1981) *Journal of the Institute of Brewing*, **87**, 303.
Müller, H-M. and Schwadorf, K. (1993) *Mycopathologia*, **121**, 115.
Niessen, L., Donhauser, S., Weideneder, A., *et al.* (1992) *Brauwelt* **132**, 702.
Niessen, L., Böhm-Schraml, M., Vogel, H., and Donhauser, S. (1993) *Mycotoxin Research*, **9**, 99.
Nip, W.K., Chang, C.C., Chu, F.S. and Prentice, N. (1975) *Applied Microbiology*, **30**, 1048.
Nummi, M., Niku-Paavola, M.-L. and Enari, T.-M. (1975) *Brauwissenschaft*, **28**, 130.
Okoye, Z.S.C. (1986) *Journal of Food Safety*, **7**, 233.
Okoye, Z.S.C. (1987) *Food Additives and Contaminants*, **4**, 57.
Okoye, Z.S.C. and Ekpenjong (1984) *Transactions of the Royal Society of Tropical Medicine and Hygiene*, **78**, 417.
Onions, A.H.S., Allsopp, D. and Eggins, H.O.W. (1981) *Smith's Introduction to Industrial Mycology*, Arnold, London.
Ouellet, T. and Seifert, K.A. (1993) *Phytopathology*, **83**, 1003.
Payen, J., Girard, T., Gaillardin, M. and Lafont, P. (1983) *Microbiologie-Aliments-Nutrition*, **1**, 143.
Pepper, E.H. and Kiesling, R.L. (1963) *Proceedings of the Association of Official Analytical Chemists*, **53**, 199.

Petters, H.I., Flannigan, B. and Austin, B. (1988) *Journal of Applied Bacteriology*, **65**, 279.
Pidoplichko, M.N., Moskovels, V.S. and Zholanova, N.M. (1960) *Mikrobiolohichnyi Zhurnal*, **22**, 15.
Pixton, S.W. and Henderson, S. (1981) *Journal of Stored Products Research*, **17**, 191.
Pixton, S.W. and Warburton, S. (1971) *Journal of Stored Products Research*, **7**, 261.
Ponchet, J. (1966) Etude des communautes mycopericarpiques du caryopse de ble, DSc Thesis, University of Paris.
Prentice, N. and Sloey, W. (1960) *Proceedings of the American Society of Brewing Chemists*, p. 28.
Rabie, C.J. and Lubben, A. (1984) *South African Journal of Botany*, **3**, 251.
Rabie, C.J., Thiel, P.G. and Marasas, W.F.O. (1983) *Third International Mycological Congress, Tokyo, Abstracts*, Mycological Society of Japan, Tokyo, p. 599.
Riddle, H.F.V., Channell, S., Blyth, W., *et al.* (1968) *Thorax*, **23**, 271.
Samson, R.A., Hoekstra, E.S., Frisrad, J.C. and Filtenborg, O. (1995) *Introduction to Food-Borne Fungi*, 4th edn, Centraalbureau voor Schimmelcultures, Baarn, The Netherlands, 299 pp.
Schapira, S.F.D. (1985) The effect of trichothecene mycotoxins on yeast and malting barley, MSc Thesis, Heriot-Watt University.
Schapira, S.F.D., Whitehead, M.P. and Flannigan, B. (1989) *Journal of the Institute of Brewing*, **95**, 415.
Schappert, K.T. and Khachatourians, G.G. (1983) *Applied and Environmental Microbiology*, **45**, 862.
Schappert, K.T. and Khachatourians, G.G. (1984) *Applied and Environmental Microbiology*, **47**, 681.
Schoental, R. (1984) *Proceedings of the Fifth Meeting on Mycotoxins in Animal and Human Health, Edinburgh, August 1984*, (eds M.O. Moss and M. Frank), University of Surrey, Guildford, pp. 62–67.
Schwabe, M., Rath, F., Golomb, A., *et al.* (1993) *Monatsschrift für Brauwissenschaft*, **46**, 408.
Schwabe, M., Fenz, R., Engels, R., *et al.* (1994) *Monatsschrift für Brauwissenschaft*, **47**, 160.
Scott, P.M., Kanhere, S.R. and Weber, D. (1993) *Food Additives and Contaminants*, **10**, 381.
Scott, P.M., Kanhere, S.R., Daley, E.F. and Farber, J.M. (1992) *Mycotoxin Research*, **8**, 58.
Sheneman, J.M. and Hollenbeck, C.M. (1960) *Proceedings of the American Society of Brewing Chemists*, p. 22.
Shlosberg, A., Zadikov, I., Perl, S. *et al.* (1991) *Mycopathologia*, **114**, 35–39.
Sinha, R.N. (1973) In *Grain Storage: Part of a System* (eds R.N. Sinha and W.E. Muir), Avi, Westport, Connecticut, p. 15.
Sinha, R.N. and Wallace, H.A.H. (1965) *Canadian Journal of Plant Science*, **45**, 48.
Skoulas, A., Williams, N. and Merriman, J.E. (1964) *Journal of Occupational Medicine*, **6**, 359.
Sloey, W. and Prentice, N. (1962) *Proceedings of the American Society of Brewing Chemists*, p. 24.
Smith, J.E., Mitchell, I. and Chiu, M.L.C. (1984) In *The Applied Mycology of Fusarium* (eds M.O. Moss and J.E. Smith), Cambridge University Press, Cambridge, p. 157.
Stankushev, K. (1969) *Mikologia i Fitopatologia*, **3**, 268.
Stones, E.L.G. (1984) In *Miscellany II* (ed. D. Sellar), The Stair Society, Edinburgh, p. 148.
Tressl, R., Hommel, E., Helak, B. (1989) *Monatsschrift für Brauwissenschaft*, **42**, 331.
Trinder, D.W. (1988) *Journal of the Institute of Brewing*, **95**, 307.
Tuite, J.F. and Christensen, C.M. (1955) *Cereal Chemistry*, **32**, 1.

Tuite, J.F. and Christensen, C.M. (1957) *Phytopathology*, **47,** 265.
Uoti, J. and Ylimaki, A. (1974) *Annales Agriculturae Fenniae*, **13,** 5.
Vaag, P. (1991) *Proceedings of the 23rd Congress of the European Brewery Convention, Lisbon*, IRL Press, Oxford, p. 553.
Vaag, P. and Pedersen, S. (1993) *Proceedings of the 24th Congress of the European Brewery Convention, Oslo*, IRL Press, Oxford, p. 111.
Valadon, L.R.G. and Lodge, E. (1970) *Transactions of the British Mycological Society*, **55,** 9.
Wallace, H.A.H. (1973) In *Grain Storage: Part of a System* (eds R.N. Sinha and W.E. Muir), Avi, Westport, Connecticut, p. 71.
Wallace, H.A.H. and Sinha, R.N. (1962) *Canadian Journal of Plant Science*, **42,** 130.
Warnock, D.W. (1971) *Journal of General Microbiology*, **67,** 197.
Warnock, D.W. (1973a) *Transactions of the British Mycological Society*, **61,** 49.
Warnock, D.W. (1973b) *Transactions of the British Mycological Society*, **61,** 547.
Warnock, D.W. and Preece, T.F. (1971) *Transactions of the British Mycological Society*, **56,** 267.
Welling, B. (1969) *Tidsskrift for Planteavl*, **73,** 291.
Whitehead, M.P. and Flannigan, B. (1989) *Journal of the Institute of Brewing*, **95,** 411.
Williams, N., Skoulas, A. and Merriman, J.E. (1964) *Journal of Occupational Medicine*, **6,** 319.
Woller, R. and Majerus, P. (1992) *Monatsschrift für Brauwissenschaft*, **35,** 88.
Wraith, D.G., Cunnington, A.M. and Seymour, W.M. (1979) *Clinical Allergy*, **9,** 545.
Ylimaki, A., Koponen, H., Hintikka, E., *et al.* (1979) *Technical Research Centre of Finland, Materials and Processing Technology, Publication 21.*
Yoshida, J., Nakagawa, A., Eto, M., Kitabatake, K. and Amaha, M. (1975) *Journal of Fermentation Technology*, **53,** 184.
Zadoks, J.C., Chang, T.T. and Konzak, C.F. (1974) *Weed Research*, **14,** 415.

CHAPTER 5

Gram-positive brewery bacteria

F.G. Priest

5.1 INTRODUCTION

Bacteria have been recognized as important spoilage agents of beer since the end of the nineteenth century. Comparison of the descriptions of bacteria prevalent in breweries and beers at that time with those of today reveals that the range of bacteria encountered is remarkably constant although the names have been changed frequently over the years. Recently, as new processes have been introduced, in particular the packaging of beers with very low oxygen contents, and microbiological techniques have been improved, so 'new' contaminants have been isolated from spoiled beers. These include the strictly anaerobic strains of *Pectinatus cerevisiiphilus* and *Megasphaera* (see Chapter 6) but in general brewers are facing today the same bacterial contaminants as they did last century.

Bacteria are traditionally divided into two categories according to their reaction to a differential staining procedure, the Gram stain. Cells that are bounded by a relatively thick, single layer of peptidoglycan, a net-like polymeric molecule that provides structural strength to the cell, generally retain the crystal violet/iodine complex of the Gram stain when washed with ethanol or acetone and are referred to as Gram positive when subjected to the Gram stain. These cells appear purple under the microscope. Gram-negative bacteria, on the other hand, possess a multilayered envelope comprising the cytoplasmic membrane, the periplasmic space, a thin layer of peptidoglycan and a hydrophobic outer membrane. Such cells do not retain the crystal violet/iodine complex when washed with ethanol or acetone and appear pink under the microscope when stained by Gram's procedure. Shimwell (1945) was first to report the particular relevance of the Gram reaction to brewery microbiology, when he noted that the growth

Brewing Microbiology, 2nd edn. Edited by F. G. Priest and I. Campbell.
Published in 1996 by Chapman & Hall, London. ISBN 0 412 59150 2

of most Gram-positive bacteria was inhibited by hop bittering substances, but Gram-negative bacteria were unaffected. This antiseptic action of hops, together with the poor nutrient status of beer, its low pH and the presence of ethanol, restricts the range of bacteria that can grow in finished beers. With regard to the Gram-positive organisms, lactic acid bacteria of the genera *Lactobacillus* and *Pediococcus* are the most common and dangerous spoilage agents and strains of some species have developed marked hop tolerance (Fernandez and Simpson, 1993). Some members of the genera *Micrococcus* and *Staphylococcus* can survive in beer and under certain conditions *Micrococcus kristinae* will grow and cause spoilage. On rare occasions aerobic, spore-forming bacteria belonging to the genus *Bacillus* have caused problems, particularly in unhopped worts (Healy and Armitt, 1980). However, like the micrococci, these bacteria are inhibited by hop components (Smith and Smith, 1993), do not appear to develop hop tolerance and are therefore not generally a serious threat. This chapter will review the physiology, taxonomy, isolation and enumeration of these various bacteria and will cover their effect on beer quality.

5.2 LACTIC ACID BACTERIA

The lactic acid bacteria are closely related to the endospore-forming bacteria of the genera *Bacillus* and *Clostridium* and, with numerous other genera, form a major division of the bacteria known as the low G + C group because they all possess a relatively low (< 55%) content of guanine plus cytosine in their chromosomal DNA. The lactic acid bacteria share considerable genetic and molecular homology and have long been recognized as a natural group (Ingram, 1975). However, a definitive description of the lactic acid bacteria cannot be agreed upon and with current developments in the taxonomy of these bacteria is probably more distant now than at any previous time. However, the typical lactic acid bacterium is a Gram- positive, non-sporulating rod or coccus. It lacks the enzyme catalase and is strictly fermentative producing either a mixture of lactic acid, CO_2, acetic acid and/or ethanol (heterofermentation) or almost entirely lactic acid (homofermentation) as the major metabolic end-product from sugar catabolism. Consequently, it will be acid tolerant and have complex nutritional requirements. Lactic acid bacteria are generally associated with habitats rich in nutrients such as milk and dairy products, vegetation and some are also members of the normal flora of the mouth, intestine and vagina of mammals.

The pioneering work of Orla-Jensen (1919), who divided the group into four genera, *Lactobacillus* for the rod-shaped organisms, *Streptococcus* for the homofermentative, facultatively anaerobic cocci, '*Betacoccus*' and '*Tetracoccus*', remains influential. The classification and identification of the lactic acid bacteria are being revolutionized however, by the application of modern molecular approaches to systematics, in particular the use of nucleic acid comparisons as indicators of evolutionary or phylogenetic relationships (Woese, 1987). These molecular taxonomies are based on sequence comparisons of the small subunit ribosomal RNA genes and are generating more detailed and reliable indications of the genera and species of all bacteria (see Priest and Austin, 1993, for details). Of the original Orla-Jensen taxa, *Streptococcus* has been divided into several genera: *Enterococcus* for '*Streptococcus faecalis*' and relatives, *Lactococcus* for the dairy streptococci, '*S. lactis*' and relatives and *Vagococcus* for some motile organisms otherwise resembling lactococci, as well as *Streptococcus sensu stricto*.

'*Betacoccus*' has been renamed *Leuconostoc* and includes heterofermentative cocci that occur in pairs or short chains. This genus has not escaped the attentions of the 'new systematicists' and rRNA gene sequence studies have shown that three phylogenetically diverse generic-ranked taxa have been contained in *Leuconostoc sensu lato*:

1. *Leuconostoc sensu stricto*;
2. *Leuc. oenos*;
3. the new genus *Weissella* which includes *Leuc. paramesenteroides* and some closely related heterofermentative lactobacilli (such as *L. confusus* and *L. kandleri*) now reclassified as *W. confusa* and *W. kandleri* (Martinez-Murcia and Collins, 1990; Collins *et al.*, 1993).

'*Tetracoccus*' has become *Pediococcus* and contains those homofermentative cocci that divide in two planes to produce pairs and tetrads. One member of this genus, *P. halophilus*, has been elevated to generic status by rRNA sequence studies as *Tetragenococcus* (Collins *et al.*, 1990).

Aerococcus is a monospecific genus which is related to *Pediococcus* and *Tetragenococcus*, and yet another genus, *Globicatella*, has been introduced recently for cocci from human clinical sources which were related to, but phylogenetically distinct from, the streptococci (Collins *et al.*, 1992). Fortunately, few of these lactic acid bacteria are commonly encountered in breweries, but the features of the principal genera are shown in Table 5.1 for reference purposes.

Orla-Jensen in his far-reaching studies, also divided *Lactobacillus* into three subgenera, mainly on the basis of temperature range for growth and mode of fermentation. Heterofermentative lactobacilli were placed in '*Betabacterium*'. Homofermentative strains that grew at 45°C but not at 15°C were placed in '*Thermobacterium*' and those with the opposite temperature relationships constituted '*Streptobacterium*'. Although these subgeneric allocations are still used, the names were not included in the Approved Lists of Bacterial Names (Skerman, McGowan and Sneath, 1980) and therefore have no standing in nomenclature.

Evidence for genetic heterogeneity in the genus can be gained from gross analysis of chromosomal DNA in the form of mol% guanine plus cytosine. Bacteria with widely different G + C ratios (> 12%) should be genetically

Table 5.1 Principal genera of lactic acid bacteria and their characteristics (data from Axelsson (1993) and Collins *et al.* (1992, 1993))

	Characteristics						
Genus	*Shape*	*Tetrads*	*CO_2 from glucose*	*Growth at 10°C*	*Growth at 45°C*	*Growth in 6.5% NaCl*	*Lactic acid formed*
Carnobacterium	Rods	–	–	+	–	•	L
Lactobacillus	Rods	–	d	d	d	d	D, L, DL†
Aerococcus	Cocci	+	–	+	–	+	L
Enterococcus	Cocci	–	–	+	+	+	L
Globicatella	Cocci	–	–	–	–	+	•
Lactococcus	Cocci	–	–	+	–	–	L
Leuconostoc	Cocci	–	+	+	–	d	D
Pediococcus	Cocci	+	–	d	d	d	L, DL†
Streptococcus	Cocci	–	–	–	d	–	L
Tetragenococcus	Cocci	–	–	+	–	+	L
Vagococcus	Cocci	–	–	+	–	–	L
Weissella	Small rods/ Cocci	–	+	d	–	d	D, DL

+, Positive; – negative; d, response varies; •, not determined.
* Configuration of lactic acid produced from sugar
† Production of D-, L-, or DL-lactic acid varies between species.

distinct and are unlikely to be members of the same genus (De Ley, 1969). That strains of *Streptococcus* (34–46% G + C), *Leuconostoc* (37–46% G + C) and *Pediococcus* (34–44% G + C) show narrow and overlapping ranges of G + C content indicates that these genera are not genetically heterogeneous and could be closely related (Bradley, 1980). Lactobacilli, on the other hand, display 32–53% G + C, suggesting that genetic diversity exists and some revision of the genus is required since this range exceeds the 10–12% maximum variation expected within a genus (Bradley, 1980; Priest, 1981a). This process has already been initiated with the establishment of the genus *Carnobacterium* (Collins *et al.*, 1987) for some species such as *L. divergens* and *L. carnis* which are commonly associated with meats and are able to grow at low temperature. A simple key for the identification of the carnobacteria has been published (Montel *et al.*, 1991).

5.2.1 Carbohydrate metabolism

Lactic acid bacteria seldom hydrolyse polysaccharides although *L. amylophilus*, which hydrolyses starch, is an exception. They do however, catabolize a range of disaccharides often including maltose which, at least in the case of lactococci, enters through a permease system and is then cleaved by a maltose phosphorylase into glucose and glucose-1-phosphate (Konings, Poolman and Driessen, 1989). Indeed, most sugars are assimilated by specific permeases driven by the proton motive force and phosphorylated in the cytoplasm by an ATP-dependent kinase. Some species use the phosphoenolpyruvate (PEP)-dependent phosphotransferase system (PTS) which phosphorylates sugars using PEP as they are transported into the cell.

The two major pathways of hexose fermentation within the lactic acid bacteria differ in the way in which the C6 skeleton is split and this yields different sets of end-products (Fig. 5.1). Homofermentation, characteristic of the streptococci, pediococci and some lactobacilli, involves the splitting of fructose-1,6-bisphosphate by the enzyme aldolase into two triosephosphates which are converted to lactate via pyruvate. Heterofermentation is practised by leuconostocs and some lactobacilli. In this process, glucose-6-phosphate is oxidized to form gluconate-6-phosphate and decarboxylated. The resultant xylulose-5-phosphate is split by phosphoketolase to yield a triosephosphate that is converted to lactate, and acetyl phosphate that is a precursor of either ethanol or acetate depending on the oxidation–reduction potential of the system. Thus equimolar amounts of CO_2 lactate and acetate/ethanol are formed from glucose. Should an additional hydrogen acceptor in the form of fructose or oxygen be available, it would be reduced to mannitol or H_2O_2 respectively and ethanol would not be formed.

There can be some confusion about the use of the terms homofermentative and heterofermentative. It is prudent to reserve the term **homofermentative** for lactic acid bacteria which use the glycolytic pathway

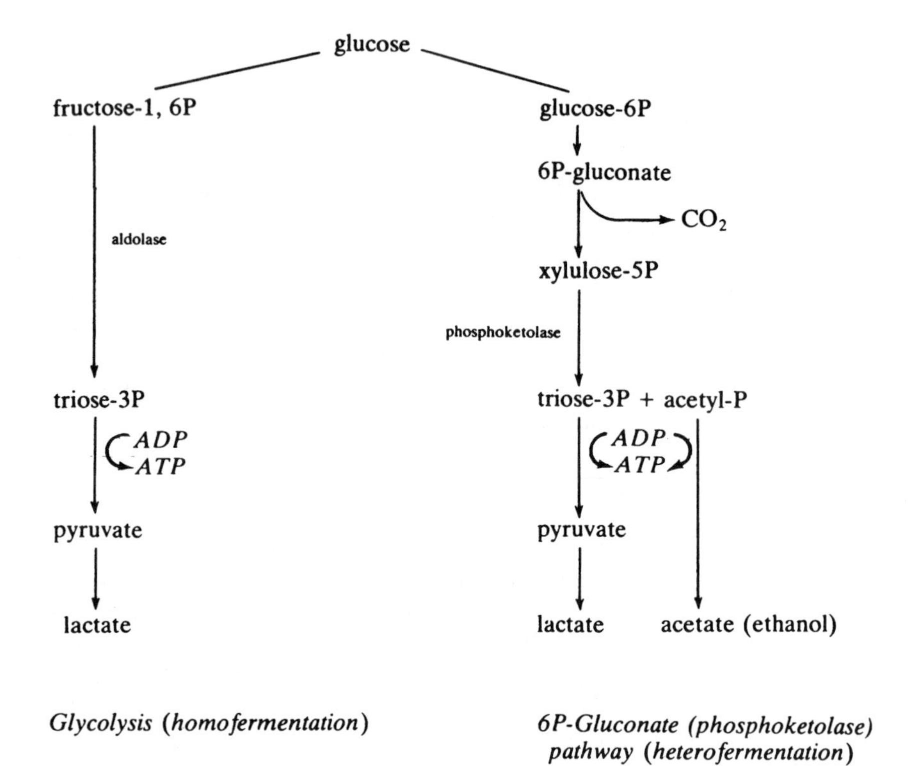

Fig. 5.1 Main pathways of hexose fermentation in lactic acid bacteria (modified from Kandler, 1983).

for glucose catabolism and therefore produce predominately lactate as end-products, and **heterofermentation** for lactic acid bacteria which use the gluconate-6-phosphate/phosphoketolase pathway. However, it should be emphasized that some lactobacilli produce various end-products from glycolysis of glucose, and others which use glycolysis for glucose catabolism switch to the gluconate-6-phosphate/phosphoketolase pathway for certain substrates, notably pentose sugars.

Heterofermentative lactic acid bacteria generally ferment pentoses, although exceptions do occur (Kandler, 1983). The various pentoses are taken up by specific permeases and induce synthesis of the relevant enzymes for conversion of the substrate into the key intermediate D-xylulose-5-phosphate, which enters the phosphoketolase pathway. Some lactic acid bacteria also ferment pentitols via D-xylulose-5-phosphate, although these sugar alcohols are taken up via the PEP-dependent PTS (London and Chace, 1977, 1979). Since a functional phosphoketolase pathway is required for pentose metabolism by heterofermentative lactic acid bacteria it might be expected that homofermentative strains would be unable to utilize these sugars. Indeed no authentic strains of the *'Thermobacterium'* group have been found that utilize pentoses (Kandler, 1983) but the streptobacteria and pediococci are often able to use such sugars. These bacteria possess a phosphoketolase that is induced by pentose sugars and the organisms thus function heterofermentatively on such substrates, although with glucose alone they would be homofermentative. For this reason these bacteria are referred to as **facultatively heterofermentative**.

Lactic acid bacteria produce either D(–)- or L(+)-lactate and sometimes both isomers. Lactate racemase was originally thought to be responsible for the production of racemic mixtures from L(+)-lactate but it has become evident that this enzyme is not widely distributed. Instead two NAD-dependent lactate dehydrogenases of different stereospecificity are generally responsible for mixtures of L- and D-lactate. There are alternative minor pathways for pyruvate dissimilation in lactic acid bacteria which are of particular importance to the brewer since they result in the formation of volatile flavour compounds such as diacetyl. These will be covered in section 5.3.3(b).

5.2.2 Nitrogen metabolism

Lactic acid bacteria are typically fastidious and require a variety of amino acids and vitamins for growth. It seems that in many environments they obtain these amino acids through the action of proteases although, compared with bacilli or pseudomonads, lactic acid bacteria are weakly proteolytic. Proteolysis has been particularly well documented in relation to the growth of lactococci in milk, where they are largely responsible for flavour development during cheese production. In these circumstances a complex system is involved comprising:

1. cell wall-associated caseinolytic protease;
2. extracellular peptidases;
3. amino acid transport systems;
4. peptide transport systems;
5. intracellular peptidases (Smid, Poolman and Konings, 1991).

Although the proteolytic activity of lactococci is low, the overall system is efficient at providing all the amino acids for growth in milk and protease-deficient mutants soon become growth-limited. Although little is known of the proteolytic systems of brewery lactic acid bacteria, El Soda and Desmazeaud (1982) and El Soda *et al.* (1982) have demonstrated proteolytic activity in both heterofermentative and homofermentative lactobacilli. Surface-bound peptidases have also been demonstrated in lactobacilli. This is particularly relevant to the brewer since beer is deficient in various amino acids and it is often assumed that this is important in restricting the growth of spoilage lactobacilli.

Some amino acids such as glutamate and histidine are degraded by lactobacilli rather than incorporated into proteins; arginine in particular can be used as an energy source by many heterofermentative strains including *L. brevis* (Rainbow, 1981; Manca de Nadre, Pesce de Ruiz Hogado and Oliver, 1988). Arginine is hydrolysed to citrulline and ammonia and the citrulline is converted to ornithine with the production of ATP.

5.2.3 Oxygen requirements

An interesting feature of the lactic acid bacteria is their relationship to oxygen. These bacteria are described as aerotolerant anaerobes in that they cannot use oxygen as a terminal electron acceptor in respiration but they contain NADH oxidases which are often induced by oxygen (Condon, 1987). The products of the reaction are either NAD^+ and H_2O_2 or NAD^+ and H_2O depending on whether two or four electrons are transferred. Unlike strict anaerobes, lactic acid bacteria are not killed by oxygen. It is thought that toxic oxygen species such as peroxide (O_2^{2-}) and particularly superoxide (O_2^-) are produced by cells in the presence of air. Aerobic bacteria have developed the enzymes catalase and superoxide dismutase respectively to remove these radicals, whereas strict anaerobes have not. Lactic acid bacteria are unable to synthesize haem and therefore cannot produce catalase unless haematin is provided. Nevertheless, they synthesize a pseudocatalase that requires a reducing substance, usually NADH, to remove peroxides. Pseudocatalase was first reported in pediococci (Delwiche, 1961) and soon after in lactobacilli, leuconostocs and streptococci (Johnston and Delwiche, 1965). The enzyme from *L. plantarum* has been purified and is a Mn^{2+}-containing molecule. Lactobacilli, leuconostocs and pediococci do not synthesize superoxide dismutase, however (Archibald and Fridovich, 1981). Instead, these bacteria contain high levels of Mn^{2+}, which is believed to be responsible for scavenging intracellular toxic oxygen species, in particular the superoxide radical. Indeed there is a good correlation between the intracellular Mn^{2+} concentration and tolerance to oxygen amongst these bacteria. Nevertheless, it is advisable to grow these bacteria under anaerobic or partially anaerobic conditions, particularly upon initial isolation.

5.3 *LACTOBACILLUS*

5.3.1 Classification

Following the initial work of Orla-Jensen (1919) the taxonomy of the lactobacilli progressed slowly despite the economic and environmental importance of the group (London, 1976). The three subgenera became firmly entrenched and formed the basis of a most useful article on the identification of these bacteria published by Sharpe, Fryer and Smith (1966). It then became apparent that the heterofermentative 'betabacteria' could be divided into two groups, one comprising metabolically inert strains that were highly ethanol tolerant and common in wines, and a group of more widely distributed, physiologically versatile species. This classification was incorporated in Sharpe's revision of her identification scheme (Sharpe, 1979) but was slightly modified in *Bergey's Manual of Systematic Bacteriology* (Kandler and Weiss, 1986). The subgeneric names were no longer used and the 'thermobacteria' were labelled Group I (obligate homofermentative strains), the 'streptobacteria' (facultative heterofermentative strains) formed Group II and the 'betabacteria', both ethanol tolerant and other types, constituted the obligately heterofermentative Group III. Hammes, Weiss and Holzapfel (1993) similarly consider these three groups for identification purposes in their comprehensive review of these bacteria in the second edition of *The Prokaryotes*.

This grouping is helpful for identification purposes but has been challenged as an appropriate approach to the classification of these bacteria. Perhaps it is useful to stress at this point that classification and identification are different disciplines with different aims. Classification is the arrangement of organisms into groups on the basis of their relationships which may be either evolutionary (phylogenetic) and/or phenetic (based on similarities between the properties of the organisms and with no evolutionary considerations). For various reasons, classifications based on these two different approaches do not always concur (see Priest and Austin, 1993). Identification, on the other hand, is the assignment of an unknown organism to its correct place in a pre-existing classification. Identification makes no assumptions about the classification processes.

With the current interest in phylogenetic classifications based on 16S rRNA gene sequences, the traditional (phenetic) classification of the lactobacilli has been challenged. Initially, certain 'atypical' lactobacilli isolated from refrigerated vacuum-packed meats including *L. divergens*, *L. piscicola*, *L. maltaromicus* and *L. carnis* were placed in a new genus *Carnobacterium*. These bacteria are not encountered in breweries but a simple scheme to aid identification of these bacteria has been published (Montel *et al.*, 1991). The remaining lactobacilli have been assigned to three groups which are similar to, but not identical with the physiological groups (Table 5.1). Group 1 (designated the *L. delbrueckii* group) contains *L. delbrueckii*, the type species of the genus, and 11 other obligately homofermentative species most of which are uncommon in breweries. Group 2 is based on the the facultatively heterofermentative *L. casei* and comprises 32 *Lactobacillus* species and five *Pediococcus* species including *P. damnosus* (see below). Some obligately homofermentative species and most of the obligately heterofermentative species including those found in breweries, such as *L. brevis*, *L. buchneri* and *L. fermentum*, were recovered in RNA Group 2. Finally, Group 3 comprises some atypical heterofermentative lactobacilli, including *L. confusus* and

Table 5.2 Some *Lactobacillus* species found in fermented beverages and related habitats and their allocation to taxonomic groups

Species	*RNA group*	*Fermentation group**	*Subgeneric allocation*	*Common habitats*
L. delbrueckii	1	I	Thermobacterium	Grain, mashes
L. yamanashiensis†	2	II	Streptobacterium	Wine, cider
L. casei		II	Streptobacterium	Milk and milk products
L. coryneformis subsp. *coryneformis*	2	II	Streptobacterium	Fermented vegetables
L. coryneformis subsp. *torquens*	2	II	Streptobacterium	Beer
L. homohiochii	2	II	Streptobacterium	Wine
L. paracasei subsp. *paracasei*‡	2	II	Streptobacterium	Beer
L. sake	2	II	Streptobacterium	Wine
L. brevis	2	III	Betabacterium	Beer, wine
L. buchneri	2	III	Betabacterium	Beer, wine
L. fructivorans§	2	III	Betabacterium	Wines, vinegar preserves
L. fermentum	2	III	Betabacterium	Fermented vegetables
L. hilgardii	2	III	Betabacterium	Wine
L. lindneri	?	III	Betabacterium	Beer
Weissella confusa¶	3	III	Betabacterium	Fermented vegetables

* Group I, obligately homofermentative; Group II, facultatively heterofermentative (homofermentative with hexose sugars); Group III, obligately heterofermentative.
† Includes *L. mali* (Carr *et al.*, 1977).
‡ Previously *L. casei* subsp. *pseudoplantarum.*
§ Includes *L. trichodes* and *L. heterohiochii.*
¶ Previously *L. confusus.*

Leuconostoc paramesenteroides; this group was designated the 'paramesenteroides' group (Collins *et al.*, 1991) initially and more recently has been raised to generic status as *Weissella* (Collins *et al.*, 1993). Although this classification describes the molecular evolutionary pathways which gave rise to the lactobacilli and relatives, it is unclear at present how it will fit into the established physiological classification.

Numerical or phenetic taxonomy involves the estimation of similarity between strains based on a large number of attributes, usually data from phenotypic tests. Strains are then allocated to clusters (generally species) of high mutual relatedness derived from these estimates (Sneath and Sokal, 1973). If each of the fermentative groups of lactobacilli contains phenotypically similar bacteria this should be revealed by numerical analysis. Unfortunately, the few studies published to date provide conflicting conclusions. Barre (1969) examined 65 strains of lactobacilli from wine together with some reference strains. The bacteria were recovered in two large clusters representing RNA Group 2 and based on *L. casei*/*L. plantarum* and *L. buchneri*/*L. fermentum*, but most of the wine isolates were identified as *L. buchneri*. Support for the three physiological groups was provided by the numerical analysis of 30 lactobacilli by Seyfried (1968). However, other studies (Wilkinson and Jones, 1977; Barbour and Priest, 1983; Shaw and Harding, 1984) have not resulted in the precise allocation of species to fermentation groups. Indeed numerical taxonomic analysis of *Lactobacillus* has been neither popular nor particularly successful, for reasons that have been discussed at length elsewhere (Priest and Barbour, 1985).

One of the principal problems in *Lactobacillus* taxonomy that has been highlighted by the numerical approach is the high phenetic similarity amongst the species. Most numerical taxonomic studies generally show 80–85% similarity between strains from different bacterial species. Amongst the lactobacilli, however, strains from separate species sometimes possess as much as 95% similarity despite being unrelated by criteria such as rRNA sequence or DNA homology. It would seem that evolutionary convergence has resulted in species that are genetically distinct but have similar phenotypes. Thus species are distinguished by few characters and such monothetic classifications lead to problems in identification since a strain need only be aberrant in one or two key features to be incorrectly identified. Since such schemes involving phenotypic attributes have proved unreliable for the identification of lactobacilli, other approaches such as serology, protein profiling and DNA probes have been explored and will be discussed in connection with brewery lactobacilli in section 5.3.5 and in Chapter 9.

In conclusion, the classification of *Lactobacillus* is emerging in a stable fashion. Classification at the genus level and above will be based on rRNA sequences supported where appropriate by chemotaxonomic considerations such as cell wall structure and physiological traits. Species will be based largely on DNA sequence homology and related techniques such as whole cell protein profiling which provide sound criteria for species distinction. However, this leaves a difficult situation for the microbiologist wishing to identify an unknown strain and for the present the traditional tables presented in *Bergey's Manual of Systematic Bacteriology* (Kandler and Weiss, 1986) and the second edition of *The Prokaryotes* (Hammes, Weiss and Holzapfel, 1992) remain the best option for the non-specialist laboratory. Fortunately, for the brewery microbiologist, the range of lactobacilli encountered is fairly restricted thus simplifying identification (see section 5.3.5).

5.3.2 Distribution of lactobacilli in beer and breweries

The frustrations associated with identifying *Lactobacillus* species have hindered ecological studies aimed at determining the range of species found in the brewing environment. Most such studies have focused on spoiled beer, in which one of the most common contaminants appears to be *L. brevis*. This obligate heterofermentative bacterium is generally tolerant towards hops and grows optimally at 30°C and pH 4–5. Various other species have been described in the past, e.g. '*L. diastaticus*' (amylolytic, responsible for superattenuation in beers) and '*L. pastorianus*' (produces extracellular polysaccharide or rope). These atypical properties are not considered sufficient to merit species status and both names are considered synonymous with *L. brevis*. It should be remembered therefore that *L. brevis* is physiologically versatile and can cause various problems in beer. *L. lindneri* was originally used to describe strains from lager that grow optimally at 19°C; it was later deemed to be synonymous with *L. brevis* (Rogosa, 1974). However, a more recent study indicates that strains of '*L. lindneri*' are sufficiently different from *L. brevis* in the spectrum of sugars fermented and DNA base composition (*L. brevis*, 44–47%; *L. lindneri* 42–46%) to warrant species status (Back, 1981, 1982). '*L. frigidus*' and '*L. parvus*' are considered to be synonyms of *L. buchneri*, a close relative of *L. brevis* that differs only in its ability to ferment melezitose and some strains have a requirement for riboflavin (Sharpe, 1979; Back, 1981).

Recently, a new lactic acid bacterium was isolated from spoiled beer which differs from *L. brevis* and *L. lindneri* in the range of sugars fermented and in DNA base composition to a sufficient extent to warrent species status, and which has been given the name *L. brevisimilis* (Back, 1987). This bacterium is difficult to culture and does not cause serious problems in bottled beer since it dies rapidly in this environment. Other lactobacilli that appear to occur less frequently in beer and breweries include *L. casei* (Eschenbecher, 1968; Dolezil and Kirsop, 1975), *L. delbrueckii* (McMurrough and Palmer, 1979), *L. fructivorans*, *L. fermentum* (Lawrence and Leedham, 1979), *L. coryneformis*, *L. coryneformis* subsp. *torquens*, *L. curvatus* (Eschenbecher, 1969; Back, 1981; Makinen, Tanner and Haikara, 1981) and *L. plantarum* (Wackerbauer and Emeis, 1968; Back, 1981).

One of the most comprehensive studies of brewery lactobacilli was that of Eschenbecher (1968, 1969), who examined 660 *Lactobacillus* cultures capable of causing beer spoilage from 81 breweries. These strains were identified to five species using sugar fermentations and other conventional tests. Of the 660 strains, 185 were identified as *L. casei*, 97 as '*L. casei* subsp. *fusiformis*', six as '*L. plantarum* var. *arabinosus*', 56 as *L. coryneformis*, four as *L. coryneformis* subsp. *torquens*, 250 as *L. brevis*, 14 as '*L. brevis* var. *lindneri*' and 48 as *L. buchneri*. Eschenbecher noted that Group II ('streptobacteria') occurred more frequently than heterofermentative (Group III) bacteria in heavily hopped beers: this is inconsistent with the general notion that only *L. brevis* is hop tolerant and suggests that more work is needed in this area. Back (1981, 1982) has provided excellent reviews of the taxonomy of brewery lactobacilli and assigned brewery isolates to species broadly similar to those of Eschenbecher with comparable frequencies. In addition to those described by Eschenbecher, Back included *L. curvatus* and two unnamed groups designated species group 8 and species group 9 (Back, 1981). The variety of lactobacilli in beer and breweries is therefore considerably more diverse than is usually believed and it becomes important to know if these species vary in their potential for beer spoilage.

5.3.3 Spoilage of beer by lactobacilli

(a) Growth of lactobacilli in beer

Since strains of at least nine species of *Lactobacillus* can be detected in beer and breweries, the traditional view that all lactobacilli are spoilage organisms may be erroneous if some of these organisms are unable to grow in beer. Unfortunately information in this area is scant. Back (1982) considered lactobacilli almost invariably responsible for beer spoilage and suggested that trace evidence of their presence in beer should be rectified immediately. Nevertheless different species vary in their ability to grow in beer and in their tolerance to hop bittering compounds. Using five different beers, Back (1981) showed that strains of *L. curvatus*, *L. brevis*, *L. lindneri* and species groups 8 and 9 could grow in virtually every beer examined whereas strains of *L. casei* grew in only three beers and *L. plantarum*, *L. buchneri* and *L. coryneformis* grew in only one. Obviously, the presence in beer of an organism that is unable to grow does not pose a serious threat to the brewer and, if these findings could be extended to other beers, then it may be that some lactobacilli could be ignored as spoilage organisms.

The growth of lactobacilli in beer appears to depend as much on the type of beer as it does on the offending organism. This has been noted above, and in an examination of 13 lactobacilli incubated in 31 different beers, Dolezil and Kirsop (1980) found that only in three beers were all the organisms able to grow. In five beers none developed, and diverse results were obtained with the remaining 23 beers. The reasons for the resistance

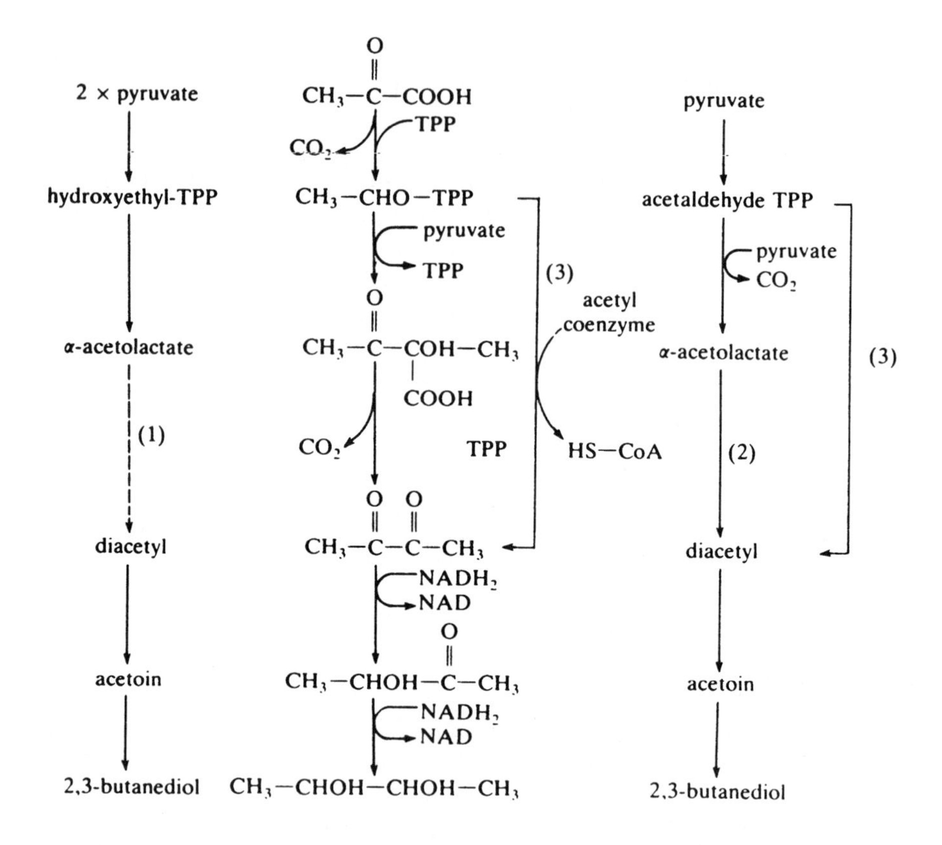

Fig. 5.2 Pathways of diacetyl and acetoin formation in yeast fermentations and *Lactobacillus*. TPP, thiamine pyrophosphate.

of some beers to infection are unclear. There was no correlation between any of the usual analytical parameters such as pH, specific gravity, nitrogen and amino nitrogen content, the quantity of fermentable carbohydrate or of sulphur dioxide and the ability of lactobacilli to proliferate. Nor was there any particular feature of the bacteria responsible for growth in a wide range of beers, although metabolic diversity, as judged by the spectrum of sugars fermented, appeared to correlate with the spoilage potential. This contradicts Back's (1981) findings, however, since *Lactobacillus* species groups 8 and 9 ferment only a few sugars and yet are potent beer spoilage organisms.

Bacterial growth in beer depends on the ability to scavenge residual nutrients left following yeast growth, and tolerance to the relatively hostile environment posed by a fermented beverage. Those nutrients that support the growth of lactic acid bacteria in beer have been examined by inoculating beers with lactic acid bacteria and following the changes in sugar and amino acid content with growth (Pfenninger *et al.*, 1979). In most cases small amounts of sucrose were metabolized along with varying amounts of maltose, maltotriose and maltotetraose. In some cases dextrins up to 14 or 15 glucose units were hydrolysed. Amongst the amino acids absorbed by lactobacilli, lysine, tyrosine and arginine predominated.

The major growth-inhibiting substances in beer are derived from hops and include *trans*-humulone and the related (–)-humulone and colupulone (Verzele, 1986). These compounds act as ionophores and dissipate the trans-membrane pH gradient of sensitive bacteria (Fernandez and Simpson, 1993). Lactobacilli vary more than 10-fold in their sensitivity to *trans*-humulone with beer spoilage organisms invariably showing high tolerance to this compound. Resistance is not associated with plasmids but seems to be chromosomally encoded (Fernandez and Simpson, 1993).

Resistance of specific beers to spoilage by lactobacilli has been tentatively attributed to the presence of a heat-labile yeast metabolite (Dolezil and Kirsop, 1980), but in a small screening exercise, Young and Talbot (1979) were unable to detect any extracellular substances from four yeasts that inhibited bacterial growth. An interesting approach was that of Munekata (1973), who examined various dipeptides for their inhibitory effect on the growth of lactic acid bacteria. Some alanyl dipeptides were found to be potent growth inhibitors. Resistance to infection has also been attributed to the cosedimentation of certain bacteria with brewing yeasts (White and Kidney, 1979). It is likely that a combination of these factors will be responsible for the susceptibility of beers to spoilage by various bacteria.

(b) *Effect of spoilage on beer quality*

Lactobacilli are most dangerous during conditioning of beer and after packaging. Spoilage is characterized by a 'silky' turbidity but sometimes before this is apparent the 'buttery' flavour of diacetyl may be noticed. Although lactic acid is the major metabolic end-product of these bacteria, its high flavour threshold in beer, which contains up to 300 ppm lactate (Hough *et al.*, 1982), compared with the low threshold for diacetyl, about 0.15 ppm (Hough *et al.*, 1982), makes the latter a potent flavour-modifying compound.

Diacetyl (and 2,3-pentanedione) are produced during normal fermentations. During growth, yeast synthesizes α-acetolactate as an intermediate in valine biosynthesis. Some α-acetolactate is excreted by the yeast and is non-enzymically converted to diacetyl in the fermenting wort. The diacetyl is then enzymically reduced by the yeast to form acetoin and finally the relatively innocuous 2,3-butanediol (Fig. 5.2, route 1). Premature removal of the yeast from the beer leaves α-acetolactate which subsequently oxidatively decarboxylates to form diacetyl which accumulates and causes an unacceptable flavour and aroma (Haukeli and Lie, 1978; Inoue, 1981). Similarly, contamination by lactobacilli results in accumulation of diacetyl but by a slightly different route (Fig. 5.2, route 3). Acetolactate synthesis followed by enzymic oxidation to diacetyl was originally thought to be the only pathway operating in lactobacilli, but direct synthesis from acetyl-CoA and acetaldehyde-TPP was demonstrated later (Speckman and Collins, 1968, 1973) and it appears that both pathways can operate in the same microorganism (Jonsson and Pettersson, 1977). Thus trace infections of beer, in the absence of yeast, can under certain conditions give rise to high levels of diacetyl accumulation. Some lactobacilli have been implicated in the production of extracellular polysaccharide or 'rope', but this is not a serious problem with lactobacilli and is more often attributable to pediococci or acetic acid bacteria (Rainbow, 1981).

5.3.4 Media for the isolation of lactobacilli

Perhaps nothing in brewery microbiology has generated as much interest and controversy as the development of culture media for the selective isolation and enumeration of lactobacilli and pediococci. In their excellent review of the topic, Casey and Ingledew (1981) list some 23 media designed for the growth of lactic acid bacteria; some of the more successful formulations are listed in Table 5.3. A more recent review by Holzapfel (1992) lists 31 media for lactobacilli and related bacteria but is not restricted to brewery varieties. Nevertheless, this is a comprehensive listing of media for lactic acid bacteria which may give some ideas for new approaches specific for brewery lactobacilli.

There have been almost as many studies of the effectiveness of *Lactobacillus* media as there have been media designed. Unfortunately they reach almost as many conclusions! It should be remembered that these media are often being tested for two criteria although this is not always made clear: recovery and selectivity. Generally, media that are effective at recovering the majority of lactobacilli from a sample have low selectivity against other brewery bacteria such as enterobacteria, and those that are highly selective for lactobacilli demonstrate lower recovery rates. Nevertheless, some general comments can be made.

For recovery of lactobacilli, maltose is a preferred carbon source. Many lactobacilli cannot use glucose as a carbon source on initial isolation and, by supplementing MRS with maltose, Lawrence and Leedham (1979) found this medium to be superior to Sucrose Agar, Universal Beer Agar, MRS

Table 5.3 Some common media suitable for the cultivation of lactic acid bacteria from breweries

Medium	*Comments*	*Reference*
Cinnamond Medium 2 (CM2)	Fructose as carbon source. Bromocresol green indicator	Harrison (1979)
KOT medium	Enhanced growth of pediococci	Taguchi *et al.* (1990)
de Man, Rogosa and Sharpe (MRS)*	Glucose as carbon source	de Man, Rogosa and Sharpe (1960)
Modified MRS†	Supplemented with maltose	Lawrence and Leedham (1979)
MRS and beer	Improvement of MRS for brewery microbiology	Back (1978)
Modified Homohiochi Medium	General medium used for sourdough, vegetables and wine	Kleynmans, Heinzl and Hammes (1989)
Homoferm–Heteroferm–Differential medium	For differential recovery of homoferm./heteroferm. lactic acid bacteria	McDonald *et al.* (1987)
Improved Nakagawa Medium	Semi-solid or solidified agar, with glucose as carbon source	Eto and Nakagawa (1975)
Lactobacillus Growth Medium (LGM)	Glucose-based complex medium supplemented with vitamins	Van Keer *et al.* (1983)
NBB	Rapid growth of *Lactobacillus* and *Pediococcus*	Back (1980)
Modified NBB	Improved version of NBB	Nishikawa and Kohgo (1985)
Raka-Ray No. 3†	Complex medium with maltose and fructose as carbon source	Saha, Sonda and Middlekauf (1974)
Sucrose agar†‡	Sucrose-based medium similar to MRS	Boatwright and Kirsop (1976)
VLLB-L41	Beer-based medium supplemented with nutrients, maltose and xylose	Wackerbauer and Emeis (1967)
VLB-S7	Designed for isolation and identification of pediococci	Emeis (1969)

* Recommended by International Committee on Systematic Bacteriology; Subcommittee on the Taxonomy of the Genus *Lactobacillus*, for the growth of *Lactobacillus*.
† Recommended by European Brewery Convention, Analytica Microbiologia Subcommittee.
‡ Recommended by Institute of Brewing, Analysis Committee.

Agar and Wallerstein Laboratories Differential Agar. Hsu and Taparowsky (1977) favoured MRS and Improved Nakagawa Agars for recovery of lactobacilli and Hug, Schlienger and Pfenniger (1978), in a comparison of six media, concluded that MRS and Raka-Ray-3 (RR-3) were the most effective for the detection of lactobacilli. A similar conclusion was reached by Matsuzawa, Kirchner and Wackerbauer (1979), but they showed that these media had low selectivity (wort bacteria and *Escherichia coli* could grow), which would limit their application in the enumeration of lactic acid bacteria in yeast and beer samples. To improve the selectivity, phenylethanol and cycloheximide were added but growth of lactobacilli was then restricted and when compared with VLB-S7, the latter was found to be superior. Van Keer *et al.* (1983) concluded that RR-3 was a valuable general-purpose medium for lactic acid bacteria and described *Lactobacillus* Growth Medium (LGM) which was as efficient as RR-3 in recovery but promoted better growth. A recent general-purpose medium that promoted good growth of lactobacilli and pediococci is NBB Agar (Back, 1980; Dachs, 1981) which has been modified by Nishikawa and Kohgo (1985). When compared with MRS and VLB-S7, development of lactic acid bacteria occurred very much more rapidly on NBB Agar (Wackerbauer and Rinck, 1983; Back, Durr and Anthes, 1984). However, a recent inter-laboratory evaluation exercise concluded that no significant differences in colony counts for *Lactobacillus* or *Pediococcus* strains were detected for three media, NBB, modified NBB or Universal Beer Medium, and that no preference for one medium was reported (Crumplen, 1988).

Thus, depending on the purpose of the medium and desirability for selectivity, it would seem that MRS, modified MRS, RR-3, VLB-S7 and NBB are probably the most favoured media for the isolation and enumeration of lactobacilli. With regard to the incubation conditions, 30°C in an anaerobic or semi-anaerobic environment is most suitable.

5.3.5 Identification of lactobacilli

Virtually any non-sporulating, rod-shaped bacterium that is Gram positive and catalase negative and has been isolated from yeast, beer or brewery plant will be a *Lactobacillus*, and since these bacteria are usually responsible for beer spoilage it is generally unnecessary to identify it further. Nevertheless, for some purposes it may be desired to identify an isolate to species level, particularly since different species have been attributed with different spoilage potential (Back, 1981). Information for the identification of brewery lactobacilli based on the results of Eschenbecher (1968, 1969), Sharpe (1979), Back (1981), Kandler and Weiss (1986) and Hammes, Weiss and Holzapfel (1993) is given in Table 5.4. Because of the phenotypic similarity of many of these bacteria, accurate identification is often a problem. General methods for conducting these tests are given by Harrigan and McCance (1976) and most practical microbiology manuals.

Table 5.4 Phenotypic features of *Lactobacillus* species of brewery origin

Test	L. casei	L. delbrueckii	L. coryneformis *subsp.* coryneformis	L. coryneformis *subsp.* torquens	L. curvatus	L. plantarum	L. brevis	L. brevisimilis	L. buchneri	L. fermentum	L. fructivorans	L. lindneri
Acid from:												
arabinose	–	–	–	–	–	d	+	+	+	d	–	–
cellobiose	+	d	–	–	+	+	–	–	–	d	–	–
dextrin	+	•	d	d	d	d	–	•	–	•	•	–
glucose	+	+	+	+	+	+	+	+	+	+	+	+
lactose	d	–	d	+	d	+	d	–	d	+	–	–
maltose	+	d	+	+	+	+	+	+	+	+	d	+
mannitol	+	–	+	+	–	+	–	•	–	–	–	–
mannose	+	+	+	d	+	+	–	•	–	w	–	–
melezitose	+	–	–	–	–	d	–	–	+	–	–	–
melibiose	–	–	d	–	–	+	+	–	+	+	–	–
rhamnose	–	–	+	–	–	–	–	–	–	–	–	•
ribose	+	–	–	–	+	+	+	+	+	+	w	–
sorbitol	+	–	d	–	–	+	–	–	–	–	–	•
sucrose	+	+	+	+	–	+	d	–	d	+	d	•
trehalose	+	d	–	–	–	+	–	–	–	d	–	–
xylose	–	–	–	–	–	d	d	+	d	d	–	–
Gas from glucose	–	–	–	–	–	–	+	+	+	+	+	+
Growth at 15°C	+	–	+	+	+	+	+	+	+	–	+	•
Arginine dihydrolase	–	d	–	–	–	–	+	–	+	+	+	–
Lactate configuration	L	D	D(L)	D	DL	DL	DL	DL	DL	DL	DL	DL

+, 90% or more strains positive; –, 90% or more strains negative; d, 11–89% strains positive; w, positive to weak reaction; • not determined.

(a) *Serological identification of lactobacilli*

Serological detection and identification of lactobacilli is attractive due to the specificity and rapidity of the antibody/antigen reaction but this approach has not been exploited to any extent. This probably stems from the taxonomic complexity of the group and problems due to cross-reactions of multivalent sera (Wackerbauer and Emeis, 1968). Dolezil and Kirsop (1975) demonstrated at least two antibody components in sera raised to lactobacilli from beers: one that reacted with a species-specific antigen and a second non-specific antibody that reacted with pediococci and yeast. By adsorbing these preparations, rapid identification of spoilage lactobacilli including *L. casei* and *L. brevis* was possible, but difficulties were still encountered with *L. plantarum*. Similarly, Nishikawa, Kohgo and Karakawa (1979) used antisera to detect *L. brevis* in beer and Rinck and Wackerbauer (1987) detected *L. lindneri* amongst other lactobacilli serologically. It is unfortunate that this promising approach has not been exploited and it is likely that the introduction of monoclonal antibodies will stimulate interest in this area (Whiting *et al.*, 1992).

(b) *Identification kits for lactobacilli*

One of the most significant recent developments in diagnostic microbiology has been the introduction of commercial miniaturized systems for the identification of bacteria. In general, these comprise disposable trays of media for the testing of about 20 phenotypic attributes (see Chapter 9). The pattern of results is then compared with patterns for reference organisms and an identification obtained (D'Amato, Holmes and Bottone, 1981). Concomitant with these kits, computerized data bases and probabilistic identification schemes have been developed (Sneath, 1978; Priest and Williams, 1993). Such systems were developed for medically important bacteria such as Enterobacteriaceae and have been used to identify enteric bacteria from breweries and related environments (Van Vuuren *et al.*, 1978, 1981; Ingledew, Sivaswamy and Burton, 1980). The phenotypic homogeneity of the lactobacilli necessitates the use of at least 50 tests (e.g. the API 50 tray) to obtain an identification and still the results may not be entirely reliable. This is often due to strain variability on storage of the organism. In one example, upon initial isolation a strain was identified as *L. brevis* but reidentified after 15 weeks as *L. plantarum*. This transition required the change of five test results which occurred independently of experimental error (D.R. Lawrence, personal communication). In this connection it should be remembered that lactobacilli harbour numerous plasmids, many of which code for carbohydrate utilization pathways (McKay, 1983), and it may be that plasmid loss is responsible for the altered phenotype.

A second reason for the failure of commercial identification kits is simply that they have been used for brewery microbiology so infrequently that results for brewery bacteria such as *L. lindneri* or *L. brevisimilis* are not included in the data bases. Nevertheless, the Minitek system has been used successfully for the identification of lactobacilli (Gilliland and Speck, 1977) and a simple modification of this system was claimed to produce results for lactobacilli that were consistent with those obtained by traditional procedures (Benno and Mitsuoka, 1983). As the ecology of brewery bacteria is understood in more detail and the benefits of the identification of dangerous spoilage bacteria are acknowledged, identification kits will probably be used more extensively in brewery quality control and research laboratories.

5.4 *PEDIOCOCCUS*

5.4.1 Classification

Pediococci are homofermentative cocci that occur in pairs and tetrads through division in two planes. They have a long association with brewing microbiology and were originally known as 'sarcinae' because their cell morphology was confused with the cubical packets of eight cells of true sarcinae. Shimwell and Kirkpatrick (1939) first showed that the brewery cocci were lactic acid bacteria but assigned them to the genus *Streptococcus* as '*Str. damnosus*' (Shimwell, 1941). This classification was not accepted, although the close relationships of pediococci and streptococci are often stressed (Whittenbury, 1978), and these bacteria were placed in the genus *Pediococcus* which had been used by Claussen for strains he had earlier isolated from European beers. The classification and nomenclature of the pediococci continued to cause confusion however, largely because Nakagawa and Kitahara (1959) used the name '*P. cerevisiae*' to describe the common cocci from beer and breweries rather than *P. damnosus* as used by Gunther and White (1961) and Coster and White (1964). Solberg and Claussen (1973a) noted that these two names were being used for the same bacterium and in response to Garvie's (1974) request, the Judicial Commission of the International Committee on Systematic Bacteriology ruled that '*P. cerevisiae*' had not been validly published and the name *P. damnosus* was conserved. Thus *P. damnosus* and '*P. cerevisiae*' are synonyms and the former is used for the common brewery cocci (Sharpe, 1979; Priest, 1981b).

The close relationship of the pediococci with *Lactobacillus* has been emphasized by the rRNA studies in which most pediococci form a small cluster with the *L. casei* group of obligately homofermentative and heterofermentative bacilli. The closest phylogenetic relatives include *L. kefir* and *L. buchneri* (Collins *et al.*, 1991). The rRNA studies have also clarified the boundaries of the genus *Pediococcus* and reveal it as a homogeneous

Table 5.5 Phenotypic features of *Pediococcus* species of brewery origin

Test	P. damnosus	P. pentosaceus	P. pentosaceus *subsp.* intermedius	P. inopinatus	P. dextrinicus	P. parvulus	P. acidilactici
Acid from:							
arabinose	–	+	–	–	–	+	+
cellobiose	+	+	+	+	+	+	+
dextrin	–	–	–	–	+	–	–
lactose	–	+	+	+	d	+	–
maltose	+	+	+	+	+	–	+
maltotriose	–	–	–	+	+	–	+
melibiose	–	d	–	–	–	–	–
raffinose	–	d	–	–	–	–	–
ribose	–	+	+	–	–	+	+
sorbitol	–	–	–	–	–	–	–
starch	–	–	–	–	+	–	–
sucrose	d	d	d	–	+	–	+
trehalose	+	+	+	+	–	d	+
xylose	–	d	–	–	–	+	–
Growth at:	–	+	+	+	+	+	+
35°C							
50°C	–	–	–	–	–	+	–
6.5% NaCl	–	+	+	+	–	+	+
15% NaCl	–	–	–	–	–	–	+
pH 4.5	+	+	+	+	+	+	–
pH 7.5	–	+	+	+	+	+	+
Arginine dihydrolase	–	+	+	–	–	–	+
Pseudocatalase	–	+	•	•	–	–	+
Lactate configuration	DL	DL	DL	DL	L(+)	DL	L(+)

All species grow at 10°C and at pH 5.5, produce acid from glucose, fructose and mannose and split aesculin.
For key see footnote to Table 5.4.

taxon with *P. dextrinicus* as the most distantly related true *Pediococcus* forming an independent lineage. Within the *Pediococcus* taxon, *P. damnosus* and *P. parvulus* form one group while *P. acidilactici* and *P. pentosaceus* form a second (Collins *et al.*, 1991) and this distinction is also evident from physiological tests such as the arginine dihydrolase reaction (Table 5.5). The non-aciduric, microaerophilic species *P. urinae-equi* is synonymous with *'Aerococcus viridans'* and the non-aciduric, salt-requiring species, *P. halophilus* has been removed to a separate genus as *Tetragenococcus* (Collins, Williams and Wallbanks, 1990).

A comprehensive study of 840 cocci from various sources including wine, beer and brewery equipment was based on phenotypic tests and DNA sequence homology and supported the integrity of the six established species, *P. acidilactici, P. damnosus, P. dextrinicus, P. parvulus, P. pentosaceus* as well as *T. halophilus* (Back, 1978; Back and Stackebrandt, 1978). Some strains, largely derived from plant material, which had previously been placed in *P. damnosus* because of their inability to ferment pentoses, showed high DNA homology with typical strains of *P. pentosaceus* and were assigned to the subspecies '*P. pentosaceus* subsp. *intermedius*'. Furthermore, some brewery strains showed only 40% reassociation with DNA from *P. damnosus* or *P. parvulus*, their closest relatives, and were placed in the new species, *P. inopinatus* (Back, 1978). A numerical taxonomic analysis of 96 Gram-positive cocci from beer and breweries and reference strains confirmed the species *P. damnosus, P. dextrinicus, P. parvulus* and *T. halophilus*. No strains of *P. inopinatus* were recovered in this study, however; virtually all the beer isolates were identified as *P. damnosus* (Lawrence and Priest, 1981).

5.4.2 Distribution of pediococci in beer and breweries

Pediococcus damnosus is undoubtedly the most common, and feared, *Pediococcus* found in beer (Solberg and Claussen, 1973a; Back, 1978; Lawrence and Priest, 1981). It is particularly interesting that this organism is apparently only found in beer, brewing yeast and wines and not in brewing raw materials or plant materials. This suggests considerable adaptation to this particular habitat. *P. inopinatus* has been isolated from wine, beer and beer yeast but is also present in fermented vegetables and milk (Back, 1978). *P. dextrinicus* and *P. pentosaceus* can be isolated from beer and brewery plant but more rarely than *P. damnosus*.

The distribution of pediococci within the brewery has been examined in a study of some 200 isolates (McCaig and Weaver, 1983). *P. damnosus* was common in beer and late fermentations but seldom encountered in pitching yeast. *P. damnosus* var. I (probably *P. inopinatus*) and *P. pentosaceus* were frequently detected in pitching yeasts (44% and 35% respectively of those cocci detected) but rarely found in late fermentations or the finished product. With regard to those pediococci found in fermented foods such as silage, cucumbers or olives, *P. parvulus, P. inopinatus* and *P. dextrinicus* are commonly isolated. The biotechnological applications of the pediococci in the processing and preservation of meat and plant foods have been reviewed recently (Raccach, 1987).

5.4.3 Spoilage of beer by pediococci

(a) Growth of pediococci in beer

Only *P. damnosus* and, to a lesser extent, *P. inopinatus* can grow in beer although other species will survive in beer for long periods (Back, 1978; Lawrence and Priest, 1981). Not only does the low pH inhibit these other species but they are markedly less hop tolerant than *P. damnosus*, which is particularly resistant to the antiseptic action of hop constituents (Fernandez and Simpson, 1993). As with the lactobacilli, there is evidence that some beers are more resistant to spoilage by *P. damnosus* than others. This may result from availability of growth factors. There is some controversy concerning the vitamins required for the growth of these bacteria, but biotin and riboflavin promote the growth of most *P. damnosus* strains and calcium pantothenate is essential for the growth of this organism (Solberg and Claussen, 1973b). Calcium pantothenate can be replaced by a β-glucoside of pantothenate, a compound that occurs in tomato juice and is essential for the growth of *Leuconostoc oenos* (Eto and Nakagawa, 1974, 1975).

(b) Effect of pediococci on beer quality

Pediococci are traditionally associated with 'sarcina sickness', characterized by acid formation and the 'buttery' aroma of diacetyl. The pathways for diacetyl synthesis conform to those shown in Fig. 5.2 for lactobacilli but the amount synthesized varies between species. *P. pentosaceus* does not produce diacetyl and *P. inopinatus* does not make it as prolifically as *P. damnosus* (McCaig and Weaver, 1983). In trial fermentations, contamination by either *P. damnosus* or *P. inopinatus* extended the fermentation time, depressed the amount of yeast in suspension by some 30% and produced high diacetyl contents in the beer. The diacetyl concentration was much greater in the case of *P. damnosus*. Ropiness in beers has been attributed to pediococci (Shimwell, 1945) but is dependent on the presence of some fermentable sugar. The extracellular slime produced by these bacteria is thixotropic and comprises a complex heteropolymer containing glucose, mannose and nucleic acid (Hough *et al.*, 1982).

5.4.4 Media for the isolation of pediococci

Those media designed for the cultivation of lactobacilli can generally be used for pediococci. However, the importance of pediococci in relation to beer spoilage and their poor growth rates has led to the development of specific media for their isolation. Nakagawa's medium based on unhopped beer is a semi-solid agar in which pediococci appear as granular colonies after 5 days at 25°C. This has since been modified for use for all lactic acid bacteria (Table 5.3). Medium VLB-S7 (Emeis, 1969) was introduced for the isolation of beer pediococci and found to be superior to several general-purpose media. However, NBB medium and modified NBB had no advantages for the growth and recovery of pediococci when tested against Universal Beer Agar (Crumplen, 1988).

5.4.5 Identification of pediococci

Pediococci are typically catalase-negative organisms which can be distinguished from leuconostocs by virtue of their homofermentative mode of sugar metabolism and characteristic formation of tetrads. However, the catalase test is not always reliable since many strains of *P. pentosaceus*, when grown on a low sugar-containing medium, produce a 'pseudocatalase' which decomposes H_2O_2 but differs from catalase in being sensitive to azide and does not contain a haem group (Whittenbury, 1978). Thus there is the possibility of confusing these strains with the catalase-positive micrococci, but only the latter will grow at pH 7.0 on nutrient agar. A scheme for the identification of the genus *Pediococcus* is given at the end of this chapter; this section is devoted to identification at the species level.

Since pediococci vary in their potential for beer spoilage, accurate identification may sometimes be important. The information in Table 5.5 has been derived from the results of Back (1978, 1982), Sharpe (1979), Lawrence and Priest (1981) and Weiss (1992). Only *P. damnosus*, *P. inopinatus*, *P. pentosaceus*, *P. pentosaceus* subsp. *intermedius* and *P. dextrinicus* are likely to be encountered in the brewery but data for *P. acidilactici*, *P. parvulus* and *T. halophilus* have been included for completeness. Methods for these tests are the same as those for lactobacilli and are outlined at the end of this chapter. Other approaches to the identification of pediococci have successfully used the API identification kit for lactobacilli (Dolezil and Kirsop, 1977).

Serological methods for the identification of pediococci were first introduced by Dolezil and Kirsop (1976) and have received added interest from the introduction of monoclonal antibodies. A membrane filter-based immunofluorescent antibody test has been developed which can detect contaminating pediococci at 0.001% of pitching yeasts in less than 4 h (Whiting *et al.*, 1992). This is, of course, a major improvement on traditional plating tests.

5.5 *LEUCONOSTOC*

Heterofermentative cocci that are sometimes oval or even short rods and occur in pairs or short chains are classified in the genus *Leuconostoc*. These organisms are found on vegetables and fruit and in fermenting vegetable matter but apparently occur rarely in breweries. Leuconostocs and obligately heterofermentative lactobacilli have similar nutritional requirements, they grow on the same media, and can often be confused since some lactobacilli can occur as coccobacilli. Nevertheless, rRNA studies have clearly demonstrated the distinction between the leuconostocs and *Lactobacillus* (Martinez-Murcia, Harland and Collins, 1993). Leuconostocs can be distinguished from most lactobacilli by their inability to produce ammonia from arginine and by forming D(–)- rather than DL-lactate from glucose (Sharpe, 1979). Leuconostocs form three major phylogenetic lineages:

- *Leuconostoc sensu stricto*, which includes *Leuc. mesenteroides* and five other *Leuconostoc* species;
- the *Leuc. paramesenteroides* group now reclassified as *Weissella* and which includes the single *Leuconostoc* species (as *W. paramesenteroides*) and *W. confusa* (previously *Lactobacillus confusus*) and some other heterofermentative lactobacilli;
- *Leuc. oenos*, which forms a single membered taxon (Collins *et al.*, 1991, 1993; Martinez-Murcia, Harland and Collins, 1993).

Leuconostocs occur in the early stages of Scotch whisky fermentations (Bryan-Jones, 1975) and a recent numerical taxonomic study of these organisms classified the majority of isolates into two groups: *Leuc. mesenteroides* and a possible new species (Priest and Barbour, 1985; Priest and Pleasants, 1988). There was no indication that *Leuc. oenos*, the bacterium responsible for the malolactic fermentation in wines, is present in distilleries. The only species to be found in breweries appears to be *Leuc. mesenteroides* (Back, 1982) but it is not responsible for beer spoilage. Exceptionally acid-tolerant strains of *Leuc. mesenteroides* have been isolated from fruit mashes and they are also associated with traditional African beverages (Sanni and Oso, 1988). Data for the identification of *Leuconostoc* species are given in Table 5.6.

5.6 HOMOFERMENTATIVE COCCI

Gram-positive cocci that are homofermentative and occur in pairs or chains are classified in the genera *Enterococcus*, *Lactococcus*, *Streptococcus* and *Vagococcus*. These bacteria are widely distributed in raw milk and dairy products, on plant material and in the mouth and intestines of humans and animals (Sharpe, 1979). The only member of this group likely to be encountered in the brewery is *Lactococcus lactis*, a common organism on plants, but best known for the production of diacetyl from citrate and its role in butter manufacture. Although *Lactococcus lactis* occurs rarely in breweries (Back, 1982) it could be confused with other catalase-negative, Gram-positive cocci since it would grow on media designed for the cultivation of lactic acid bacteria. It can be distinguished from the leuconostocs by its homofermentative metabolism of sugars and from the brewery pediococci by its cell morphology (chains rather than tetrads) and growth at pH 9.2

(pediococci are unable to grow at high pH). A more definite identification could be obtained using commercially available Group N antiserum for the typing of lactococci but it should be remembered that cross-reactions between this and pediococci have been noted (Dolezil and Kirsop, 1976). There have been no reports of *Lactococcus lactis* growing in beer.

5.7 *MICROCOCCUS* AND *STAPHYLOCOCCUS*

The genera *Micrococcus* and *Staphylococcus* contain Gram-positive cocci which form catalase and do not have a lactic fermentative mode of metabolism. Microscopically these bacteria form clusters of cells but these are not always apparent and then the bacteria are virtually indistinguishable from lactic cocci. Nevertheless, they can be readily differentiated by their ability to grow at pH 7.0 on nutrient agar and the catalase test (see section 5.9). Although *Micrococcus* and *Staphylococcus* differ widely in DNA base composition (30–40 and 66–75 mol%, respectively) and have very different rRNA sequences, there are few reliable tests that distinguish them. The most useful criterion is that staphylococci are facultative anaerobes although *Staph. saprophyticus* and related species grow poorly or not at all under anaerobic conditions (Kloos and Schleifer, 1975a,b) whilst micrococci, with few exceptions, are strict aerobes. Moreover, staphylococci are erythromycin resistant and not lysed by lysozyme, but most micrococci are erythromycin and lysozyme sensitive. Finally, staphylococci are sensitive to lysostaphin whereas most micrococci are resistant.

These bacteria are not generally considered to be important brewery bacteria but recent studies have shown that they are widely distributed in beer and breweries and in some instances have been responsible for beer spoilage. A numerical taxonomic study of 96 cocci from the brewing environment revealed that the skin commensals *Staph. saprophyticus*, *Staph. epidermidis* and the widely distributed *M. varians* were common in beer and breweries, with the latter sometimes being isolated from pitching yeast (Lawrence and Priest, 1981). However, these bacteria grew poorly, if at all, at pH 4.5 and below, were susceptible to hop constituents, and in the case of *M. varians* were obligately aerobic. They are therefore unable to spoil beer and are probably adventitious contaminants arising from humans or from raw materials. Many can survive in beer for long periods, however, and could be confused with spoilage organisms in quality control procedures, particularly since some strains of *M. kristinae* display tetrad-like packets of cells in Gram stains.

Micrococcus kristinae is an atypical *Micrococcus* in that it is facultatively anaerobic (Kloos, Tornabene and Schleifer, 1974). Strains of this organism are relatively acid tolerant and hop resistant and have been responsible for beer spoilage particularly in relatively high-pH, low-bitterness products (Back, 1981; Lawrence and Priest, 1981). The intensity of growth is related to the oxygen content of the beer, but even traces of growth under anaerobic conditions can yield a fruity aroma and an atypical taste (Back, 1981). It is therefore important to be able to distinguish *M. kristinae* from the harmless *M. varians* and this is most readily achieved by its ability to grow anaerobically. Phenotypic descriptions of *M. kristinae* have been published by Back (1981) and Kocur, Kloos and Schleifer (1992).

5.8 ENDOSPORE-FORMING BACTERIA

Aerobic endospore-forming bacteria belonging to the genus *Bacillus* have on occasion caused problems in breweries. Spores from these bacteria are present in malt and cereal adjuncts and will survive wort boiling but are subsequently unable to germinate because of the low pH of fermenting wort and beer. Moreover, all bacilli are susceptible to hops so there is no problem from these bacteria in beers (Smith and Smith, 1993). It is a sobering thought, however, that a spore-forming variant of *L. brevis* would create havoc in the modern brewing industry!

Thermophilic, endospore-forming bacilli have been isolated from brewery plant, malt and sweet worts (McMurrough and Palmer, 1979; Healy and Armitt, 1980; Smith and Smith, 1992). Identified as *B. coagulans* and *B. coagulans–B. stearothermophilus* intermediates, these bacteria produce copious quantities of lactic acid in sweet wort held at 55–70°C for more than 2 h. Although usually considered detrimental, controlled acidification of the wort can have benefits both in beer preparation and for the beer in general, including improvements in flavour balance and microbiological stability (Back, 1988). However, one of the more disturbing features of the *B. coagulans* strains is their ability to contribute to nitrosamine formation through the reduction of nitrate to nitrite (Smith and Smith, 1992).

5.9 IDENTIFICATION OF GENERA OF GRAM-POSITIVE BACTERIA OF BREWERY ORIGIN

This scheme (Table 5.6) is designed to enable rapid identification, to the generic level, of a Gram-positive bacterium isolated from the brewing environment. This will generally be sufficient to predict the spoilage potential of the organism but in some instances it may be desirable to pursue the identification to the species. Tables and keys for this purpose have been included in the relevant sections of this chapter.

5.9.1 Test procedures

The tests involved in the identification key use common microbiological procedures but details of the methods are given below for those readers

Table 5.6 A key for the identification of genera of Gram-positive bacteria of brewery origin

(1) Catalase*	Positive → (5)
	Negative → (2)
(2) Cell morphology	Rods: *Lactobacillus*
	Cocci† → (3)
(3) Gas production from glucose	Positive: *Leuconostoc/Weissella*‡
	Negative → (4)
(4) Cells in tetrads	*Pediococcus*
Cells in pairs/chains	*Lactococcus*
(5) Endospore-forming rods	Positive: *Bacillus*
	Negative → (6)
(6) Anaerobic growth	Positive → (7)
	Negative: *Micrococcus*
(7) Lysis by lysostaphin	Positive: *Staphylococcus*
	Negative: *M. kristinae*

* Some pediococci may be catalase positive on media of low sugar content. If doubtful, repeat test using culture from medium with at least 1% glucose or test reaction for sensitivity to 0.01 M azide; catalase is inhibited, pseudocatalase is not (Whittenbury, 1978).
† Some leuconostocs are elongate and can be confused with heterofermentative lactobacilli and vice versa: see section 5.5.
‡ Differentiation of *Leuconostoc* from *Weissella* is problematic and requires combinations of characters for the particular species (see Collins *et al.*, 1993).

unfamiliar with bacterial identification. For further information the reader should consult *Cowan and Steele's Manual for the Identification of Medical Bacteria* (Barrow and Feltham, 1993) for general tests, Sharpe (1979) and Hammes, Weiss and Holzapfel (1993) for tests involving lactic acid bacteria and Kloos, Tornabene and Schleifer (1974), Baird Parker (1979) and Kocur, Kloos and Schleifer (1992) for those involving micrococci and staphylococci.

(a) Catalase test

Method 1

Grow the organism on a plate or slope of a suitable medium. Add 1 ml of 3% (v/v) H_2O_2 and examine immediately for evolution of gas which indicates catalase activity. Anaerobically incubated cultures should be exposed to air for at least 30 min before adding the H_2O_2.

Method 2

Grow the organism in a suitable broth; after incubation add 1 ml of 3% (v/v) H_2O_2 and examine for gas evolution.

Controls

Positive: *Staphylococcus epidermidis* or *Micrococcus varians*; negative: *Lactobacillus brevis*.

(b) Gas production from glucose

Since the amount of gas produced is often slight, the method of Gibson and Abd-el-Malek (1945) is used. A suitable semi-solid medium comprises: meat extract (Lab-Lemco), 0.5 g; tryptone, 0.5 g; yeast extract, 0.5 g; Tween 80, 0.05 ml; agar, 0.2 g; water to 100 ml. Glucose, sterilized separately, is added to 5% (w/v) and the medium adjusted to pH 6.0. Cultures are stab inoculated into the medium in tubes using a vigorous heavy inoculum. The cultures are overlayered with a seal of sterile molten agar and, after incubation, gas bubbles collecting beneath the agar seal denote heterofermentative growth.

Controls
Positive: *Lactobacillus brevis*; negative: *Pediococcus damnosus*.

(c) Anaerobic growth

A modification of the Hugh and Leifson (1953) oxidation/fermentation (O/F) test is used. Duplicate tubes containing tryptone, 1 g; yeast extract, 0.1 g; glucose, 1 g; bromocresol purple, 0.004 g; agar, 0.2 g; in 100 ml of water, pH 7.0 are stab inoculated with the organism and one tube is overlayered with paraffin oil to provide anaerobic conditions. Growth and acid (yellow) production in both tubes denote fermentative metabolism; growth and acid in only the open tube indicate strictly oxidative metabolism.

Controls
Fermentative: *Staphylococcus epidermidis*; oxidative: *Micrococcus varians*.

(d) Lysis by lysostaphin and lysozyme

Sterile filtered lysostaphin (final concentration 200 μg ml^{-1}) or lysozyme (final concentration 25 μg ml^{-1}) is added to sterile, molten peptone (1%), yeast extract (0.1%), glucose (1.0%), sodium chloride (0.5%), agar medium at pH 7.0. Plates are streaked with the organism and incubated for 2 days.

Controls
Lysostaphin-sensitive, lysozyme-resistant, *Staphylococcus epidermidis*. Lysostaphin-resistant, lysozyme-sensitive, *Micrococcus varians*.

5.10 CONCLUDING REMARKS

In this article I have emphasized the diversity of Gram-positive bacteria that can be recovered from beer. It is evident that rather than simply 'lactic rods', at least ten species of *Lactobacillus* can be the cause of beer spoilage.

Similarly 'lactic cocci' encompasses several *Pediococcus* species and probably *M. kristinae*. The time may be ripe to consider some of the rapid detection and identification methods for bacteria described in later chapters in a quality control context and on a routine basis. Only then will we become aware of the full spectrum of brewery bacteria and their role in beer spoilage. The cost might be high, but savings would accrue for example if it could be demonstrated unequivocally that a beer contaminated with *Lactobacillus* X could be sent to trade rather than destroyed or reprocessed because strains of that species were unable to grow in and consequently spoil that particular product. Beyond such immediate interests a more detailed picture of brewery microbiology would emerge that would be of interest and value to all those involved in the technical side of the industry.

REFERENCES

Archibald, F.S. and Fridovich, L. (1981) *Journal of Bacteriology*, **146**, 928.
Axelsson, L.T. (1993) In *Lactic Acid Bacteria* (eds S. Salminen and A. van Wright), Marcel Dekker, New York, p. 1.
Back, W. (1978) *Brauwissenschaft*, **31**, 237, 312, 336.
Back, W. (1980) *Brauwelt*, **120**, 1562.
Back, W. (1981) *Monatsschrift für Brauerei*, **34**, 267.
Back, W. (1982) *Brauwelt*, **45**, 1562, 2090.
Back, W. (1987) *Monatsschrift für Brauwissenschaft*, **40**, 484.
Back, W. (1988) *Monatsschrift für Brauwissenschaft*, **41**, 152.
Back, W. and Stackebrandt, E. (1978) *Archives of Microbiology*, **118**, 79.
Back, W., Durr, P. and Anthes, S. (1984) *Monatsschrift für Brauwissenschaft*, **37**, 126.
Baird-Parker, A.C. (1979) In *Identification Methods for Microbiologists*, 2nd edn (eds F.A. Skinner and D.W. Lovelock), Academic Press, London, p. 201.
Barbour, E.A. and Priest, F.G. (1983) In *Current Developments in Malting, Brewing and Distilling* (eds F.G. Priest and I. Campbell), Institute of Brewing, London, p. 289.
Barre, P. (1969) *Archiv für Mikrobiologie*, **68**, 74.
Barrow, G.I. and Feltham, R.K.A. (eds) (1993) *Cowan and Steel's Manual for the Identification of Medical Bacteria*, 3rd edn. Cambridge University Press, Cambridge.
Benno, Y. and Mitsuoka, T. (1983) *Systematic and Applied Microbiology*, **4**, 123.
Boatwright, J. and Kirsop, B.H. (1976) *Journal of the Institute of Brewing*, **82**, 343.
Bradley, S.G. (1980) In *Microbiological Classification and Identification* (eds M. Goodfellow and R.G. Board), Academic Press, London, p. 11.
Bryan-Jones, G. (1975) In *Lactic Acid Bacteria in Beverages and Food* (eds J.G. Carr, C.V. Cutting and G.C. Whiting), Academic Press, London, p. 165.
Carr, J.G., Davies, P.A., Dellaglio, F. *et al.* (1977) *Journal of Applied Bacteriology*, **42**, 219.
Casey, G.P. and Ingledew, W.M. (1981) *Brewers' Digest*, **55**, (March), 38.
Collins, M.D., Williams, A.M. and Wallbanks, S. (1990) *FEMS Microbiology Letters*, **70**, 255.
Collins, M.D., Farrow, J.A.E., Philips, B.A. *et al.* (1987) *International Journal of Systematic Bacteriology*, **97**, 310.
Collins, M.D., Rodrigues, U., Ash, C. *et al.* (1991) *FEMS Microbiology Letters*, **77**, 5–12.

Collins, M.D., Aguirre, M., Facklam, R.R. *et al.* (1992) *Journal of Applied Bacteriology*, **73**, 433.
Collins, M.D., Samelis, J., Metaxopoulos, J. and Wallbanks, S. (1993) *Journal of Applied Bacteriology*, **75**, 593.
Condon, S. (1987) *FEMS Microbiology Reviews*, **46**, 269.
Coster, E. and White, H. (1964) *Journal of General Microbiology*, **37**, 15.
Crumplen, R. (1988) *Journal of American Society of Brewing Chemists*, **46**, 129.
Dachs, E. (1981) *Brauwelt*, **121**, 1778.
D'Amato, R.F., Holmes, B. and Bottone, E.J. (1981) *Critical Reviews in Microbiology*, **9**, 1.
De Ley, J. (1969) *Journal of Theoretical Biology*, **22**, 89.
Delwiche, E.A. (1961) *Journal of Bacteriology*, **81**, 416.
de Man, J.C., Rogosa, M. and Sharpe, M.E. (1960) *Journal of Applied Bacteriology*, **23**, 130.
Dolezil, L. and Kirsop, B.H. (1975) *Journal of the Institute of Brewing*, **81**, 281.
Dolezil, L. and Kirsop, B.H. (1976) *Journal of the Institute of Brewing*, **82**, 93.
Dolezil, L. and Kirsop, B.H. (1977) *Journal of Applied Bacteriology*, **42**, 213.
Dolezil, L. and Kirsop, B.H. (1980) *Journal of the Institute of Brewing*, **86**, 122.
El Soda, M. and Desmazeaud, M.J. (1982) *Canadian Journal of Microbiology*, **28**, 1181.
El Soda, M., Zeyada, N., Desmazeaud, M.J., Mashaly, R. and Ismail, A. (1982) *Science Alimentaire*, **2**, 261.
Emeis, C.C. (1969) *Monatsschrift für Brauerei*, **22**, 8.
Eschenbecher, F. (1968) *Brauwissenschaft*, **21**, 424, 464.
Eschenbecher, F. (1969) *Brauwissenschaft*, **22**, 14.
Eto, M. and Nakagawa, A. (1974) *Bulletin of Brewing Science*, **20**, 37.
Eto, M. and Nakagawa, A. (1975) *Journal of the Institute of Brewing*, **81**, 232.
Fernandez, J.L. and Simpson, W.J. (1993) *Journal of Applied Bacteriology*, **75**, 369.
Garvie, E.I. (1974) *International Journal of Systematic Bacteriology*, **24**, 301.
Gibson, T. and Abd-el-Malek, Y. (1945) *Journal of Dairy Research*, **14**, 35.
Gilliland, S.E. and Speck, M.L. (1977) *Applied and Environmental Microbiology*, **30**, 1289.
Gunther, H.L. and White, H. (1961) *Journal of General Microbiology*, **26**, 199.
Hammes, W.P., Weiss, N. and Holzapfel, W. (1993) In *The Prokaryotes*, 2nd edn (eds A. Balows, H.G. Trüper, M. Dworkin *et al.*), Springer-Verlag, New York, p. 1535.
Harrigan, W.F. and McCance, M.E. (1976) *Laboratory Methods in Food and Dairy Microbiology*, Academic Press, London.
Haukeli, A.D. and Lie, S. (1978) *Journal of the Institute of Brewing*, **84**, 85.
Healy, P. and Armitt, J. (1980) *Journal of the Institute of Brewing*, **86**, 169.
Holzapfel, W.H. (1992) *International Journal of Food Microbiology*, **17**, 113.
Hough, J.S., Briggs, D.E., Stevens, R. and Young, T.W. (1982) *Malting and Brewing Science*, Vol. 2, 2nd edn, Chapman & Hall, London.
Hsu, W.P. and Taparowsky, J.A. (1977) *Brewers' Digest*, **52**, (February), 48.
Hug, H., Schlienger, E. and Pfenniger, H. (1978) *Schweizer Brauerei-Rundschau*, **89**, 145.
Hugh, R. and Leifson, E. (1953) *Journal of Bacteriology*, **66**, 24.
Ingledew, W.M., Sivaswamy, G. and Burton, J.D. (1980) *Journal of the Institute of Brewing*, **86**, 165.
Ingram, M. (1975) In *Lactic Acid Bacteria in Beverages and Food* (eds J.G. Carr, C.V. Cutting and G.C. Whiting), Academic Press, London, p. 1.
Inoue, T. (1981) *MBAA Technical Quarterly*, **18**, 62.
Johnston, M.A. and Delwiche, E.A. (1965) *Journal of Bacteriology*, **90**, 347.
Jonsson, H. and Pettersson, H.-E. (1977) *Milchwissenschaft*, **32**, 587.
Kandler, O. (1983) *Antonie van Leeuwenhoek*, **49**, 209.

Kandler, O. and Weiss, N. (1986) In *Bergey's Manual of Systematic Bacteriology*, Vol. 2 (eds P.H.A. Sneath, N.S. Mair, M.E. Sharpe and J.G. Holt), Williams and Wilkins, Baltimore, p. 1208.

Kleynmans, U., Heinzl, H. and Hammes, W.P. (1989) *Systematic and Applied Microbiology*, **11,** 267.

Kloos, W.E. and Schleifer, K.-H. (1975a) *Journal of Clinical Microbiology*, **1,** 82.

Kloos, W.E. and Schleifer, K.-H. (1975b) *International Journal of Systematic Bacteriology*, **25,** 62.

Kloos, W.E., Tornabene, T.G. and Schleifer, K.-H. (1974) *International Journal of Systematic Bacteriology*, **24,** 79.

Kocur, M., Kloos, W.E. and Schleifer, K.H. (1992) In *The Prokaryotes*, 2nd edn (eds A. Balows, H.G. Trüper, M. Dworkin *et al.*), Springer-Verlag, New York, p. 1301.

Konings, W.N., Poolman, B. and Driessen, A.J.M. (1989) *CRC Critical Reviews in Microbiology*, **16,** 419.

Lawrence, D.R. and Leedham, P.A. (1979) *Journal of the Institute of Brewing*, **85,** 119.

Lawrence, D.R. and Priest, F.G. (1981) *Proceedings of the 18th Congress of the European Brewery Convention, Copenhagen*, IRL Press, Oxford, p. 217.

Lawrence, R.C. and Thomas, T.D. (1979) *Symposium of the Society for General Microbiology*, **29,** 187.

London, J. (1976) *Annual Review of Microbiology*, **30,** 279.

London, J. and Chace, N.M. (1977) *Proceedings of the National Academy of Sciences of the United States of America*, **74,** 4296.

London, J. and Chace, N.M. (1979) *Journal of Bacteriology*, **140,** 949.

Makinen, V., Tanner, R. and Haikara, A. (1981) *Brauwissenschaft*, **34,** 173.

Manca de Nadre, M.C., Pesce de Ruiz Holgado, A.A. and Oliver, G. (1988) *Biochemie*, **70,** 367.

Martinez Murcia, A.J. and Collins, M.D. (1990) *FEMS Microbiology Letters*, **70,** 73.

Martinez-Murcia, A.J., Harland, N.M. and Collins, M.D. (1993) *Journal of Applied Bacteriology*, **74,** 532.

Matsuzawa, K., Kirchner, G. and Wackerbauer, K. (1979) *Monatsschrift für Brauerei*, **32,** 312.

McCaig, R. and Weaver, R.L. (1983) *MBAA Technical Quarterly*, **20,** 31.

McDonald, L.C., McFeeters, R.F., Daeschiel, M.A. and Fleming, H.P. (1987) *Applied and Environmental Microbiology*, **53,** 1382.

McKay, L.L. (1983) *Antonie van Leeuwenhoek*, **49,** 259.

McMurrough, I. and Palmer, V. (1979) *Journal of the Institute of Brewing*, **85,** 11.

Montel, M.C., Talon, R., Fournoud, J. and Champomier, M.C. (1991) *Journal of Applied Bacteriology*, **70,** 469.

Munekata, M. (1973) *Bulletin of Brewing Science*, **19,** 27.

Nakagawa, A. and Kitahara, K. (1959) *Journal of General and Applied Microbiology*, **5,** 95.

Nishikawa, N. and Kohgo, M. (1985) *Technical Quarterly of the Master Brewers Association of the Americas*, **22,** 61.

Nishikawa, N., Kohgo, M. and Karakawa, T. (1979) *Bulletin of Brewing Science*, **25,** 13, 17, 23.

Orla-Jensen, S. (1919) *Lactic Acid Bacteria*, Andre Fred Host, Copenhagen.

Pfenninger, H.B., Schur, F., Anderegg, P. and Schlienger, E. (1979) *Proceedings of the 17th Congress of the European Brewery Convention, Berlin*, IRL Press, London, p. 491.

Priest, F.G. (1981a) In *The Aerobic Endospore-forming Bacteria: Classification and Identification* (eds R.C.W. Berkeley and M. Goodfellow), Academic Press, London, p. 33.

Priest, F.G. (1981b) *Journal of the Institute of Brewing*, **87,** 279.

Priest, F.G. and Austin, B. (1993) *Modern Bacterial Taxonomy*, 2nd edn, Chapman & Hall, London.
Priest, F.G., and Barbour, E.A. (1985) In *Computer-Assisted Bacterial Systematics* (eds M. Goodfellow, D. Jones and F.G. Priest), Academic Press, London, p. 137.
Priest, F.G. and Pleasants, J.G. (1988) *Journal of Applied Bacteriology*, **64,** 379.
Priest, F.G. and Williams, S.T. (1993) In *Handbook of New Bacterial Systematics* (eds M. Goodfellow and A.G. O'Donnell), Academic Press, London, p. 361.
Raccach, M. (1987) *CRC Critical Review of Microbiology*, **14,** 291.
Rainbow, C. (1981) In *Brewing Science*, Vol. 2 (ed. J.R.A. Pollock), Academic Press, London, p. 491.
Rinck, M. and Wackerbauer, K. (1987) *Monatsschrift für Brauwissenschaft*, **40,** 324.
Rogosa, M. (1974) In *Bergey's Manual of Determinative Bacteriology*, 8th edn (eds R.E. Buchanan and N.E. Gibbons), Williams and Wilkins, Baltimore, p. 576.
Saha, R.B., Sonda, R.J. and Middlekauf, J.E. (1974) *Proceedings of the American Society of Brewing Chemists*, **32,** 9.
Sanni, A.I. and Oso, B.A. (1988) *Nahrung*, **32,** 319.
Seyfried, P.L. (1968) *Canadian Journal of Microbiology*, **14,** 313.
Sharpe, M.E. (1979) In *Identification Methods for Microbiologists*, 2nd edn (eds F.A. Skinner and D.W. Lovelock), Academic Press, London, p. 233.
Sharpe, M.E., Fryer, T.F. and Smith, D.G. (1966) In *Identification Methods for Microbiologists* Part A (eds B.M. Gibbs and F.A. Skinner), Academic Press, London, p. 65.
Shaw, B.G. and Harding, C.D. (1984) *Journal of Applied Bacteriology*, **56,** 25.
Shimwell, J.L. (1941) *Wallerstein Laboratories Communications*, **4,** 41.
Shimwell, J.L. (1945) *Wallerstein Laboratories Communications*, **8,** 23.
Shimwell, J.L. and Kirkpatrick, W.F. (1939) *Journal of the Institute of Brewing*, **45,** 137.
Skerman, V.B.D., McGowan, V. and Sneath, P.H.A. (1980) *International Journal of Systematic Bacteriology*, **30,** 225.
Smith, N.A. and Smith, P. (1992) *Journal of the Institute of Brewing*, **98,** 409.
Smith, N.A. and Smith, P. (1993) *Journal of the Institute of Brewing*, **99,** 43.
Smid, E.J., Poolman, B. and Konings, W.N. (1991) *Applied and Environmental Microbiology*, **57,** 2447.
Sneath, P.H.A. (1978) In *Essays in Microbiology* (eds J.R. Norris and M.H. Richmond), John Wiley, London, p. 10/1.
Sneath, P.H.A. and Sokal, R.R. (1973) *Numerical Taxonomy*, W.H. Freeman, San Francisco.
Solberg, O. and Claussen, O.G. (1973a) *Journal of the Institute of Brewing*, **79,** 227.
Solberg, O. and Claussen, O.G. (1973b) *Journal of the Institute of Brewing*, **79,** 231.
Speckman, R.A. and Collins, E.B. (1968) *Journal of Bacteriology*, **95,** 174.
Speckman, R.A. and Collins, E.B. (1973) *Applied Microbiology*, **26,** 744.
Taguchi, H., Ohkochi, M., Vehara, H. *et al.* (1990) *Journal of the American Society of Brewing Chemists*, **48,** 72.
Van Keer, C., Van Melkebeke, L., Vertriest, W. *et al.* (1983) *Journal of the Institute of Brewing*, **89,** 361.
Van Vuuren, H.J.J., Kersters, K., De Ley, J. *et al.* (1978) *Journal of the Institute of Brewing*, **84,** 315.
Van Vuuren, H.J.J., Kersters, K., De Ley, J. and Toerien, D.F. (1981) *Journal of Applied Bacteriology*, **51,** 51.
Verzele, M. (1986) *Journal of the Institute of Brewing*, **92,** 32.
Wackerbauer, K. and Emeis, C.C. (1967) *Monatsschrift für Brauerei*, **20,** 160.
Wackerbauer, K. and Emeis, C.C. (1968) *Monatsschrift für Brauerei*, **22,** 3.
Wackerbauer, K. and Rinck, M. (1983) *Monatsschrift für Brauwissenschaft*, **36,** 392.
Weiss, N. (1992) In *The Prokaryotes*, 2nd edn (eds A. Balows, H.G. Trüper, M. Dworkin *et al.*), Springer-Verlag, New York, p. 1502.

White, F.H. and Kidney, E. (1979) *Proceedings of the 17th Congress of the European Brewery Convention, Berlin*, IRL Press, Oxford, p. 801.
Whiting, M., Crichlow, M., Ingledew, W.M. and Ziola, B. (1992) *Applied and Environmental Microbiology*, **58,** 713.
Whittenbury, R. (1978) In *Streptococci* (eds F.A. Skinner and L.B. Quesnel), Academic Press, London, p. 51.
Wilkinson, B.J. and Jones, D. (1977) *Journal of General Microbiology*, **98,** 399.
Woese, C.R. (1987) *Microbiological Reviews*, **51,** 221.
Young, T.W. and Talbot, N.P. (1979) *Proceedings of the 17th Congress of the European Brewery Convention, Berlin*, IRL Press, Oxford, p. 817.

CHAPTER 6

Gram-negative spoilage bacteria*

H.J.J. Van Vuuren

6.1 INTRODUCTION

Brewing bacteriology was born when microorganisms responsible for the spoilage of beer were investigated by Louis Pasteur in his classic nineteenth century study. He was called upon to determine why French beer was inferior to German beer. He isolated a number of bacterial contaminants from French beer and malt wort and in 1876 published his famous book *Études sur la Bière, ses Maladies, Causes qui les Provoquent. Procédés pour la Rendre Inaltérable, avec une Théorie Nouvelle de la Fermentation* (Studies of Beer, its Diseases and the Causes That Provoke Them. Procedures for Making it Unalterable, with a New Theory of Fermentation). Today, it is generally believed that the presence of bacteria in the beer brewing process is undesirable.

The Gram stain was developed in 1884 by the Danish physician Christian Gram. Many modifications and improvements have subsequently been made (Society of American Bacteriologists, 1957). The Gram reaction is governed by certain structural properties of the bacterial cell wall. The distinction between Gram-positive and Gram-negative organisms is thus of prime importance in the taxonomy of eubacteria. Both Gram-positive and Gram-negative bacteria are involved in beer spoilage, but this chapter is confined to the latter.

A diagnostic key for the identification of beer spoilage bacteria was devised by the eminent brewery microbiologist Shimwell (1947). The key was simplified by Ault (1965), who restricted it to bacteria likely to be encountered in the beer brewing process. Van Vuuren (1976) modified Ault's key further by only considering beer spoilers. A number of

*Manuscript submitted December 1991

Brewing Microbiology, 2nd edn. Edited by F. G. Priest and I. Campbell.
Published in 1996 by Chapman & Hall, London. ISBN 0 412 59150 2

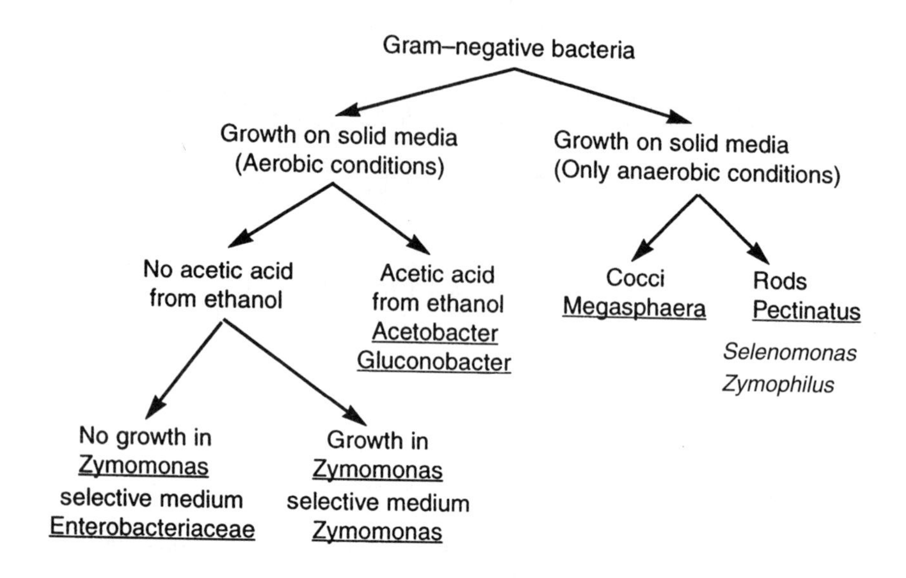

Fig. 6.1 Diagnostic scheme for the identification of Gram-negative beer spoilage bacteria.

diagnostic keys have subsequently been published (Back, 1980; Ingledew, Sivaswamy and Burton, 1980; Casey and Ingledew, 1981; Harper, 1981; Campbell, 1983). The simple key shown in Fig. 6.1 can be used to identify Gram-negative contaminants in beer, including the recently isolated genera *Selenomonas* and *Zymophilus*.

Enterobacteriaceae isolated from beer breweries can be identified further by the API 20E system (Ingledew, Sivaswamy and Burton, 1980; Van Vuuren *et al.*, 1981).

The occurrence of Gram-negative bacteria in the beer brewing process is generally regarded as being undesirable. Important Gram-negative spoilage bacteria considered in this chapter include acetic acid bacteria, certain members of the family Enterobacteriaceae, *Zymomonas, Pectinatus cerevisiiphilus, Pectinatus frisingensis, Selenomonas lacticifex, Zymophilus raffinosivorans, Zymophilus paucivorans* and *Megasphaera*. The occurrence of a few non-fermentative bacteria occasionally found in beer is also discussed.

This brings me to the term infection, often used incorrectly by brewery microbiologists. Bacteria infect humans, animals and plants but they contaminate substrates such as beer, other beverages and food. They are therefore contaminants of beer and the term contamination should be used instead of infection.

6.2 ACETIC ACID BACTERIA

The acetic acid bacteria comprise an industrially important group of Gram-negative, rod-shaped bacteria capable of converting ethanol to acetic acid. They spoil food, soft drinks and alcoholic beverages such as beer, wine and cider (for a review see Swings and De Ley, 1981). The acetic acid bacteria are subdivided into the genera *Acetobacter* and *Gluconobacter* (*Acetomonas*). In the eighth edition of *Bergey's Manual of Determinative Bacteriology* (Buchanan and Gibbons, 1974), *Gluconobacter* was included in the family Pseudomonadaceae, whereas *Acetobacter* was regarded as a genus of uncertain affiliation among the aerobic Gram-negative rods. However, recent taxonomic studies have shown that *Acetobacter* and *Gluconobacter* are closely related and share little affinity with the pseudomonads. They are therefore placed in the new family Acetobacteriaceae (Gillis and De Ley, 1980).

6.2.1 *Acetobacter*

(a) General characteristics

Acetobacter cells are Gram negative or Gram variable, ellipsoidal to rod-shaped, straight or slightly curved, 0.6–0.8 μm × 1.0–4.0 μm, occurring singly, in pairs or chains. Pleomorphic forms occur which may be spherical, elongated, swollen, club shaped, curved or filamentous. They are non-motile or motile; if motile, peritrichous or lateral flagella are present. *Acetobacter* spp. are obligately aerobic with a respiratory metabolism (O_2 is the terminal electron acceptor), catalase positive, oxidase negative, and oxidize ethanol to acetic acid and acetate to CO_2 and H_2O. They are often referred to as the 'overoxidizers'. In liquid media, *Acetobacter* forms a ring, film or pellicle; uniform turbidity of the medium and a cell deposit is sometimes observed (De Ley, Swings and Gossele, 1984).

(b) Taxonomy

There have been extensive taxonomic studies on the genus *Acetobacter*. The eighth edition of *Bergey's Manual of Determinative Bacteriology* (Buchanan and Gibbons, 1974) contained three species with nine subspecies. At present, only *Acetobacter aceti*, *Acetobacter liquefaciens*, *Acetobacter pasteurianus* and *Acetobacter hansenii* are recognized (De Ley, Swings and Gossele, 1984b). Features differentiating these species are given in Table 6.1.

(c) Metabolism

Acetobacter spp. possess a respiratory metabolism (De Ley, 1961; Leisinger, 1965) and, like *Gluconobacter*, directly oxidize sugars, alcohols and steroids

Table 6.1 Characteristics differentiating species of the genus *Acetobacter* (data from *Bergey's Manual of Systematic Bacteriology* (Krieg and Holt, 1984))

Characteristics	A. aceti	A. liquefaciens	A. pasteurianus	A. hansenii
Formation of:				
water-soluble brown pigments on GYC*	–	+	–	–
γ-pyrones from D-glucose	–	d	–	–
γ-pyrones from D-fructose	–	+	–	–
5-ketogluconic acid from D-glucose	+	d	–	d
2,5-diketogluconic acid from D-glucose	–	+	–	–
Ketogenesis from glycerol	+	+	–	+
Growth on carbon sources:				
ethanol	+	+	d	–
dulcitol	–	–	–	d
Na acetate	+	d	d	–
Growth on L-amino acids in the presence of D-mannitol as carbon source:				
L-glycine, L-threonine, L-tryptophan	–	d	–	–
L-asparagine, L-glutamine	d	+	–	+
Growth in the presence of 10% ethanol	–	–	d	–
G + C mol% (T_m)	55.9–59.5	62.3–64.6	52.8–62.5	58.1–62.6

* GYC: 5% D-glucose + 1% yeast extract + 3% $CaCO_3$ + 2·5% agar.
–, 0–15% strains positive; d, 16–84% strains positive; +, 85–100% strains positive.

(for reviews see De Ley and Kersters, 1964; Asai, 1968). For example, glucose is oxidized to gluconic and 2- and 5-oxogluconic acids, glycerol to dihydroxyacetone and sorbitol to sorbose (for a review see Rainbow, 1966). They also metabolize sugars via the hexose monophosphate pathway and the tricarboxylic acid cycle (Asai, 1968; Rainbow, 1981). In certain *A. pasteurianus* strains, the Entner–Doudoroff pathway appears to be more active than the hexose monophosphate shunt (White and Wang, 1964a,b). Strains which utilize ethanol as sole carbon source, and ammonium sulphate as the only source of nitrogen, utilize enzymes of the glyoxylate by-pass (Stouthamer, Van Boom and Bastiaanse, 1963; Leisinger, 1965).

Acetobacter grows well on media containing ethanol, glycerol and sodium DL-lactate as sole carbon source. Several strains are able to grow using NH_4^+ and ethanol as sole nitrogen and carbon sources respectively. Only a few strains are able to use single amino acids as sole nitrogen source. Growth factor requirements depend on the carbon source supplied. Some *A. pasteurianus* and *A. hansenii* strains produce extracellular cellulose.

6.2.2 Gluconobacter

(a) General characteristics

The morphology of *Gluconobacter* is very similar to that of *Acetobacter*. However, motile strains have between three and eight polar flagella; a single flagellum is rarely observed. They are obligately aerobic, catalase positive, oxidase negative, do not reduce nitrate to nitrite but reduce ethanol to acetic acid. Furthermore, *Gluconobacter* does not oxidize acetate or lactate to CO_2 and H_2O. A strong ketogenesis occurs and acid formation from D-glucose and D-xylose is pronounced ($pH \leq 4.5$). Some strains produce a pink, non-diffusible pigment whereas others may produce a soluble, dark brown pigment and γ-pyrone. The pathway for γ-pyrone formation has been elucidated (Asai, 1968) and it is believed that the production of brown pigments is related to γ-pyrone synthesis (Rainbow, 1981). Acetic acid formed in Frateur's (1950) ethanol medium dissolves the calcium carbonate, which is not re-deposited afterwards. This method, as well as that of Carr (1968), can be used to distinguish between *Acetobacter* and *Gluconobacter*. In beer, wort or glucose-containing liquid media, a pellicle or film forms at the surface. Some strains produce viscous growth in beer, due to formation of dextrans and levans (De Ley and Swings, 1984). *Gluconobacter oxydans* is the most important species of this genus for the brewer.

(b) Metabolism

Gluconobacter possesses a respiratory metabolism with oxygen as the terminal electron acceptor. The hexose monophosphate shunt constitutes the most important route for the metabolism of sugars to CO_2 and H_2O. The complete glycolytic and tricarboxylic acid cycles do not function in *Gluconobacter* (De Ley, 1961; Williams and Rainbow, 1964; Greenfield and Claus, 1972). Enzymes of the Entner–Doudoroff pathway are present (Leisinger, 1965; Kersters and De Ley, 1968). *Gluconobacter* is able to oxidize sugars, steroids and aliphatic and cyclic alcohols. *Gluconobacter* utilizes a number of sugars and alcohols as sole source of carbon and good growth is obtained on D-mannitol, sorbitol, glycerol, D-fructose and D-glucose. Ammonium ions as well as single amino acids are used as sole nitrogen source by many *Gluconobacter* strains. A detailed description of *Acetobacter* and *Gluconobacter* spp. is given in Krieg and Holt (1984).

6.2.3 Beer spoilage

Acetification of beer was studied by pioneers of microbiology such as Persoon (see Rainbow, 1981), Pasteur, Hansen, Henneberg and Beijerinck (see Swings and De Ley, 1981). *Acetobacter* and *Gluconobacter* spp. have been isolated from breweries (Van Vuuren, 1976; Ploss, Erber and Eschenbecher, 1979; Hough *et al.*, 1982). Acetomonads are resistant to the bacteriostatic activity of hops, acid and ethanol (Masschelein, 1973; Hough *et al.*, 1982) and are therefore capable of growing and spoiling beer. Being strict aerobes, they should not grow in wort or beer once anaerobic conditions develop. However, beer strains probably grow under microaerophilic conditions and have been isolated from beer with a low oxygen content (Harper, 1980). Laboratory strains of acetic acid bacteria fail to grow on media incubated under carbon dioxide but comparable strains isolated from public houses grow successfully (Harper, 1980).

Acetic acid bacteria are particularly prevalent in public houses and draught beer is rapidly acetified if air is allowed to enter casks. These bacteria spoil beer by causing acid, off-flavours, turbidity and ropiness (Shimwell, 1936; Cosbie, Tosic and Walker, 1942, 1943; Tosic and Walker, 1944; Walker and Tosic, 1945; Shimwell, 1948a; Masschelein, 1973). The production of off-flavours without acetification is probably caused by the oxidation of polyalcohols, such as glycerol, to dihydroxyacetone. Ropiness is often observed as a pellicle or greasy-looking covering on the surface of beer. The extent of beer spoilage by acetic acid bacteria is strain dependent and the alcohol content of beer also affects spoilage. Comrie (1939) and Shimwell (1936) found that the bacteria were completely inhibited by alcohol concentrations higher than 6%. However, some strains grow in media containing 10% ethanol (De Ley, Swings and Gossele, 1984). *Acetobacter* sp. BS05 and *Gluconobacter oxydans* subsp. *oxydans* strain NCIB 9013 are able to survive 12–13% v/v ethanol produced in high gravity brews (Magnus, Ingledew and Casey, 1986). Gilliland and Lacey (1966) showed that an *Acetobacter* sp. produced a compound toxic to yeasts.

Ploss, Erber and Eschenbecher (1979) failed to find acetic acid bacteria on barley and hops but a few malt, wort, yeast and air samples were contaminated. Storage vessels, pressure equilibration, stability and filtration samples were highly contaminated. However, the low incidence of acetomonads in air samples is rather surprising since they are ubiquitous in breweries and beer exposed to air can be rapidly acetified. Acetomonads occur on many plant species and contamination from plant materials and air should not be ignored (Ingledew, 1979; Hough *et al.*, 1982). *Acetobacter* thrives in alcohol-enriched niches whereas *Gluconobacter* prefers sugars as carbon source. Acetomonads are resistant to hop antiseptics and acid conditions and are ethanol tolerant. They can therefore grow and spoil beer in the presence of oxygen.

6.3 ENTEROBACTERIACEAE

6.3.1 General characteristics

The family Enterobacteriaceae comprises a large number of bacterial species isolated from different ecological niches. Some species are free living whereas others are pathogenic to humans, animals, insects and plants. The following genera are included in the family: *Escherichia, Shigella, Enterobacter, Klebsiella, Citrobacter, Salmonella, Erwinia, Kluyvera, Serratia, Cedecea, Morganella, Hafnia, Edwardsiella, Providencia, Proteus, Yersinia, Obesumbacterium, Xenorhabdus, Rahnella* and *Tatumella* (Brenner, 1984).

Enterobacteriaceae are Gram-negative, non-acid-fast, straight rods, 0.3–1.0 μm × 1.0–6.0 μm. Motile as well as non-motile species occur. Motile strains have peritrichous flagella. They do not form endospores or microcysts, are facultatively anaerobic, resistant to bile salts and capable of growing in a mineral salts medium containing glucose as sole carbon and energy source. However, some require vitamins or amino acids or both. Acid and often detectable gas are produced during fermentation of D-glucose. All species important in brewing are catalase positive and oxidase negative. Nitrate is reduced to nitrite except by some *Erwinia* strains and *Yersinia*, and this has implications in the synthesis of *N*-nitrosamines in beer (Smith, 1994). The G + C content of the DNA of members of the family is 38–60 mol%.

6.3.2 Metabolism

Enterobacteria ferment D-glucose via the Embden–Meyerhof–Parnas and hexose monophosphate pathways to yield varying amounts of formic acid (or CO_2 and H_2), acetic acid, lactic acid, succinic acid, ethanol, acetoin and 2,3-butanediol (Fig. 6.2). The concentration of these metabolites depends to a large extent on the bacterial species and is also affected by growth conditions (Van Vuuren, 1978; Tracey, 1984). Two main fermentation types occur. In the mixed acid fermentation typified by *Escherichia coli*, ethanol and acids are formed with little or no acetoin and 2,3-butanediol. In contrast, relatively large amounts of acetoin and 2,3-butanediol but smaller amounts of acids are produced by *Klebsiella* spp. via the 2,3-butanediol pathway. The methyl red and Voges–Proskauer tests are often used as an indication of the mixed acid and 2,3-butanediol fermentations respectively. The results obtained are affected by the ratio of end-products produced and care should be taken when the tests are used for identification purposes.

Gas chromatographic analyses of metabolic end-products of 60 enterobacterial contaminants from beer showed that *Citrobacter freundii, Proteus mirabilis* and *Klebsiella oxytoca* ferment D-glucose via the mixed acid fermentation pathway. *Rahnella aquatilis* (previously *Enterobacter agglomerans*), *Enterobacter cloacae, Klebsiella pneumoniae, Serratia marcescens* and *Hafnia alvei*

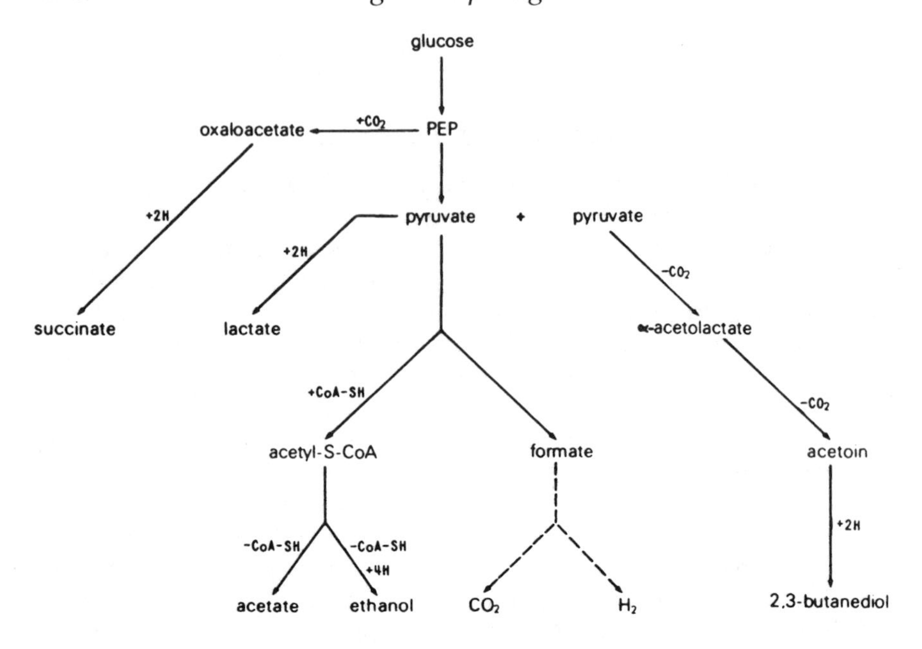

Fig. 6.2 Metabolism of glucose by Enterobacteriaceae.

produced large but varying amounts of 2,3-butanediol (Van Vuuren, 1978). Although *H. alvei* is regarded to ferment D-glucose via the 2,3-butanediol pathway, Tracey (1984) found that the organism is metabolically diverse; some *H. alvei* strains ferment D-glucose via the mixed acid fermentation pathway but others produce large amounts of 2,3-butanediol. *Obesumbacterium proteus* has the mixed acid fermentation pathway.

In the eighth edition of *Bergey's Manual of Determinative Bacteriology* (Buchanan and Gibbons, 1974), the family Enterobacteriaceae was divided into tribes mainly on the basis of D-glucose fermentation via the mixed acid or 2,3-butanediol pathways. The tribe concept has been abandoned in *Bergey's Manual of Systematic Bacteriology* (Brenner, 1984) as it was felt that 'the use of tribes is of no diagnostic significance and of questionable taxonomic significance'.

6.3.3 Beer spoilage

Enterobacterial contaminants in beer breweries have been referred to as termobacteria (Lindner, 1895), wort bacteria (Ault, 1965; Cowbourne, Priest and Hough, 1972; Mäkinen, Tanner and Haikara, 1981) and as *coli-aerogenes* or coliform bacteria (Ingledew, 1979; Keevil, Hough and Cole, 1979; McCaig and Morrison, 1984). The term termobacteria has no scientific standing; wort bacteria include non-enterobacterial contaminants whereas *coli-aerogenes* or coliform bacteria exclude important beer spoilers such as

O. proteus. Enterobacteriaceae, enterobacterial contaminants or enterobacteria are the correct terms to describe these contaminants.

Lindner in 1895 was probably the first to report contamination and spoilage of beer by Enterobacteriaceae. Several enterobacterial species such as *Enterobacter aerogenes, E. cloacae, Rahnella aquatilis, H. alvei, O. proteus, C. freundii, K. pneumoniae, K. oxytoca, Serratia* strains and *P. mirabilis* have subsequently been isolated from beer breweries (Thorwest, 1965; Bernstein, Blenkinship and Brenner, 1968; Von Riemann and Scheible, 1969; Prücha and Scheible, 1970; Anderson, Howard and Hough, 1971; Niefind and Späth, 1971; Cowbourne, Priest and Hough, 1972; Priest and Hough, 1974; Eschenbecher and Ellenrieder, 1975; Van Vuuren and Toerien, 1981; Van Vuuren *et al.*, 1981; McCaig and Morrison, 1984; Hamze *et al.*, 1991).

Pathogenic species occurring in beer have not been described. Ault (1965), Shimwell (1948b) and Seidel (1980) considered members of the family Enterobacteriaceae of little importance to the brewing industry. It is now well known that enterobacterial species retard or accelerate the fermentation process and significantly influence the flavour and aroma of the final product. It is generally believed that apart from *O. proteus* and *R. aquatilis*, other enterobacterial contaminants do not survive the adverse conditions during fermentation (Priest, Cowbourne and Hough, 1974; Hamze *et al.*, 1991). However, bacteria may exist in a 'non-recoverable' stage (Xu *et al.*, 1982), i.e. where no growth is observed on standard laboratory media. They are thus viable and exist in a 'dormant' stage. 'Non-recoverable' bacterial contaminants could have serious implications for the brewing industry and question the often-held view that many enterobacteria do not survive the brewing process. When present in recycled pitching yeast, they will not be detected by standard culture media and methods; contamination and spoilage of beer might therefore occur when they develop under favourable conditions in a new fermentation. Investigations should be undertaken to determine the efficacy of media and methods to detect possible 'non-recoverable' bacterial contaminants. Enriched or more selective culture media and prolonged incubation should be considered.

(a) Obesumbacterium proteus

O. proteus is the best known enterobacterial contaminant. This bacterium was probably first described by Lindner as *'Termobacterium lutescens'* (Priest, 1974). Shimwell (1936) named it 'Bacterium Y', Shimwell and Grimes (1936) renamed it *'Flavobacterium proteum'* (later *'F. proteus'*) and finally Shimwell (1963, 1964) created a new genus *Obesumbacterium*, with a single species *O. proteus*. The extensive studies of Priest *et al.* (1973) and Priest, Cowbourne and Hough (1974) indicated that *O. proteus* is closely related to *Hafnia* and thus a member of the family Enterobacteriaceae. They proposed that the genus *Obesumbacterium* be discontinued and that

O. proteus be transferred to *Hafnia* as '*Hafnia protea*'. Furthermore, they recognized two genetic groups of '*H. protea*'. This has been confirmed by Brenner (1981) and Prest, Hammond and Stewart (1994). However, *O. proteus* rather than '*H. protea*' was included in the Approved Lists of Bacterial Names (Skerman, McGowan and Sneath, 1980) and subsequently in *Bergey's Manual of Systematic Bacteriology* (Brenner, 1984).

O. proteus has thus far only been isolated from the brewery environment. It is often found in pitching yeast and referred to as the 'short fat rod of pitching yeast'. *O. proteus* is highly pleomorphic, grows well in hopped or unhopped wort and tolerates ethanol concentrations of up to 6% v/v. However, the inhibitory concentration of ethanol appears to be pH dependent. *O. proteus* can grow in unhopped wort with pH ranging from 4.4 to 9.0. Brown (1916) considered it a harmless yeast contaminant whereas Strandskov (1965) concluded that it could spoil beer only when present in pitching yeast at a concentration exceeding 1.5% of the yeast count. However, *O. proteus* is an important beer spoilage organism; it is able to grow alongside yeast in a brewery fermentation, retard the fermentation process and result in beer with a high final specific gravity and pH (Case, 1965; Priest, Cowbourne and Hough, 1974). *O. proteus* is also responsible for increased levels of dimethyl sulphide, dimethyl disulphide, n-propanol, isobutanol, isopentanol, 2,3-butanediol and diacetyl (Cowbourne, Priest and Hough, 1972; Priest, Cowbourne and Hough, 1974). The final concentration of these compounds is often strain dependent. Beer produced with yeast contaminated with *O. proteus* has a parsnip-like or fruity odour and flavour. Pitching yeast is the main source of contamination in breweries.

(b) Rahnella aquatilis

Enterobacter agglomerans was first recognized as an important contaminant in beer breweries by Van Vuuren *et al.* (1978). This finding was confirmed by Mäkinen, Tanner and Haikara (1981) and McCaig and Morrison (1984). However, at that stage *E. agglomerans* comprised a large and heterogeneous group of bacteria and its nomenclature and classification was in a state of flux (Van Vuuren, 1987). The 18 aerogenic strains of *E. agglomerans* isolated from breweries (Van Vuuren *et al.*, 1981) have been reclassified as strains of *Rahnella aquatilis* (Hamze *et al.*, 1991). The taxonomic position *E. agglomerans* strains studied by Mäkinen, Tanner and Haikara (1981) and McCaig and Morrison (1984) have not been clarified. For the purpose of this review all strains of *E. agglomerans* previously isolated from breweries will be regarded as strains of *R. aquatilis*.

R. aquatilis cells are small rod-shaped cells, 0.5–0.7 μm × 2.0–3.0 μm. They are motile when grown at 25°C but motility is lost at 37°C. D-Glucose is fermented with the production of acid and most strains form gas. Lactose, maltose, L-rhamnose, raffinose and salicin are fermented. *R. aquatilis* is negative for lysine and ornithine decarboxylases and for arginine dihydrolase. All strains are Voges–Proskauer positive and most strains are methyl red positive. Strains are weakly positive for phenylalanine deaminase. The mol% G + C of the DNA is 51–56 (determined by melting temperature, T_m) (Izard *et al.*, 1979, 1981; Brenner, 1984). *R. aquatilis* occurs mainly in fresh water but has occasionally been isolated from human clinical specimens.

R. aquatilis grows well in hopped or unhopped wort, in the presence or absence of yeast. It survives the brewing process when normal gravity wort is used and, like *O. proteus*, accumulates in recycled pitching yeast (Van Vuuren, Cosser and Prior, 1980; Magnus, Ingledew and Casey, 1986). In high-gravity brews, however, *R. aquatilis* is killed by ethanol concentrations of 11–12% v/v (Magnus, Ingledew and Casey, 1986).

In ale fermentations, *O. proteus* rises to the surface of the wort with the top yeasts and remains viable for 60–70 h; other enterobacteria remain suspended in the wort and are killed by adverse conditions, i.e. a low pH and a high ethanol concentration (Priest, Cowbourne and Hough, 1974). During lager beer fermentations, however, bacterial contaminants such as *R. aquatilis* settle or flocculate with the yeast. Bacteria accumulated in harvested pitching yeast are a serious source of contamination. It also becomes impossible to obtain accurate counts of bacteria during fermentation.

R. aquatilis greatly affects the brewing process as well as beer flavour and aroma when present in fermenting wort (Van Vuuren, Cosser and Prior, 1980; McCaig and Morrison, 1984). According to Cowbourne, Priest and Hough (1972), enterobacterial contaminants retard the fermentation process; this effect might, however, be yeast strain dependent. The presence of *R. aquatilis* during lager beer fermentations increases the rate of fermentation during the initial stages but the final specific gravity of contaminated beer is higher than that of control beer (Van Vuuren, Cosser and Prior, 1980). In ale brewing, however, the specific gravity of the final beer is lower when fermentations are contaminated by *R. aquatilis* (McCaig and Morrison, 1984). The final pH of fermenting wort contaminated with *R. aquatilis* is higher than usual, probably due to neutral metabolites produced during fermentation. *R. aquatilis* ferments D-glucose via the 2,3-butanediol fermentation pathway (Van Vuuren, Cosser and Prior, 1980). However, in ale fermentations the final pH of the contaminated beer is reportedly similar to that of control beer (McCaig and Morrison, 1984).

Abnormally high levels of diacetyl (0.7 mg l^{-1}) and dimethyl sulphide (142.8 μg l^{-1}) were detected in beer produced from wort contaminated with 10^6 *R. aquatilis* cells ml^{-1}. The concentration of these compounds in the uncontaminated control beer was considerably lower (0.04 mg l^{-1} and 54.6 μg l^{-1} respectively). The production of these compounds was affected by the level of contamination. Significant increases were only found with a contaminant population of 10^6 cells ml^{-1}; a slight increase in the concentration of both compounds was observed at 10^3 cells ml^{-1} (Van Vuuren, Cosser and Prior, 1980). The amount of dimethyl sulphide produced by *R. aquatilis* is strain dependent (McCaig and Morrison, 1984). The production of excess vicinal diketones by *R. aquatilis* in beer was confirmed by McCaig and Morrison (1984). It has been suggested that other enterobacteria are indirectly responsible for synthesis of diacetyl by producing higher levels of <F128Ma-acetolactate (Priest and Hough, 1974), a heat-labile precursor of diacetyl (Inoue *et al.*, 1968; Suomalainen and Ronkainen, 1968). The mechanism by which *R. aquatilis* produces diacetyl during fermentation has not yet been elucidated. *R. aquatilis* contamination also leads to increased levels of acetaldehyde and methyl acetate; however, the concentration of ethyl acetate, n-propanol, isobutanol, isoamyl acetate and isoamyl alcohol is lower (Van Vuuren, Cosser and Prior, 1980). Beer contaminated with *R. aquatilis* is unpalatable, with a fruity, milky and sulphury taste and aroma. Barley, malt and hops as well as recycled pitching yeast could well be the source of contamination (Van Vuuren, Cosser and Prior, 1980).

(c) Citrobacter freundii

C. freundii has been reported as an occasional contaminant of pitched wort (Cowbourne, Priest and Hough, 1972; Priest, Cowbourne and Hough, 1974; Eschenbecher and Ellenrieder, 1975; Van Vuuren *et al.*, 1981). This bacterium is closely related to *E. coli*. The cells are straight rods, ~1.0 μm in diameter and 2.0–6.0 μm long. They occur singly and in pairs and are usually motile by peritrichous flagella. *C. freundii* is facultatively anaerobic, oxidase negative and catalase positive. Citrate can be used as sole carbon source; however, citrate negative strains have been found in beer breweries (Van Vuuren *et al.*, 1981). D-Glucose is fermented via the mixed acid fermentation pathway and gas is produced. The G + C content of the DNA is 50–51 mol%. *C. freundii* has been found in humans, animals, soil, water, sewage, clinical specimens and wound swabs. It is regarded as an opportunistic or secondary pathogen (Krieg and Holt, 1984).

Keevil, Hough and Cole (1979) studied the effect of *C. freundii* on beer fermentation and flavour. The organism reduces yeast viability and affects the redox potential of fermenting wort. Contamination of fermenting wort with *C. freundii* leads to faster fermentations. Similar results were obtained with *R. aquatilis* (Van Vuuren, Cosser and Prior, 1980). The presence of *C. freundii* during fermentation greatly increases concentrations of lactate, pyruvate, succinate, isocitrate and dimethyl sulphide in the final beer. Even a low number of *C. freundii* cells, viable for only a restricted period during fermentation, may spoil beer (Keevil, Hough and Cole, 1979). *C. freundii* is, however, sensitive to high concentrations of ethanol and does not survive the brewing process (Magnus, Ingledew and Casey, 1986).

(d) Klebsiella

Klebsiella spp. have been isolated from different stages of the brewing process and are associated with beer spoilage (Anderson, Howard and Hough, 1971; Masschelein, 1973). *Klebsiella* cells are straight rods, 0.3–1.0 μm × 0.6–6.0 μm, occurring singly, in pairs or short chains. They are non-motile, capsulated, facultatively anaerobic and oxidase negative. D-Glucose is fermented by brewery isolates via the mixed acid or 2,3-butanediol fermentation pathways (Van Vuuren, 1978). Acid and gas are produced but anaerogenic strains occur. The G + C content of the DNA is 53–58 mol%. *Klebsiella* spp. have also been isolated from intestinal contents, clinical specimens, soil, water and plants (Krieg and Holt, 1984).

K. pneumoniae is a pathogen of humans and causes inflammation of the lungs. It is most unlikely that this pathogen occurs in samples from beer breweries. A new species, *Klebsiella terrigena*, was recently described (Izard *et al.*, 1981). It occurs mainly in aquatic and soil environments and can be distinguished from *K. pneumoniae* by its ability to grow at 10°C and inability to produce gas from lactose at 44.5°C. Its ability to ferment melezitose distinguishes *K. terrigena* from *Klebsiella planticola* (Brenner, 1984). *K. pneumoniae* strains isolated from beer breweries conform to criteria laid down for the new species (Van Vuuren *et al.*, 1981). Therefore, *K. terrigena*, rather than *K. pneumoniae*, occurs as a contaminant in beer breweries. This information will certainly be received with relief by brewers.

Beer produced from wort contaminated with an indole-negative *Klebsiella* strain (possibly *K. terrigena*) had a typical phenolic off-flavour. Only indole-negative enterobacteria produce phenolic off-flavours in beer (West, Lautenbach and Brumsted, 1963; Von Riemann and Scheible, 1969; Prücha and Scheible, 1970). Cowbourne, Priest and Hough (1972) suggested that bacteria produce these off-flavours by deaminating tyrosine (Fig. 6.3).

Abnormally high concentrations of 4-ethylguaiacol have been detected in beer having a phenolic off-flavour (Halcrow *et al.*, 1966). *Klebsiella* spp. decarboxylate ferulic acid, normally present in beer, to 4-vinylguaiacol (Fig. 6.4). According to Finkle *et al.* (see Halcrow *et al.*, 1966) hydrogenation of the vinyl group occurs as a side reaction. Both 4-vinylguaiacol (Dadic, van Gheluwe and Valyi, 1971) and 4-ethylguaiacol (Halcrow *et al.*, 1966) impart a phenolic taste to beer.

Indole-positive *K. pneumoniae* strains have been isolated from beer breweries (Van Vuuren, 1978). Indole-positive *K. pneumoniae* strains have been reclassified as *K. oxytoca* (Brenner, 1984) and strains of this species have been isolated from fermenting wort in two different breweries (Van Vuuren and Toerien, 1981). The organism is now considered a contaminant of beer but it is not known how it affects the fermentation process, flavour or aroma. *K. oxytoca* is sensitive to ethanol and does not survive the brewing process (Magnus, Ingledew and Casey, 1986).

Klebsiella spp. have also been associated with production of excessive

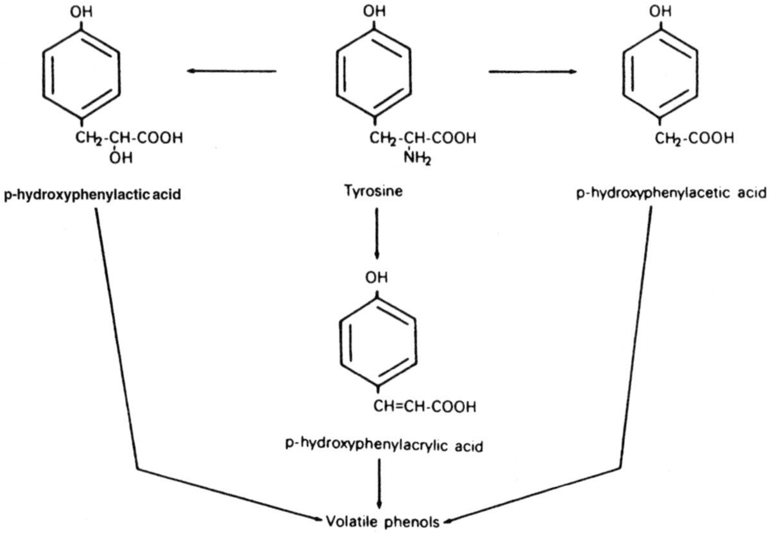

Fig. 6.3 Production of volatile phenols from tyrosine.

volatile organo-sulphur compounds such as dimethyl sulphide (Anderson, Howard and Hough, 1971) which affect beer quality adversely. Use of ^{35}S-labelled sulphur showed that less than 2% of sulphur produced by *Klebsiella* is derived from methionine, cysteine, cystine or sulphate (Wainwright, 1972). It has been shown that *R. aquatilis* produces dimethyl sulphide mainly by reduction of dimethyl sulphoxide (McCaig and Morrison, 1984).

(e) Other enterobacterial contaminants

Other enterobacterial contaminants isolated from brewery worts include *E. cloacae, E. aerogenes, H. alvei, Serratia* strains and *P. mirabilis*. Two genetic

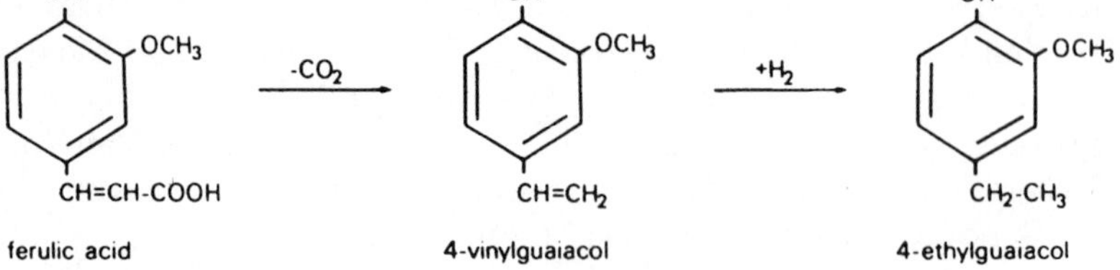

Fig. 6.4 Production of 4-vinylguaiacol and 4-ethylguaiacol from ferulic acid.

groups of *E. cloacae* have been isolated from beer breweries (Van Vuuren, 1978). *E. cloacae* and *E. aerogenes* produce phenolic compounds and dimethyl sulphide respectively in beer. *H. alvei* contamination leads to increased concentrations of n-propanol and dimethyl sulphide (for a review see Priest, Cowbourne and Hough, 1974). Spoilage of beer by *Serratia* and *P. mirabilis* has not been reported.

Enterobacterial contaminants occur most commonly in pre-fermentation wort and are associated with early stages of fermentation. However, a much wider distribution of most types occurs (Van Vuuren *et al.*, 1978) than was recognized previously. They continue to grow in fermenting wort and produce metabolites which have a detrimental effect on beer quality. Sulphury or phenolic off-flavours in beer might indicate enterobacterial contamination. Another feature of enterobacteria is their ability to reduce nitrate to nitrite. Unacceptable levels of nitrite may be formed in beer if nitrate-containing water is used for brewing and contamination with enterobacteria occurs (Weiner, Rolph and Taylor, 1975; Hough *et al.*, 1982). It has recently been shown that *O. proteus* contains the enzyme nitrate reductase which converts nitrate to nitrite (Calderbank and Hammond, 1989). In experimental fermentations, levels of apparent total *N*-nitroso compounds (ATNC) at the end of fermentation were dependent on the initial wort nitrate levels as well as the initial number of bacteria present in the pitching yeast (Smith, 1994).

Enterobacteriaceae occurring in beer breweries are regarded as free-living forms normally encountered in water, soil and plant material. Water, barley, malt and hops are probably the main sources of contamination. It is obvious that members of the Enterobacteriaceae are actively involved in beer spoilage. Many brewery microbiologists have so far underestimated the significance of these contaminants.

6.4 *ZYMOMONAS*

6.4.1 General characteristics

Zymomonas cells are short, plump, Gram-negative rods, 1.0–1.4 μm × 2.0–6.0 μm. They occur mostly in pairs but single cells are usually present. Some strains form chains, rosettes, and in many strains filamentous cells are present. Endospores are not formed. Most strains are non-motile. Motile strains have one to four polar flagella. Glucose and fructose are fermented to nearly equimolar amounts of ethanol. They do not ferment maltose, are oxidase negative and anaerobic but tolerate some oxygen. No growth is observed on nutrient agar. *Zymomonas* has a G + C content of 47.5–49.5 mol% (Swings and De Ley, 1984).

6.4.2 Taxonomy

Swings and De Ley (1977) studied a large collection of strains from different ecological niches. They suggested that only a single species, *Zymomonas mobilis*, be recognized with two subspecies: *Z. mobilis* subsp. *mobilis* and *Z. mobilis* subsp. *pomacii*. The latter was created specifically for the cider sickness zymomonads. This classification system was followed in *Bergey's Manual of Systematic Bacteriology* (Swings and De Ley, 1984). Synonyms assigned to *Zymomonas* are listed in Table 6.2.

6.4.3 Metabolism

The almost quantitative fermentation of glucose and fructose to ethanol and CO_2 via a modified Entner–Doudoroff pathway (Fig. 6.5) is a distinctive characteristic of the genus *Zymomonas*. Small amounts of acetaldehyde, acetylmethylcarbinol, acetic acid, lactic acid and glycerol are formed (Schreder, Brunner and Hampe, 1934).

The ability of *Zymomonas* to grow and metabolize in the presence of high ethanol concentrations is unusual for a bacterium and emphasizes its importance as a contaminant in the alcoholic beverage industry. Comrie

Table 6.2 Synonyms of *Zymomonas* (data from Swings and De Ley, 1977)

Zymomonas mobilis subsp. *mobilis* (Lindner) De Ley & Swings (1976)

Synonyms
- *Termobacterium mobile* Lindner (1928)
- *Pseudomonas lindneri* Kluyver & Hoppenbrouwers (1931)
- *Saccharomonas lindneri* (Kluyver & Hoppenbrouwers) Shimwell (1950)
- *Zymomonas mobile* (sic) (Lindner) Kluyver & Van Niel (1936)
- *Zymomonas mobilis* (Lindner) Kluyver & Van Niel (1936)
- *Zymomonas mobilis* var. *anaerobia* Richards & Corbey (1974)
- *Zymomonas congolensis* Van Pee & Swings (1971)
- *Achromobacter anaerobium* Shimwell (1937)
- *Saccharobacter* Shimwell (1937)
- *Saccharomonas anaerobia* Shimwell (Shimwell) (1950)
- *Zymomonas anaerobia* Shimwell (Kluyver) (1957)
- *Zymomonas anaerobia* var. *anaerobia* (Shimwell) Carr (1974)
- *Saccharomonas anaerobia* var. *immobilis* Shimwell (1950)
- *Zymomonas anaerobia* var. *immobilis* (Shimwell) Carr (1974)
- *Zymomonas mobilis* var. *recifensis* Goncalves de Lima, De Araujo, Schumacher & Cavalcanti Da Silva (1970)

Zymomonas mobilis subsp. *pomacii* (Millis) De Ley & Swings (1976)

Synonyms
- Cider sickness organism Barker & Hillier (1912)
- *Zymomonas anaerobia* var. *pomaceae* Millis (1956)

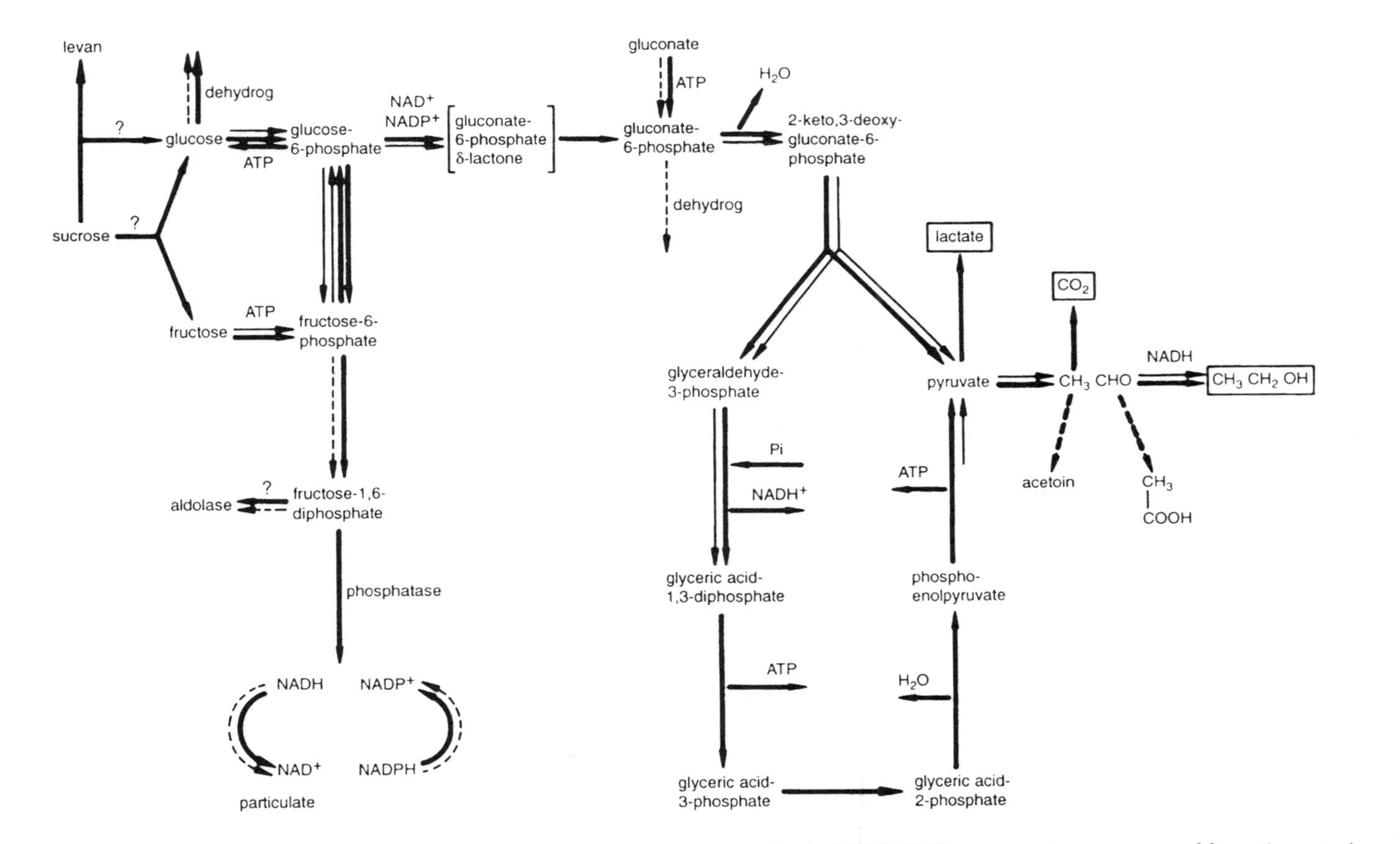

Fig. 6.5 Mechanism of carbohydrate catabolism in two *Zymomonas* strains. Strain NCIB 8938: ▬▬, active, ▬ ▬ ▬, weakly active; strain NCIB 8777: ——, active, – – –, weakly active. (From Swings and De Ley, 1977, reproduced with permission of *Bacteriological Reviews*.)

(1939) reported that growth of *Zymomonas* is inhibited by 8% v/v ethanol. However, Swings and De Ley (1977) found that almost half of their strains grew at 10% v/v ethanol. *Z. mobilis* strain BS057 survived ethanol concentrations of 12–13% v/v produced during high gravity brewing (Magnus, Ingledew and Casey, 1986). Although growth of some *Zymomonas* strains may be arrested by 8–10% v/v ethanol, they remain metabolically active and can produce up to 15% v/v ethanol. *Zymomonas* is an attractive organism for the industrial production of ethanol (Swings and De Ley, 1977).

6.4.4 Beer spoilage

Shimwell (1937) first isolated *Zymomonas* from beer and named it *'Achromobacter anaerobium'*. *Zymomonas* strains studied by Shimwell produced hydrogen sulphide and acetaldehyde in primed beer. This was confirmed by Anderson, Howard and Hough (1971), Dadds, MacPherson and Sinclair (1971), Dadds and Martin (1973) and Wainwright (1972). Trace amounts of dimethyl sulphide and dimethyl disulphide are also produced (Dadds, Martin and Carr, 1973). Shimwell considered *Zymomonas* the most malignant beer spoilage bacterium recorded at that time. Beer contaminated with *Zymomonas* has a rotten apple (Shimwell, 1950) or fruity (Dadds and Martin, 1973) odour. *Zymomonas* is an important contaminant in ale breweries, mainly in the UK (for a review see Dadds and Martin, 1973).

Zymomonas has not been reported in lager breweries. Its relatively stringent carbohydrate requirements probably restrict the organism to ale breweries (Dadds, MacPherson and Sinclair, 1971). However, according to Swings and De Ley (1977) its possible presence in lager breweries should not be ignored. They suggested that the low temperatures (8–12°C) of lager fermentations are unfavourable for the development of *Zymomonas*.

Zymomonas has also been isolated from soil near breweries and from brushes of cask-washing machines (Shimwell, 1937, 1950), well water (Stevens, according to Dadds and Martin, 1973), fermented apple juice (cider; Barker and Hillier, 1912; Millis, 1951, 1956), pear juice (perry; Millis, 1951) and ripening honey (Ruiz-Argueso and Rodriquez-Navarro, 1975). *Zymomonas* strains are also responsible for the spontaneous alcoholic fermentation of agave sap (pulque) (Lindner, 1928), palm wine (Swings and De Ley, 1977) and sugar-cane juice (Goncalves de Lima *et al.*, 1970).

6.5 ANAEROBIC GRAM-NEGATIVE RODS

Brewery microbiologists have generally accepted that beer spoilage bacteria are restricted to certain aerobic and facultatively anaerobic species. However, spoilage of packaged beer by strictly anaerobic rods has been reported (Lee, Mabee and Jangaard, 1978; Lee *et al.*, 1980). Lee, Mabee and Jangaard (1978) named a new genus and species for these obligately anaerobic rods, viz. *Pectinatus cerevisiiphilus*. Bacterial contaminants similar to *P. cerevisiiphilus* have since been isolated from spoilt beer in Germany (Back, Weis and Seidel, 1979; Seidel-Rüfer, 1990), Scandinavia (Haikara, 1985; Haikara *et al.*, 1980) and Japan (Takahashi, 1983). In an extensive taxonomic study of anaerobic rods isolated from breweries, Schleifer *et al.* (1990) identified a new species of the genus *Pectinatus, P. frisingensis*, as well as a new species of the genus *Selenomonas*, *S. lacticifex* and a new genus, *Zymophilus*, comprising two species *Z. raffinosivorans* and *Z. paucivorans*.

6.5.1 *Pectinatus cerevisiiphilus*

The description of *P. cerevisiiphilus* below is based on the amended description by Schleifer *et al.* (1990). *P. cerevisiiphilus* cells are slightly curved rods, 0.7–0.9 μm × 2.0–30 μm; they occur singly, in pairs, and only rarely in short chains. The occurrence of longer, helical filaments is characteristic for older cells. They are motile by flagella emanating from only one side of the cell (comb-like). This organism is an obligately anaerobic, non-spore-forming mesophile. No growth is obtained on agar plates under aerobic conditions or under a CO_2 atmosphere. They can, however, be cultured on thioglycollate agar by the roll-tube technique of Hungate (1969), or in a modified MRS medium under anaerobic conditions (Schleifer *et al.*, 1990). Colonies are circular, entire, beige to white, glistening and opaque. Glucose is fermented to acetic and propionic acids. The G + C content of the DNA is 38–41 mol%.

6.5.2 *Pectinatus frisingensis*

Morphologically *P. frisingensis* is similar to *P. cerevisiiphilus*. Colonies on modified MRS agar medium are shiny, opaque, circular, slightly yellow, and 1–2 mm in diameter after 3 days at 30°C. In addition to acetic and propionic acids, acetoin and sometimes minor amounts of succinic acid are produced from glucose under fermentative conditions. The G + C content of the DNA is 38–41 mol%. *P. frisingensis* can be distinguished from *P. cerevisiiphilus* by its ability to ferment cellobiose, inositol and *N*-acetyl-glucosamine and its inability to utilize xylose and melibiose (Schleifer *et al.*, 1990).

6.5.3 *Selenomonas lacticifex*

Four strains of *S. lacticifex* have been isolated from pitching yeast. These strains were obligately anaerobic, usually motile and non-spore-forming rods, 0.6–0.9 μm × 5.0–15 μm. The cells are curved and crescent shaped and occur predominantly as single cells or (rarely) in pairs. The optimum temperature for growth is 30°C. Colonies on modified MRS agar are smooth, opaque, circular, yellowish and 2–3 mm in diameter after 3 days at 30°C. D-Glucose is fermented to acetic, lactic and propionic acids. The G + C content of the DNA is 51–52 mol% (Schleifer *et al.*, 1990).

6.5.4 *Zymophilus raffinosivorans*

Z. raffinosivorans was isolated from pitching yeast and brewery waste. Cells are straight to slightly curved rods, 0.7–0.9 μm × 3.0–15 μm, and occur predominantly as single cells; pairs or short chains are sometimes observed (Schleifer *et al.*, 1990). They are non-spore-forming motile rods but motility can be lost after several subcultivations. Colonies on modified MRS agar are circular, smooth, opaque, slightly yellow and 1–2 mm in diameter after 3 days at 30°C. D-Glucose is fermented to acetic and propionic acids. The G + C content of the DNA is 38–41 mol%.

6.5.5 *Zymophilus paucivorans*

Z. paucivorans was isolated from pitching yeast. In contrast to *Z. raffinosivorans*, *Z. paucivorans* cells are curved or helically shaped. The rod-shaped cells are 0.8–1.0 μm × 5.0–30 μm and occur singly, in pairs or in short chains. They are motile but motility may disappear after subculturing. Colonies on modified MRS agar are circular, smooth, slightly yellow, and 1–2 mm in diameter after 3 days at 30°C. D-Glucose is fermented to acetic and propionic acids and trace amounts of lactic acid. *Z. raffinosivorans* and *Z. paucivorans* differ in their ability to utilize xylose, rhamnose, xylitol, raffinose, melibiose and inositol (Schleifer *et al.*, 1990). The G + C content of *Z. paucivorans* DNA is 39–41 mol%.

6.5.6 Beer spoilage

It has become clear that strictly anaerobic rod-shaped bacteria constitute an important group of spoilage bacteria of packaged beer. Bacterial isolates considered to be *Pectinatus* strains produced a considerable amount of acetic and propionic acids as well as acetoin in wort and packaged beers (Seidel, Back and Weis, 1979; Lee *et al.*, 1980). High concentrations of H_2S were also detected in beer contaminated by *Pectinatus* strains. The beer became turbid with an odour of rotten eggs. However, strains previously thought to be similar to *P. cerevisiiphilus* have now been reclassified; they belong to at least three different genera (Schleifer *et al.*, 1990). *Selenomonas lacticifex* is considered to be a spoilage organism and *Zymophilus raffinosivorans* a potential spoilage organism of beer (Seidel-Rüfer, 1990). *Z. paucivorans*, however, is not regarded as a beer spoilage organism by Seidel-Rüfer (1990).

Pectinatus strains have been isolated from lubrication oil mixed with beer and water, and from a drainage system (Lee *et al.*, 1980). However, the source of contamination by anaerobic rods in breweries is not clear. Transmission of anaerobic rod-shaped bacteria may well be by contaminated pitching yeast.

6.6 *MEGASPHAERA*

Weiss, Seidel and Back (1979) isolated strictly anaerobic Gram-negative cocci from finished beer spoilt by cloudiness and unpleasant odours. The isolates were phenotypically similar to *Megasphaera elsdenii*. However, the G + C values differed from that of the *M. elsdenii* type species and they were regarded as *Megasphaera* species I and II. Engelmann and Weiss (1985) studied the taxonomic position of 12 *Megasphaera* strains isolated from beer. These strains were genetically closely related and they suggested a new species, *M. cerevisiae*, in the family Veilonellaceae.

M. cerevisiae cells are Gram-negative, slightly elongated cocci, 1.3–1.6 μm in diameter, in pairs and occasionally in short chains. They are non-motile and non-spore forming. They are strictly anaerobic and colonies on a basal medium containing fructose or lactate are whitish, smooth and flat, 2–3 mm in diameter. Growth occurs between 15°C and 37°C with an optimum of 28°C. *M. cerevisiae* is negative for catalase and benzidine; it produces H_2S. Nitrate is not reduced to nitrite and it is indole negative. Fructose is fermented by all strains but only some strains ferment arabinose. The mol% G + C of the DNA is 42.4–44.8 (T_m). Their habitat is unknown (Engelmann and Weiss, 1985).

Megasphaera strains produce considerable amounts of butyric acid in beer. Acetic, isovaleric, valeric and caproic acids as well as acetoin are also produced in smaller quantities (Seidel, Back and Weiss, 1979). In wort, large quantities of butyric and caproic acids and smaller amounts of acetic, isovaleric and valeric acids are produced. No acetoin is produced in wort (Seidel, Back and Weiss, 1979). *Megasphaera* is considered a true beer spoiler by Seidel, Back and Weiss, (1979). *M. cerevisiae*, however, is sensitive to alcohol and low pH (Haikara and Lounatmaa, 1987). Growth of this contaminant in beer is also restricted at an alcohol content of 2.8% w/v.

6.7 MISCELLANEOUS NON-FERMENTATIVE BACTERIA

Alcaligenes, Acinetobacter, Flavobacterium and *Pseudomonas* spp. have been isolated from worts and brewing liquors (Masschelein, 1973; Eschenbecher and Ellenrieder, 1975; Priest, 1981a). These bacteria have a strict respiratory type of metabolism. Spoilage of beer by these occasional contaminants has not been reported.

6.8 DETECTION, ENUMERATION AND ISOLATION

The ability of certain Gram-negative bacteria to spoil beer is now widely accepted. To prevent or minimize spoilage, effective microbiological controls are essential. A number of methods exist for the detection of bacterial contaminants in beer breweries. Traditionally, bacteria are detected by direct microscopic observation and by cultivation on suitable nutrient media. The need for more rapid detection has been recognized and several other methods have been employed. These include the ATP assay, pH change assay, measurement of the uptake of radioactive ^{14}C, and use of the Coulter counter, fluorescence microscopy, serology, optical brighteners and staining techniques. These alternative methods are discussed in Chapter 8. Therefore, only microscopic techniques and some methods of culturing Gram-negative spoilage bacteria are considered in this chapter.

6.8.1 Microscopic examination

Direct microscopic examination of a suspension or stained smears can be used to detect bacterial contaminants. These rapid methods yield valuable information concerning the presence, morphology and motility of bacteria. When used in conjunction with the Gram stain, a distinction can be made between Gram-negative and Gram-positive isolates. However, these methods can only be used when the numbers of bacteria exceed 30 000 ml^{-1}. Furthermore a distinction between live and dead cells cannot be made and it is sometimes difficult to distinguish non-motile bacteria from insoluble matter.

The Petroff–Hausser counting chamber can be used to determine bacterial cell numbers in suspensions. The disadvantages of the method are threefold: distinguishing dead cells from living cells; low bacterial populations cannot be calculated unless the sample is centrifuged beforehand; and the actual counting procedure is tiresome. Despite these shortcomings, it is a good practice to examine samples from breweries regularly under the microscope.

6.8.2 Culture media

Many Gram-negative bacterial species grow well on simple laboratory media. Others require specific nutrients such as vitamins and various growth-promoting substances. Furthermore, special-purpose media are available to facilitate recognition, enumeration and isolation of certain types of bacteria.

The proper application of media in brewing bacteriology has been thoroughly reviewed by Casey and Ingledew (1981). No attempt has been made to repeat that review. I will rather concentrate on media that I have found suitable for rapid culture of acetic acid bacteria, Enterobacteriaceae and *Zymomonas*. In addition, the detection of the strictly anaerobic contaminants, *Pectinatus*, *Selenomonas*, *Zymophilus* and *Megasphaera*, will also be discussed.

For the selective inhibition of yeasts, actidione (cycloheximide) can be added to culture media at 20–100 μg ml^{-1}. The lower concentration will inhibit brewing yeasts and the higher concentration will prevent the growth of most 'wild' yeasts.

Membrane filtration is most useful for enumerating low levels of organisms in suspensions. A relatively large volume of fluid can be passed through the filter which is then placed on a suitable agar medium. Colonies developing on the surface of the filter are counted after a suitable incubation period.

(a) *Acetic acid bacteria*

The nutrient requirements of acetic acid bacteria depend on the carbon source supplied (De Ley, Gillis and Swings, 1984; De Ley, Swings and Gossele, 1984). Williamson (1959) developed a medium for the detection and isolation of acetic acid bacteria in breweries. However, *Acetobacter* grows well on a standard medium containing 0.5% Bacto yeast extract (Difco), 1.5% ethanol and 2.5% agar (Suomalainen, Keränen and Kangasperko, 1965). De Ley and Swings (1984) described enrichment and isolation procedures for *Gluconobacter*.

(b) *Enterobacteriaceae*

Enterobacteriaceae comprise a large and heterogeneous family of bacteria with differing nutritional requirements. Numerous media, some selective, have been developed to sustain growth of these organisms (for a review see Casey and Ingledew, 1981). Lactose broth can be used to detect *Citrobacter*, *Enterobacter*, *Rahnella* and *Klebsiella* strains whereas Universal Liquid Medium supports good growth of most bacterial contaminants, including enterobacteria, encountered in beer breweries (Van Vuuren *et al.*, 1977). Furthermore, enterobacterial contaminants can readily be cultured at 30°C on selective media such as MacConkey agar. The lactose-positive *Citrobacter*, *Enterobacter* and *Klebsiella* strains form red colonies whereas those of *O. proteus* and *H. alvei* are pink to colourless. A prolonged incubation period (36–48 h) is needed for growth of *O. proteus* (Priest, 1981b).

(c) Zymomonas

Dadds (1971) described a suitable medium for the detection of *Zymomonas*. The presence of *Zymomonas* is indicated by abundant gas production at 30°C after 2–6 days. False positive results may be due to the growth of lactobacilli or wild yeasts. Swings and De Ley (1977, 1984) list suitable media for the isolation and maintenance of *Zymomonas*.

(d) Pectinatus, Selenomonas *and* Zymophilus

A selective differential medium (LL Agar; lactate-lead acetate) was developed for isolation and differentiation of *Pectinatus* (Lee, Moore and Mabee, 1981). Lee tubes (Ogg, Lee and Ogg, 1979) containing LL agar can be used for the detection and isolation of *Pectinatus*, a strict anaerobe. *Pectinatus*, *Selenomonas* and *Zymophilus* spp. can be grown under anaerobic conditions at 30°C in modified MRS medium (Schleifer *et al.*, 1990).

(e) Megasphaera

Megasphaera strains can be cultured under anaerobic conditions in a GasPak system (Seidel and Back, 1979) on beer agar enriched with 1% glucose and 1% peptone, Micro-assay culture agar and nutrient agar (Merck AG, 7881).

6.8.3 Summary

Many breweries rely on their own media and methods. The nature of this review precludes the possibility of considering these applications, of which many undoubtedly have merit.

The nutritional requirements of Gram-negative bacterial contaminants vary considerably and I doubt if any single medium meets all their requirements. The use of more than one medium type is therefore advisable. Furthermore, it is good practice occasionally to include other medium types to ensure that all bacterial contaminants are detected.

6.9 CONCLUSIONS

A wide range of Gram-negative bacterial contaminants including acetic acid bacteria, certain enterobacteria, *Zymomonas, Pectinatus, Selenomonas* and *Megasphaera* are able to spoil beer. The presence of these contaminants during different stages of the brewing process or in the final beer affects the concentration of certain volatile compounds to some degree depending on the level of contamination. A complex and delicate balance of flavour compounds determines flavour, aroma and quality of beer. Increased concentrations of certain volatiles affect beer quality adversely; a reduction thereof could also lead to off-flavours by allowing other compounds present at normal concentrations to dominate flavour and aroma. To ensure flavour consistency, bacterial contaminants in the brewing process either should be eliminated or their numbers should be kept as low as possible. It is not possible to operate a commercial brewery under aseptic conditions, but special attention should be given to ensure that cleaning procedures are adequate.

Since the previous edition of this book appeared in 1987, knowledge concerning Gram-negative spoilage bacteria, particularly the anaerobic rods and cocci, has increased considerably. *P. frisingensis*, *S. lacticifex*, *Z. raffinosivorans* and *Z. paucivorans* have been described as contaminants in beer. *P. frisingensis* and *S. lacticifex* are now regarded as spoilage organisms and *Z. raffinosivorans* as a potential spoilage bacterium. The taxonomic position of *Megasphaera* species I and II and that of *E. agglomerans* has been clarified. However, it seems highly likely that hitherto undescribed Gram-negative spoilage bacteria exist in breweries. Brewing microbiologists should accept the challenge to isolate and classify such strains and determine their effect on beer fermentation.

ACKNOWLEDGEMENTS

I thank Professor M.J. Hattingh, Department of Plant Pathology, University of Stellenbosch, for a critical review of this manuscript and The South African Breweries (Pty) Ltd for cooperation, financial assistance and access to their breweries.

REFERENCES

Aho, P.E., Seidler, R.J., Evans, H.J. and Raju, P.N. (1974) *Phytopathology*, **64**, 1413.

Anderson, R.J., Howard, G.A. and Hough, J.S. (1971) *Proceedings of the 13th Congress of the European Brewery Convention, Estoril*, IRL Press, Oxford, p. 253.

Asai, T. (1968) *Acetic Acid Bacteria. Classification and Biochemical Activities*, University of Tokyo Press. Tokyo and University Park Press, Baltimore, p. 103.

Ault, R.G. (1965) *Journal of the Institute of Brewing*, **71**, 376.

Back, W. (1980) *Brauwelt*, **43**, 1562.

Back, W., Weis, N. and Seidel, H. (1979) *Brauwissenschaft*, **32**, 233.

Barker, B.T.P. and Hillier, V.F. (1912) *Journal of Agricultural Science*, **5**, 67.

Bernstein, L., Blenkinship, B.K. and Brenner, M.W. (1968) *Proceedings of the Annual Meeting of the American Society of Brewing Chemists*. St Paul, MN, p. 150.

Brenner, D.J. (1981) In *The Prokaryotes*. Vol. 2 (eds M.P. Starr, H. Stolp, H.G. Trüper *et al.*). Springer, Berlin. p. 1105.

Brenner, D.J. (1984) In *Bergey's Manual of Systematic Bacteriology*, 9th edn, Vol. 1 (eds N.R. Krieg and J.G. Holt), Williams and Wilkins, Baltimore, p. 408.

Brown, H.T. (1916) *Journal of the Institute of Brewing*, **22**, 265.

Buchanan, R.E. and Gibbons, N.E. (eds) (1974) *Bergey's Manual of Determinative Bacteriology*, 8th edn, Williams and Wilkins, Baltimore, p. 291.

Calderbank, J. and Hammond, J.R.M. (1989) *Journal of the Institute of Brewing*, **95**, 277.

Campbell, I. (1983) *The Brewer* (October) 414.

Carr, J.G. (1968) In *Identification Methods for Microbiologists, Part B* (eds B.M. Gibbs and D.A. Shapton), Academic Press, London, p. 1.

Carr, J.G. (1974) In *Bergey's Manual of Determinative Bacteriology*, 8th edn (eds R.E. Buchanan and N.E. Gibbons), Williams and Wilkins, Baltimore, p. 352.

Case, A.C. (1965) *Journal of the Institute of Brewing*, **71,** 250.
Casey, G.P. and Ingledew, W.M. (1981) *Brewers' Digest* (February), 26.
Comrie, A.A.D. (1939) *Journal of the Institute of Brewing*, **45,** 342.
Cosbie, A.J.C., Tosic, J. and Walker, T.K. (1942) *Journal of the Institute of Brewing*, **48,** 82.
Cosbie, A.J.C., Tosic, J. and Walker, T.K. (1943) *Journal of the Institute of Brewing*, **49,** 88.
Cowbourne, M.A., Priest, F.G. and Hough, J.S. (1972) *Brewers' Digest*, **47,** 76.
Dadds, M.J.S. (1971) In *Isolation of Anaerobes, Technical Series No. 5* (eds D.A. Shapton and R.G. Board), Academic Press, London, p. 219.
Dadds, M.J.S. and Martin, P.A. (1973) *Journal of the Institute of Brewing*, **79,** 386.
Dadds, M.J.S., MacPherson, A.L. and Sinclair, A. (1971) *Journal of the Institute of Brewing*, **77,** 453.
Dadds, M.J.S., Martin, P.A. and Carr, J.G. (1973) *Journal of Applied Bacteriology*, **36,** 531.
Dadic, M., van Gheluwe, J.E.A. and Valyi, Z. (1971) *Wallerstein Laboratories Communications*, **34,** 5.
De Ley, J. (1961) *Journal of General Microbiology*, **24,** 31.
De Ley, J. and Kersters, K. (1964) *Bacteriological Reviews*, **28,** 164.
De Ley, J. and Swings, J. (1976) *International Journal of Systematic Bacteriology*, **26,** 146.
De Ley, J. and Swings, J. (1984) In *Bergey's Manual of Systematic Bacteriology*, 9th edn, Vol. 1 (eds N.R. Krieg and J.G. Holt), Williams and Wilkins, Baltimore, p. 275.
De Ley, J., Gillis, M. and Swings, J. (1984) In *Bergey's Manual of Systematic Bacteriology*, 9th edn, Vol. 1 (eds N.R. Krieg and J.G. Holt), Williams and Wilkins, Baltimore, p. 267.
De Ley, J., Swings, J. and Gossele, F. (1984) In *Bergey's Manual of Systematic Bacteriology*, 9th edn, Vol. 1 (eds N.R. Krieg and J.G. Holt), Williams and Wilkins, Baltimore, p. 268.
Engelmann, U. and Weiss, N. (1985) *Systematic and Applied Microbiology*, **6,** 287.
Eschenbecher, F. and Ellenrieder, M. (1975) *Proceedings of the 15th Congress of the European Brewery Convention, Nice*, IRL Press, Oxford, p. 497.
Ewing, W.H. and Fife, M.A. (1972) *International Journal of Systematic Bacteriology*, **22,** 10.
Frateur, J. (1950) *La Cellule*, **53,** 287.
Gilliland, R.B. and Lacey, J.P. (1966) *Journal of the Institute of Brewing*, **72,** 291.
Gillis, M. and De Ley, J. (1980) *International Journal of Systematic Bacteriology*, **30,** 7.
Goncalves de Lima, O., De Araujo, J.M., Schumacher, E. and Cavalcanti Da Silva, E. (1970) *Revista do Instituto de Antibioticos Universidade do Recife*, **10,** 3.
Graham, D.C. and Hodgkiss, W. (1967) *Journal of Applied Bacteriology*, **30,** 175.
Greenfield, S. and Claus, G.W. (1972) *Journal of Bacteriology*, **112,** 1295.
Haikara, A. (1985) *Brauwissenschaft*, **38,** 239.
Haikara, A. and Lounatmaa, K. (1987) *Proceedings of the 21st Congress of the European Brewery Convention, Madrid*, IRL Press, Oxford, p. 473.
Haikara, A., Enari, T.M. and Lounatmaa, K. (1981) *Proceedings of the 18th Congress of the European Brewery Convention, Copenhagen*, IRL Press, Oxford, p. 229.
Haikara, A., Penttilä, L., Enari, T.M. and Lounatmaa, K. (1980) *Applied and Environmental Microbiology*, **41,** 511.
Halcrow, R.M., Glenister, P.R., Brumsted, D.D. and Lautenbach, A.F. (1966) *Proceedings of the 9th Convention of the Institute of Brewing (Australian Section)*, Institute of Brewing, London, p. 273.
Hamze, M., Mergaert, J., Van Vuuren, H.J.J. *et al.* (1991) *International Journal of Food Microbiology*, **13,** 63.
Harper, D.R. (1980) *Process Biochemistry*, (December/January), 2.
Harper, D.R. (1981) *Brewers' Guardian* (August), 23.

Hough, J.S., Briggs, D.E., Stevens, R. and Young, T.W. (1982) In *Malting and Brewing Science*, Vol. 2, Chapman & Hall, London, p. 741.
Hungate, R.E. (1969) In *Methods in Microbiology*, Vol. 3B (eds J.R. Norris and D.W. Ribbons), Academic Press, London, p. 117.
Ingledew, W.M. (1979) *Journal of the American Society of Brewing Chemists*, **37,** 145.
Ingledew, W.M., Sivaswamy, G. and Burton, J.D. (1980) *Journal of the Institute of Brewing*, **86,** 165.
Inoue, T., Masuyama, K., Yamamoto, Y. *et al.* (1968) *Proceedings of the Annual Meeting of the American Society of Brewing Chemists*, St Paul, MN, p. 158.
Izard, D., Gavini, F., Trinel, P.A. and Leclerc, H. (1979) *Annales de Microbiologie (Paris)*, **130,** 163.
Izard, D., Ferragut, C., Gavini, F. *et al.* (1981) *International Journal of Systematic Bacteriology*, **31,** 116.
Joyner, Jr, A.E. and Baldwin, R.L. (1966) *Journal of Bacteriology*, **92,** 1321.
Keevil, W.J., Hough, J.S. and Cole, J.A. (1979) *Journal of the Institute of Brewing*, **85,** 99.
Kersters, K. and De Ley, J. (1968) *Antonie van Leeuwenhoek*, **34,** 393.
Kirchner, G., Lurz, R. and Matsuzawa, K. (1980) *Brauwissenschaft*, **33,** 461.
Kluyver, A.J. (1957) In *Bergey's Manual of Determinative Bacteriology*, 7th edn (eds R.S. Breed, E.G.D. Murray and N.R. Smith), Williams and Wilkins, Baltimore, p. 199.
Kluyver, A.J. and Hoppenbrouwers, W.J. (1931) *Archiv für Mikrobiologie*, **2,** 245.
Kluyver, A.J. and Van Niel, C.B. (1936) *Zentralblatt für Bakteriologie, Parasitenkunde, Infektionskrankheiten und Hygiene*, Abt. II, **94,** 369.
Krieg, N.R. and Holt, J.G. (eds) (1984) *Bergey's Manual of Systematic Bacteriology*, 9th edn, Vol. 1, Williams and Wilkins, Baltimore.
Lee, S.Y., Mabee, M.S. and Jangaard, N.O. (1978) *International Journal of Systematic Bacteriology*, **28,** 582.
Lee, S.Y., Moore, S.E. and Mabee, M.S. (1981) *Applied and Environmental Microbiology*, **41,** 386.
Lee, S.Y., Mabee, M.S., Jangaard, N.O. and Horiuchi, E.K. (1980) *Journal of the Institute of Brewing*, **86,** 28.
Leisinger, T. (1965) *Zentralblatt für Bakteriologie, Parasitenkunde, Infektionskrankheiten und Hygiene*, Abt. II, **119,** 329.
Lindner, P. (1895) *Mikroskopische Betriebskontrolle in den Gärungsgewerben mit einer Einführung in die technische Biologie, Hefenreinkultur und Infektionslehre*, 1st Aufl., Parey, Berlin, p. 1.
Lindner, P. (1928) *Bericht des Westpreussischen botanisch-zoologischen Vereins*, **50,** 253.
Magnus, C.A., Ingledew, W.M. and Casey, G.P. (1986) *Journal of the American Society of Brewing Chemists*, **44,** 158.
Mäkinen, V., Tanner, R. and Haikara, A. (1981) *Brauwissenschaft*, **34,** 173.
Masschelein, C.A. (1973) *Brewers' Digest*, **48,** 54.
McCaig, R. and Morrison, M. (1984) *Journal of the American Society of Brewing Chemists*, **42,** 23.
Millis, N.F. (1951) PhD Thesis, University of Bristol, Bristol, UK.
Millis, N.F. (1956) *Journal of General Microbiology*, **15,** 521.
Niefind, H.J. and Späth, G. (1971) *Proceedings of the 13th Congress of the European Brewery Convention, Estoril*, IRL Press, Oxford, p. 459.
Ogg, J.E., Lee, S.Y. and Ogg, B.J. (1979) *Canadian Journal of Microbiology*, **25,** 987.
Olsen, E. (1939) *Svenska Bryggareforeningens Monadsblad*, **54,** 8.
Ploss, M., Erber, J. and Eschenbecher, F. (1979) *Proceedings of the 17th Congress of the European Brewery Convention, West Berlin*, IRL Press, Oxford, p. 39.
Prest, A.G., Hammond, J.R.M. and Stewart, G.S.A.B. (1994) *Applied and Environmental Microbiology*, **60,** 1635.
Priest, F.G. (1981a) *Journal of the Institute of Brewing*, **87,** 299.

Priest, F.G. (1981b) In *An Introduction to Brewing Science and Technology, Part II, Contamination*, Institute of Brewing, London, p. 23.
Priest, F.G. and Hough, J.S. (1974) *Journal of the Institute of Brewing*, **80,** 370.
Priest, F.G., Cowbourne, M.A. and Hough, J.S. (1974) *Journal of the Institute of Brewing*, **80,** 342.
Priest, F.G., Somerville, H.J., Cole, J.A. and Hough, J.S. (1973) *Journal of General Microbiology*, **75,** 295.
Prücha, J. and Scheible, E. (1970) *Brauwelt*, **110,** 1233.
Rainbow, C. (1961) In *Progress in Industrial Microbiology*, Vol. 3 (ed. D.J.D. Hockenhull), Heywood, London, p. 45.
Rainbow, C. (1966) *Wallerstein Laboratories Communications*, **29,** 5.
Rainbow, C. (1981) In *Brewing Science*, Vol. 2 (ed. J.R.A. Pollock), Academic Press, London, p. 491.
Richards, M. and Corbey, D.A. (1974) *Journal of the Institute of Brewing*, **80,** 241.
Ruiz-Argueso, T. and Rodriques-Navarro, A. (1975) *Applied Microbiology*, **30,** 893.
Schleifer, K.H., Leuteritz, M., Weiss, N. *et al.* (1990) *International Journal of Systematic Bacteriology*, **40,** 19.
Schreder, K., Brunner, R. and Hampe, R. (1934) *Biochemische Zeitschrift*, **273,** 223.
Seidel, H. (1980) *Brauwelt*, **43,** 1216.
Seidel, H., Back, W. and Weiss, N. (1979) *Brauwissenschaft*, **32,** 262.
Seidel-Rufer, H. (1990) *Brauwissenschaft*, **3,** 101.
Shimwell, J.L. (1936) *Journal of the Institute of Brewing*, **42,** 585.
Shimwell, J.L. (1937) *Journal of the Institute of Brewing*, **43,** 507.
Shimwell, J.L. (1947) *Wallerstein Laboratories Communications*, **10,** 195.
Shimwell, J.L. (1948a) *Wallerstein Laboratories Communications*, **11,** 27.
Shimwell, J.L. (1948b) *Wallerstein Laboratories Communications*, **11,** 135.
Shimwell, J.L. (1950) *Journal of the Institute of Brewing*, **56,** 179.
Shimwell, J.L. (1963) *Brewers' Journal*, **99,** 759.
Shimwell, J.L. (1964) *Journal of the Institute of Brewing*, **70,** 247.
Shimwell, J.L. and Grimes, M. (1936) *Journal of the Institute of Brewing*, **42,** 348.
Skerman, V.D.B., McGowan, V. and Sneath, P.H.A. (1980) *International Journal of Systematic Bacteriology*, **30,** 225.
Smith, N.A. (1994) *Journal of the Institute of Brewing*, **100,** 347.
Society of American Bacteriologists (1957) *Manual of Microbiological Methods*, McGraw-Hill, New York, p. 15.
Stouthamer, A.H., Van Boom, J.H. and Bastiaanse, A.J. (1963) *Antonie van Leeuwenhoek*, **29,** 393.
Strandskov, F.B. (1965) *Wallerstein Laboratories Communications*, **28,** 29.
Soumalainen, H. and Ronkainen, P. (1968) *Nature*, **220,** 792.
Soumalainen, H., Keränen, E.J.A. and Kangasperko, J. (1965) *Journal of the Institute of Brewing*, **71,** 41.
Swings, J. and De Ley, J. (1977) *Bacteriological Reviews*, **41,** 1.
Swings, J. and De Ley, J. (1981) In *The Prokaryotes*, Vol. 1 (eds M.P. Starr, H. Stolp, H.G., Trüper, A. Balows and H.G. Schlegel), Springer, Berlin, p. 771.
Swings, J. and De Ley, J. (1984) In *Bergey's Manual of Systematic Bacteriology*, 9th edn, Vol. 1 (eds N.R. Krieg and J.G. Holt), Williams and Wilkins, Baltimore, p. 576.
Takahashi, N. (1983) *Bulletin of Brewing Science*, **28,** 11.
Thorwest, A. (1965) *Brauwelt*, **105,** 845.
Tosic, J. and Walker, T.K. (1944) *Journal of the Institute of Brewing*, **50,** 296.
Tracey, R.P. (1984) MSc Thesis, University of the Orange Free State, Bloemfontein, South Africa.
Turner, J.G., Walker, S.K. and Blanko, M. (1976) *Proceedings of the American Phytopathological Society*, **3,** 268.

Van Pee, W. and Swings, J. (1971) *East African Agricultural and Forestry Journal*, **36**, 311.
Van Vuuren, H.J.J. (1976) MSc Thesis, University of Stellenbosch, Stellenbosch, South Africa.
Van Vuuren, H.J.J. (1978) PhD Thesis, Rijksuniversiteit-Gent, Gent, Belgium.
Van Vuuren, H.J.J. and Toerien, D.F. (1981) *Journal of the Institute of Brewing*, **87**, 229.
Van Vuuren, H.J.J., Cosser, K. and Prior, B.A. (1980) *Journal of the Institute of Brewing*, **86**, 31.
Van Vuuren, H.J.J., Louw, H.A., Loos, M.A. and Meisel, R. (1977) *Applied and Environmental Microbiology*, **33**, 246.
Van Vuuren, H.J.J., Kersters, K., De Ley, J. *et al.* (1978) *Journal of the Institute of Brewing*, **84**, 315.
Van Vuuren, H.J.J., Kersters, K., De Ley, J. and Toerien, D.F. (1981) *Journal of Applied Bacteriology*, **51**, 51.
Von Riemann, J. and Scheible, E. (1969) *Brauwelt*, **109**, 1074.
Wainwright, T. (1972) *Brewers' Digest*, **47**, 78.
Walker, T.K. and Tosic, J. (1945) *Journal of the Institute of Brewing*, **51**, 245.
Wallnöfer, P. and Baldwin, R.L. (1967) *Journal of Bacteriology*, **93**, 504.
Weiner, J.P., Rolph, D.J. and Taylor, R. (1975) *Proceedings of the 15th Congress of the European Brewery Convention, Nice*, IRL Press, Oxford, p. 565.
Weiss, N., Seidel, H. and Back, W. (1979) *Brauwissenschaft*, **32**, 189.
West, D.B., Lautenbach, A.F. and Brumsted, D.D. (1963) *Proceedings of the Annual Meeting of the American Society of Brewing Chemists*, St Paul, MN, p. 194.
White, G.A. and Wang, C.H. (1964a) *Biochemical Journal*, **90**, 408.
White, G.A. and Wang, C.H. (1964b) *Biochemical Journal*, **90**, 424.
Williams, P.J. le B and Rainbow, C. (1964) *Journal of General Microbiology*, **35**, 237.
Williamson, D.H. (1959) *Journal of the Institute of Brewing*, **65**, 154.
Xu, H.S., Roberts, N., Singleton, F.L. *et al.* (1982) *Microbial Ecology*, **8**, 313.

CHAPTER 7

Wild yeasts in brewing and distilling

I. Campbell

7.1 INTRODUCTION

Wild yeasts must have been a problem in beverage fermentations throughout history, to brewers and winemakers in particular, and have been the subject of numerous reviews since their microbiological nature became understood. Wiles (1953), Gilliland (1967, 1971) and Rainbow (1981) have given useful and detailed accounts of the adverse effects of wild yeasts, which have changed little over the years. Ingledew and Casey (1982) provided a comprehensive survey of culture media for detection and isolation of wild yeasts, which they defined as 'yeasts not playing a significant part in a normal fermentation'. In the review of wild yeasts in the earlier edition of this book (Campbell, 1987), it was considered more appropriate to use Gilliland's (1967) definition 'yeasts not deliberately used and under full control', which will be used again to indicate the accidental and haphazard nature of wild yeast infection.

7.2 DETECTION OF WILD YEASTS

It is good microbiological practice to test for the absence of undesirable yeasts at each sensitive stage of the brewing process. This was strongly emphasized by Wiles (1953), Gilliland (1967, 1971), Rainbow (1981) and the present author (Campbell, 1987) in their reviews. The Analysis Committees of the European Brewery Convention (Moll, 1981) and the Institute of Brewing (Baker, 1991) have published standard methods for detection of wild yeast contaminants, although with only minimal background information on the microbiological principles.

Brewing Microbiology, 2nd edn. Edited by F. G. Priest and I. Campbell.
Published in 1996 by Chapman & Hall, London. ISBN 0 412 59150 2

In breweries, the various steps prior to hop boiling are of little importance as a source of wild yeasts in the wort, since none is sufficiently heat resistant to survive even brief boiling. Wiles (1953) reported small numbers of yeasts in malt, hops and priming sugars, but even when beers were dry hopped and primed for cask conditioning there was no evidence of any harmful effect.

The unboiled worts of distillery practice might seem to be more at risk from yeast contaminants of the malt (Chapter 4). Although a proportion of bacterial contaminants may persist through mashing to the fermentation, yeasts from malt normally have no adverse effect on the fermentation. With few exceptions, yeast contaminants of malt are strictly aerobic species (Chapter 4) and therefore, even if they survive mashing temperatures, they are unable to grow during the subsequent fermentation, swamped by the great excess of fermentative culture yeast and rapid generation of CO_2 and anaerobic conditions.

In both breweries and distilleries, samples of cooled aerated wort, and swab samples of the surface of the cleaned empty fermentation vessel, should be taken routinely before every fermentation. Since no yeast should be present in either of these samples, any which are discovered should be regarded with concern: even if by chance they are not dangerous contaminants, certainly their presence indicates imperfect sterilization procedures. As indicated in Table 7.1, non-selective media have to be used for these samples. Any wild yeasts, culture yeasts or bacteria detected indicate poor hygiene and potential microbiological problems. Commercially available malt extract agar or Wallerstein Laboratories nutrient (WLN) agar are useful general non-selective media for the brewing and distilling industries, and are suitable for the detection of contaminant bacteria as well as yeasts.

Samples of pitching yeast must be examined before use to ensure satisfactory viability or vitality (Pfenninger, 1977; Baker, 1991) but it is also good practice to check each inoculum for a suitably low number, and preferably the absence, of contaminant bacteria and wild yeasts. For this purpose, a selective medium is necessary. Unfortunately there is no one medium which suppresses brewery or distillery culture yeasts and supports the growth of any possible contaminant yeast which may be present.

The most useful and versatile medium for detection of wild yeasts is lysine agar (Campbell, 1987), a synthetic medium with lysine as the sole source of nitrogen. Almost all yeast genera other than *Saccharomyces* have the ability to utilize the amino groups of lysine for biosynthesis of other amino acids, and hence proteins. Therefore on plating out contaminated pitching yeast, or a sample taken during fermentation or from early stages post-fermentation before removal of yeast, only those non-*Saccharomyces* yeasts which may be present are able to grow to colonies of normal size. Two important disadvantages of lysine agar are:

Table 7.1 Media for isolation of wild yeasts

1. Examination of wort, washings or swab samples from clean fermentors, pipework, kegs, etc., filtered or pasteurized beer and other samples which should not contain live yeasts.
   Non-selective media:
   (a) wort agar;
   (b) malt extract agar;
   (c) Wallerstein Laboratories nutrient (WLN) agar.
2. Examination of pitching yeast or samples during fermentation (Ingledew and Casey, 1982).
   (a) Inhibitory media (culture yeasts are suppressed; only certain contaminant yeasts are able to grow):
      (i) 'actidione' agar (WLN agar + actidione), suppresses culture yeasts and many sensitive wild yeasts;
      (ii) copper sulphate or crystal violet agar (wort agar or malt extract agar + either 6 $\mu g\ ml^{-1}$ $CuSO_4$ (Taylor and Marsh, 1984) or 20 $\mu g\ ml^{-1}$ crystal violet), suppress culture strains of *S. cerevisiae* but wild strains of *Saccharomyces* mat grow
      (iii) fuchsin/sulphite agar (various formulations): similar principle to $CuSO_4$ agar.
   (b) Selective media (nutrients available only for wild yeast):
      (i) lysine agar, selective for wild yeasts capable of growing on lysine as sole source of N;
      (ii) dextrin agar, or starch agar, selective for yeasts capable of utilizing dextrin or starch as sole source of C;
      (iii) citrate agar, xylose agar, etc., selective for yeasts capable of using sole C source supplied.

1. its selective effect is by starvation, and carried-over extracellular or intracellular nutrient will allow some slight growth of *Saccharomyces* yeasts to visible colonies;
2. *Saccharomyces* spp., biochemically adapted to grow well in brewery and distillery fermentations, are potentially the commonest contaminants of these situations but cannot be detected.

Therefore, although *S. bayanus*, *S. cerevisiae*, *S. exiguus*, *S. pastorianus* and *S. unisporus* are the only brewery contaminants listed in Table 7.2 which are unable to grow on lysine medium, inability to detect wild strains of *S. cerevisiae* in particular is a serious problem.

A different approach can be used to detect diastatic wild yeasts. The most important, particularly for its role as a post-fermentation contaminant of beers, was first isolated by Andrews and Gilliland (1952) and named *S. diastaticus*, but is now regarded simply as a diastatic strain of *S. cerevisiae* (Kreger-van Rij, 1984). A useful selective medium for this organism relies on its ability to utilize dextrin or starch as a sole carbon source for growth (Ingledew and Casey, 1982): synthetic medium of Wickerham's (1951) yeast

Table 7.2 Yeast species reported as brewery contaminants (Barnett, Payne and Yarrow, 1990)

Brettanomyces anomalus	*Hanseniaspora uvarum*
B. claussenii	*H. valbyensis*
B. custersianus	*H. vineae*
B. custersii	
B. lambicus	*Kluyveromyces marxianus*
Candida beechii	*Pichia anomala*
C. ernobii	*P. fabianii*
C. humilis	*P. farinosa*
C. intermedia	*P. fermentans*
C. norvegica	*P. guilliermondii*
C. oleophila	*P. membranaefaciens*
C. parapsilosis	*P. ohmeri*
C. sake	*P. onychis*
C. solani	*P. orientalis*
C. stellata	*P. subpelliculosa*
C. tenuis	
C. tropicalis	*Saccharomyces bayanus*
C. vartiovaarai	*S. cerevisiae*
C. versatilis	*S. exiguus*
	S. pastorianus
Debaryomyces hansenii	*S. unisporus*
D. marama	*Schizosaccharomyces pombe*
Dekkera bruxellensis	
D. intermedia	*Torulaspora delbrueckii*
Filobasidium capsuligenum	*Zygosaccharomyces bailii*
	Z. rouxii

When both perfect and imperfect forms of the yeast occur, only the perfect (sporing) form is named (see Table 1.2).

nitrogen base + 1% starch or dextrin allows growth of '*S. diastaticus*' and other amylolytic yeasts, while normal brewing strains of *S. cerevisiae* and other non-amylolytic yeasts of any genus are unable to grow.

The general principle of providing a selective single or mixed carbon source, which can be utilized by contaminant *Saccharomyces* strains but not brewing yeast, has been examined with a range of compounds, but with only limited success (see Ingledew and Casey, 1982). In general, contaminants recovered on media with specific carbon nutrients other than dextrin or starch are of genera other than *Saccharomyces* and would have been detected anyway on lysine agar.

Another possible medium which is occasionally effective is 'actidione agar'. Although used primarily for the detection of contaminant bacteria (Chapters 5 and 6), it is also suitable for detection of those wild yeasts which happen to be actidione resistant. Actidione (cycloheximide) is an antifungal antibiotic, but various wild yeasts are resistant to the 100 $\mu g\ ml^{-1}$ added to the commercially available medium, and can be detected, or counted if quantitative procedures are used. Most culture yeast strains are inhibited by 5 μg ml^{-1} actidione, so WLN, malt extract or any other suitable non-selective medium can be supplemented with the lowest concentration of antibiotic which will reliably inhibit culture yeast, thereby producing a medium with greater potential for isolation of wild yeast than the commercially available 100 μg ml^{-1} medium. Wild yeasts of many genera, including some strains of *Saccharomyces* spp., may be sufficiently resistant to actidione to be recovered in this way. Unfortunately, the resistant strains which can be detected may represent only a part of the wild yeast population.

Numerous ingenious attempts to provide selective media that suppress brewing strains of *S. cerevisiae* but allow growth of all wild *Saccharomyces* strains, particularly of the species *S. cerevisiae* itself, have been generally unsuccessful as routine media, despite their presumed success in their inventors' laboratories with the yeast strains at their disposal. Ingledew and Casey (1982) have provided a useful account of these media, many of which, despite their limited selective value, work well with some breweries' strains of culture and wild yeasts and may be no worse than lysine, actidione or dextrin media. Subsequent to that review, Taylor and Marsh (1984) described a medium containing $CuSO_4$ as selective agent, but with the increasing success of molecular biological methods, as indicated below, little further attention seems to have been devoted to development of new selective media.

Lysine and dextrin media exert their selective effects by starvation. Ideally, several washings by centrifugation are required to limit carry-over of nutrient from the previous medium. However, it is my impression that most microbiologists plate unwashed samples on dextrin or lysine media, and tolerate slight growth of culture yeast on the traces of nutrient transferred with the inoculum; the extra effort of washing cultures may be justified in research on yeast nutrition but probably not in routine isolation of contaminants. Actidione and other inhibitor media suppress the growth of culture yeast but it is important to remember that the cells are not killed, and colonies of wild yeasts could be contaminated with surviving culture yeast. Therefore colonies must never be inoculated directly to identification media, but must first be plated on a non-selective medium, e.g. malt extract broth, to ensure a pure culture.

Descriptions above apply to traditional culture methods on agar plates, but the same principles can be applied to rapid culture methods, e.g. growth analysers which measure change in impedance of a growing culture (Chapter 8). For example, with duplicate broth cultures in lysine and complete media, growth in the former only indicates the presence of non-*Saccharomyces* wild yeasts.

Immunological methods were described in detail in the earlier edition of this book (Campbell, 1987). Immunofluorescence (IF) is a standard recommended method (Pfenninger, 1977; Baker, 1991), useful for rapid detection and counting of microbial, including yeast contaminants, provided the appropriate specific antiserum is available. A particularly important application is the detection of contaminants of the genus *Saccharomyces* in pitching yeast and of fermentation samples, when no selective medium is possible (Campbell, 1987). Although not yet published as recommended methods for the brewing industry, applications of enzyme-linked immunosorbent assay (ELISA) and the direct epifluorescence technique (DEFT) are common in the food and drink industries (Pettipher *et al.*, 1980).

Recent developments in molecular biology have produced new methods for detection of specific contaminants. The polymerase chain reaction (PCR) for amplifying unique fragments of DNA or RNA has been suggested for detection of many different pathogens and spoilage bacteria in the food and drink industries (Chapters 3 and 8), but with equivocal results in beer (DiMichele and Lewis, 1993; Thompson *et al.*, 1995). Also, like IF which requires specific sera, PCR detection of a particular contaminant is possible only when the specific primer, reacting with the contaminant but not culture yeast, is available.

A commonly used version of DEFT based on enzymic hydrolysis of dark fluorescein di-acetate to fluorescent fluorescein (Baker, 1991) ignores dead cells. A serious disadvantage of PCR and standard IF methods is the inability to distinguish living from dead cells: both have the nucleic acid and cell wall structures which are detected in these tests. Furthermore, these methods are limited to detection and counting of specific contaminants; no culture is available for further analysis.

7.3 IDENTIFICATION OF WILD YEASTS

The principles of yeast classification and identification were discussed in Chapter 1. Many contaminant yeasts of the brewing and related industries are themselves strains of *S. cerevisiae*, but other species of *Saccharomyces*, and other fermentative and non-fermentative genera are also liable to occur. Table 7.2 shows the list of yeast species recorded by Barnett, Payne and Yarrow (1990) as contaminants of brewery fermentations, but is almost certainly incomplete as many instances of contamination are not followed through to full identification, and even then the information would not necessarily be published in the scientific literature. In distillery fermentations, few of these listed contaminants could compete with distiller's yeast, and then only if introduced as a substantial proportion of the culture. Therefore, provided the distilling yeast is of acceptable quality, the likely causes of wild yeast contamination would be either improperly cleaned fermentation vessels from a previous contaminated fermentation, or massively contaminated brewer's yeast used as supplementary pitching yeast.

How important is it to identify fully all yeast contaminants? It has been suggested (e.g. Gilliland, 1971) that it is helpful to determine the genus and

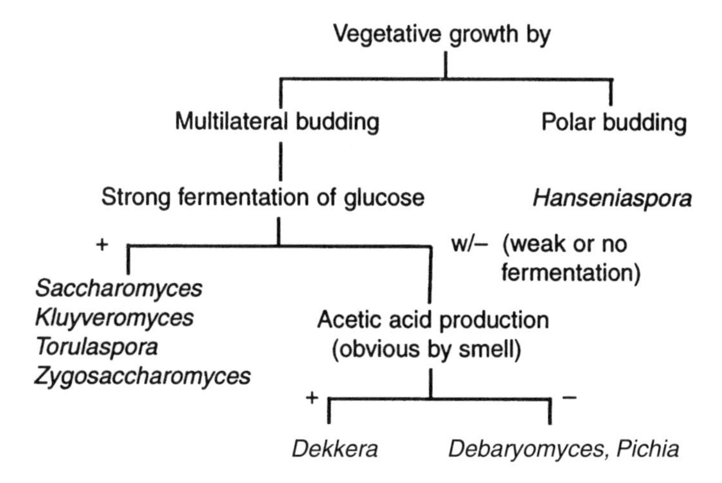

Fig. 7.1 Simple identification scheme for common perfect yeast genera of the brewing and related industries.
The genera above are 'perfect' yeasts which form spores. Equivalent non-sporing 'imperfect' genera are given below:

Spore-forming:	*Dekkera*	*Hanseniaspora*	Other genera above
Non-sporing:	*Brettanomyces*	*Kloeckera*	*Candida*

Yeast spoilage effects

Hanseniaspora/Kloeckera, *Saccharomyces/Kluyveromyces/Torulaspora/Zygosaccharomyces* and equivalent *Candida* spp:
Unintended fermentation, producing turbidity and off-flavours

Dekkera / Brettanomyces:
Acetic acid spoilage, turbidity.

*Debaryomyces / Pichia** and equivalent *Candida* spp:
Turbidity, yeasty or estery off-flavour, often with production of surface film/pellicle fragmenting to flaky particles or deposit.
**Hansenula* is no longer recognized as a separate genus: former *Hansenula* spp. are now regarded as nitrate-utilizing spp. (or, in some cases, strains) of *Pichia.*

species of contaminants for full access to the literature on their adverse effects and their elimination. On the other hand, a simple scheme such as Fig. 7.1 for determination of the principal types of yeasts should provide sufficient information to suggest possible sources of the contamination and appropriate remedial action, as suggested below (section 7.5). Colony morphology and colour (Hall, 1971) or resistance to antibiotics and other inhibitors with a multi-agent 'yeastar' (Simpson, Fernandez and Hammond, 1992) are simple labour-saving methods which are sufficient to recognize similarity or difference of separate isolates, but give no information on identity. Colony morphology and colour are difficult to describe well enough for reliable exchange of information between laboratories, but the extent of inhibition by the agents on the 'yeastar' is more easily expressed quantitatively. However, either exchange of cultures or identification to genus and preferably to species level is advisable for operations involving more than one laboratory.

In recent years the API identification kits (supplied by bioMérieux UK, Basingstoke or API, Montalieu-Vercieu, France) have simplified identification of bacteria and yeasts. The API 20C kit for yeasts, with 20 test media, is primarily intended for the identification of pathogenic *Candida* spp. and distinction from non-pathogenic yeasts which colonize body surfaces. The system is equally applicable to identification of industrial yeasts, and the manual available from API (Anon., 1994) will identify culture yeasts and the commoner contaminants of the brewing and distilling industries. The suppliers maintain a database of properties of all known yeasts, and provide a consultancy service for identification of unusual isolates.

Where identification of substantial numbers of isolates is planned, it may be more economical to use traditional identification media used according to the methods of Kreger-van Rij (1984) or Barnett, Payne and Yarrow (1990). A computer program is available as a supplement or alternative to the latter publication to assist with identification.

If a yeast isolate from pitching yeast or a fermentation is identified as *S. cerevisiae*, it is almost certain to be the pitching yeast itself. If isolated from beer post-fermentation, further testing is necessary to determine whether it is the culture yeast or a contaminant. Small-scale fermentations in tall-tube fermentors (Pfenninger, 1977) will definitively distinguish culture yeasts from each other and from contaminants, by demonstrating the characteristic and constant brewing properties of the culture yeasts (Campbell, 1987). As an alternative to the full range of tests in Table 7.3, gas chromatographic analysis of the characteristic 'flavour profile' of individual strains will suffice, but also requires several days' incubation. In cases of doubtful identity of a batch of pitching yeast, it is impracticable to wait for results of such a test, but even for less urgent situations, a lengthy and labour-intensive small-scale fermentation is an unsatisfactory method of identification.

Although all strains of *S. cerevisiae*, by definition, ferment glucose, sucrose and maltose but fail to ferment lactose, strains vary with regard to fermentation of other sugars, e.g. galactose, melibiose, melezitose, α-methyl glucoside and trehalose (Kreger-van Rij, 1984). Even after only 6 h incubation in these sugars, fermentative ability is obvious and difference from, or identity with, known culture yeast can be established. Alternatively, the 'yeastar' test can be used, as above. It is important to have a small number of rapid tests which can reliably distinguish the culture strains of the various brands of a brewery.

Molecular biology provides an alternative, but not necessarily more

Table 7.3 Tests for distinguishing strains of *Saccharomyces cerevisiae* (after Gilliland, 1971, amended in accordance with current taxonomy)

1. Laboratory fermentation tests:
   (a) all strains of *S. cerevisiae* ferment glucose, sucrose and maltose, may or may not ferment galactose, but do not ferment lactose (Kreger-van Rij, 1984);
   (b) further differentiation can be achieved by testing fermentation of maltotriose, melibiose, raffinose, trehalose, dextrin (and galactose, used in initial identification tests).
2. Small-scale fermentation tests, e.g. EBC tall-tube fermentors:
   (a) flocculation, fining;
   (b) head formation;
   (c) final pH of beer;
   (d) uptake of nitrogen (especially amino N), isohumulones;
   (e) rate of fermentation, attenuation limit;
   (f) total yeast crop;
   (g) liability to autolysis;
   (h) volatiles produced, taste of final beer.
3. Additional laboratory tests: for distinguishing strains:
   (a) giant colony morphology (and colour, if on WLN agar);
   (b) sensitivity to antibiotics (e.g. different concentrations of cycloheximide, nystatin, etc.);
   (c) vitamin requirements (very labour intensive, and largely abandoned).

rapid possibility. The concept of 'DNA fingerprinting' pioneered by Pedersen (1983) has not yet become an established analytical technique but is expected to supplant cultural methods for rapid identification of different culture yeasts, even in the most difficult situation of recognizing two different lager strains (Meaden, 1990). The 'fingerprint' can be matched with those of known cultures, but is unsuitable for identification of an unknown culture. Recent restoration of the former specific names *S. bayanus* and *S. pastorianus* has been possible only by DNA analysis: traditional, i.e. cultural, identification tests can no longer distinguish them from *S. cerevisiae* (Barnett, 1992; Martini, Martini and Cardinali, 1993): electrophoretic analysis or 'DNA fingerprinting' is necessary.

7.4 EFFECTS OF WILD YEASTS IN THE BREWERY

Gilliland's (1967) definition 'yeasts not deliberately used and under full control' can now be seen to cover four main types of wild yeast infection:

1. fermentative yeasts, which could also occur as
2. 'killer' yeasts and

3. the wrong type of culture yeast, and
4. 'aerobic' (non-fermentative) yeasts.

In the following pages we examine these types in more detail.

7.4.1 Fermentative yeasts

Yeasts of the genera *Saccharomyces, Kluyveromyces, Torulaspora* and *Zygosaccharomyces* are biochemically similar and, therefore, potentially able to compete with culture strains of *S. cerevisiae.* With the exception of 'killer' yeasts (see below), contaminant yeasts have little or no direct adverse effect on culture yeasts, but if capable of growth at even a slightly faster rate will increase relative to the culture yeast over successive fermentations. Wild *Saccharomyces* contaminants with a more efficient and therefore faster rate of nutrient uptake, with simpler vitamin requirements or with simpler metabolic activity (e.g. inability to utilize maltotriose) are all potentially capable of slightly faster growth than the culture yeast and of gradually increasing in proportion.

Contamination of pitching yeast, even if initially pure, is probably inevitable after a number of successive fermentations and repitchings in traditional open rectangular vessels. The risk is reduced, but certainly not eliminated, by the use of closed vessels. Whether chance contamination increases or disappears over successive fermentations depends on various process factors and the characteristics of the culture and wild strains. In ale fermentations, the traditional head-forming yeast is skimmed at intervals from the surface of the fermenting beer. Contaminants which rise with the head will therefore be incorporated in the skimmed yeast and continue as contaminants of the pitching yeast. If only the middle of three skimmings is used as pitching yeast, it should be possible to control the proportion of contaminants which happen to rise to the head early or late in the fermentation.

Flocculation, the ability of yeast to adhere in clumps and sediment rapidly from the medium in which they are suspended (Stratford, 1992), is an important property of brewing yeast. It is important that yeasts remain in suspension during the fermentation, but to avoid expensive clarification by filtration or centrifugation it is important that brewing yeasts spontaneously clump and settle out of the beer late in the fermentation. A good brewing yeast does this, but many wild yeasts are non-flocculent and so, in addition to any flavour faults they cause, the beer remains turbid.

Detection of contaminants or mutants of unsuitably strong flocculence or non-flocculence is difficult or impossible by the use of selective media alone, but the changed flocculation properties are themselves selective and assist recovery of the contaminant. The correct degree of flocculation is an important property of brewing yeast (Stewart, 1975; Speers *et al.*, 1992):

small-scale fermentations will confirm the changed properties, and comparison with the authentic original culture will show whether the change is due to a contaminant, in which cultural properties will be different, or a mutant, whose properties other than flocculence should be unchanged. As above, simple tests for fermentation of sugars or resistance to inhibitors give a rapid indication of a different yeast strain.

Changes in flocculence, although genetically controlled (Johnston and Reader, 1983; Stratford, 1992), are expressed as changes in cell wall structure. Other changes of the yeast cell wall also affecting brewing behaviour are fining and head-forming ability; the effect of mutations in these properties was discussed by Campbell (1987). For beers treated with isinglass or other fining agents, the presence of non-fining contaminants creates haze problems. Haze may also result from the fact that many of these contaminants are smaller than the culture yeasts which filtration or centrifugation procedures have been developed to remove.

Although fermentative diastatic yeasts are possible contaminants of the fermentation, they are also possible post-fermentation contaminants, having available a rich nutrient source by their ability to utilize dextrin (Andrews and Gilliland, 1952). Low levels, possibly only one viable cell, of post-fermentation contamination can grow to spoilage levels as turbidity, phenolic and other off-flavours and a reduction of the viscosity of the beer by hydrolysis of dextrins (Gilliland, 1971; Campbell, 1987).

'*S. diastaticus*' strains are often flocculent, and it is possible that cells in the centre of flocs could be insulated from pasteurization conditions. Although such a situation has been postulated to explain failure of pasteurization (Brightwell, 1972) it is unlikely that flocs would pass the preliminary filtration stage. More likely as a cause of failure of pasteurization is the heat resistance of yeast spores. Although brewing yeasts seldom sporulate, and certainly are unlikely to do so in wort or beer (Chapter 3), many of the potential contaminants of the brewery are spore formers (Table 7.2). Yeast spores are slightly more heat resistant than vegetative cells (Put and de Jong, 1980) but since pasteurization programmes are designed primarily to destroy those non-sporing culture yeasts which pass the filtration stage, spore-forming contaminants may survive.

7.4.2 Killer yeast

Normally in fermentations, contaminant yeasts or bacteria are in competition with the culture yeast and, at worst, increase in proportion over successive fermentations with each reuse of the pitching yeast. Killer yeast (Chapter 3) is an extreme form of contaminant, killing sensitive culture yeast, as almost all are, and leaving the killer strain itself as the dominant yeast of the fermentation. Maule and Thomas (1973) gave a detailed case history of one of the earliest confirmed examples of an outbreak of killer yeast infection replacing the culture yeast and producing an objectionable off-flavour in addition to poor attenuation and other defects. Although various zymocins, the preferred name for killer factors, are known and yeasts of many genera produce them, under the conditions of brewery or distillery fermentations, *Saccharomyces* spp. are the most likely killer strains. Therefore with little chance of a suitable selective medium for early detection, the contamination is unlikely to be recognized until the killer strain has become the dominant organism and detectable on non-selective medium, e.g. by different colony colour or morphology. It is possible to engineer resistant production yeast strains by transferring specific resistance factors, or even killer factors, into cultures (Chapter 4), but so many different killer factors exist that anticipating necessary resistance factors is impracticable. Furthermore, such manipulation risks changing the valuable properties of the culture yeast, and also with current popular aversion to 'genetic engineering', customers would react adversely.

7.4.3 Different culture yeasts

For most beers the specific yeast strain, or mixture of known strains, is an essential component of the recipe. A different culture yeast, however well it may perform in one of the other products of the brewery, is likely to give an unsatisfactory beer when used in error. Breweries where different types or brands are produced are permanently at risk of such accidents. Rate of fermentation, final attenuation, nature and amount of the flavour by-products of fermentation, flocculation and head formation are all important properties of individual strains (Nykänen and Suomalainen, 1983; Molzahn, 1993). Obviously the 'correct' yeast is completely absent after the accidental substitution of the wrong culture, so some difference from normal is possible for all of the above properties of the beer. Careful documentation is the most effective safeguard, but either rapid cultural methods or rapid tests with genetic probes, as described above (section 7.2), will confirm the identity of the strain.

7.4.4 Aerobic yeasts

Yeasts of the genus *Pichia*, and particularly *P. membranaefaciens*, are the commonest of the non-fermentative spoilage yeasts. The literature of both brewing (e.g. Wiles, 1953; Rainbow, 1980) and wine microbiology (e.g. Sponholz, 1993) treats these non- or weakly fermenting yeasts as aerobic organisms, but this is not entirely true. In pure culture in wort, spoilage *Pichia* spp. are capable of growth under completely anaerobic conditions (Campbell and Msongo, 1991). In the acid and ethanol concentrations of beer, however, these yeasts are no longer able to grow anaerobically and show their generally recognized behaviour as strict aerobes.

Acetic acid-forming *Brettanomyces* and *Dekkera* spp., although fermentative, require oxygen for growth and fermentation. They form an important component of the yeast flora of fermenting Belgian lambic beer (van Oevelen *et al.*, 1977) but also, like the acetic acid bacteria *Acetobacter* and *Gluconobacter*, are liable to grow in beer post-fermentation, if air accidentally gains access.

The nature of the spoilage flora of beer is greatly affected by access of air. Such air may itself be the source of the yeast or bacterial contaminants, but even if not, access of air to packaged beer allows growth of any aerobic yeasts which have escaped filtration and pasteurization, and encourages growth of surviving facultative anaerobes. Of the possible contaminants listed in Table 7.2, all are stimulated by oxygen, but yeasts of the genera *Debaryomyces*, *Filobasidium* and *Pichia*, the last including the former genus *Hansenula*, grow only under aerobic conditions in beer (Campbell and Msongo, 1991), as do many spoilage species of *Candida*. The effects of 'aerobic' yeast growth are turbidity, often but not always associated with a surface film or thick pellicle, which if present, fragments to a flaky suspension or deposit. Production of acetic acid by *Brettanomyces* and *Dekkera* spp. often leads to excessive production of ethyl and other acetate esters, but unpleasantly high levels of ester production are particularly associated with *Hansenula* spp., now recognized as nitrate-utilizing species of *Pichia*, e.g. *P. subpelliculosa*.

7.5 ELIMINATION OF WILD YEASTS

It is obvious that to eliminate wild yeasts from brewery fermentations or beer, the brewer must know the source of the contamination. While the original source must obviously be from outside the brewery, and reinfection from that external source must always be a possibility, subsequent instances of contamination are likely to be one or more of the internal sources:

1. pitching yeast;
2. equipment surfaces;
3. air;
4. water.

Raw materials added directly (priming sugar, dry hopping) are a possible, but rare, cause of infection (Wiles, 1953). The nature of the wild yeast, fermentative or oxidative, can be a clue to its source, e.g. aerobic yeasts are often associated with airborne infection from grain dust or growth on the dilute medium created by wort or beer rinsings.

It is unlikely that boiled and cooled wort is the initial source of contamination, except perhaps in the unfortunate combination of a faulty heat-exchanger plate and a contaminated water supply. The most likely result would be bacterial contamination, although introduction of a wild yeast is possible. Pipework leading to the fermentor, or the walls of the fermentation vessel itself, are possible sources of infection from the previous fermentation. Plant must be of perfectly smooth internal construction, and both mains and vessels must be cleaned and sterilized immediately after every fermentation (Chapter 10), subsequently confirmed by the standard microbiological tests (Baker, 1991).

With routine microbiological examination of each batch of pitching yeast, any development of microbial contaminants is obvious long before reaching unacceptable levels. The concept of 'unacceptable level' of wild yeast contamination is interpreted differently between breweries, and must also take into account the potential of the contaminant to outgrow the culture yeast. Normally the yeast is pitched before microbiological results are available, but with the exception of killer yeast, contaminants develop only slowly from one fermentation to the next. Most breweries routinely replace yeast cultures after a fixed number of successive fermentations, to prevent any substantial development of contaminants. If no new culture is ready to replace unacceptably contaminated yeast, washing with phosphoric or other acid to reduce the pH to 2.8–3.0 for 2 h is often a satisfactory remedy. Bacterial contaminants are less resistant than *S. cerevisiae* to low pH, but unfortunately many wild yeasts are at least as tolerant of low pH as brewing yeasts. Indeed there is a risk that their proportion may increase as a result of acid washing, which is therefore not necessarily an effective treatment for removal of wild yeast contamination from pitching yeast. It is often preferable to tolerate a low but stable level of contamination rather than to undertake potentially troublesome elimination of wild yeast. Yet even with that attitude, it is essential that contamination is kept to an acceptably low level by good hygiene.

Post-fermentation contamination is possible from various sources; routine microbiological sampling at each stage of post-fermentation processing is essential to ensure satisfactory quality (Hjortshøj, 1984). Pasteurization is primarily to eliminate the few surviving culture yeasts rather than spoilage organisms, but excessive numbers of either in the filtered beer presented to the pasteurizer will overcome the system and a proportion of these organisms will survive as contaminants. Microbiological quality of beer leaving filters and pasteurizers must be checked by membrane filtration; kegs, bottles and cans, and lids, crown caps and jet water must be free of spoilage organisms, and the quality of the packaged product must be monitored by membrane filtration and forcing tests. Standard microbiological tests required have been described by Moll (1981) and Baker (1991). At the final stage, the dispense system is also a possible source of contamination requiring careful surveillance (Harper, 1981).

Elimination of detected spoilage often requires extensive investigation for possible sources. Ultimately the elimination of wild yeast contamination depends on the experience and good sense of the brewer and microbiologist.

REFERENCES

Andrews, J. and Gilliland, R.B. (1952) *Journal of the Institute of Brewing*, **58**, 189.
Anonymous (1994) *API 20C Analytical Profile Index*, BioMérieux UK, Basingstoke.
Baker, C.D. (1991) *Recommended Methods of Analysis*, Institute of Brewing, London.
Barnett, J.A. (1992) *Yeast*, **8**, 1.
Barnett, J.A., Payne, R.W. and Yarrow, D. (1990) *Yeasts, Characteristics and Identification*, 2nd edn, Cambridge University Press, Cambridge.
Brightwell, R. (1972) *Proceedings of the 12th Convention of the Australian and New Zealand Section of the Institute of Brewing*, Institute of Brewing, London, p. 209.
Campbell, I (1987) In *Brewing Microbiology* (eds F.G. Priest and I. Campbell), Elsevier, London.
Campbell, I. and Msongo, H.S. (1991) *Journal of the Institute of Brewing*, **97**, 279.
DiMichele, L.J. and Lewis, M.J. (1993) *Journal of the American Society of Brewing Chemists*, **51**, 63.
Gilliland, R.B. (1967) *Brewers' Guardian*, **96** (December), 37.
Gilliland, R.B. (1971) *Journal of the Institute of Brewing*, **77**, 276.
Hall, J.F. (1971) *Journal of the Institute of Brewing*, **77**, 513.
Harper, D.R. (1981) *Brewers' Guardian*, **110** (July), 23.
Hjortshøj, B. (1984) *European Brewery Convention Symposium on Quality Assurance, Zoeterwoude*, IRL Press, Oxford, p. 65.
Ingledew, W.M. and Casey, G.P. (1982) *Brewers' Digest*, **57** (March), 18.
Johnston, J.R. and Reader, H.P. (1983) In *Yeast Genetics: Fundamental and Applied Aspects* (eds J.F.T. Spencer, D.M. Spencer and A.R.W. Smith), Springer, New York, p. 205.
Kreger-van Rij, N.J.W. (1984) *The Yeasts, a Taxonomic Study*, 3rd edn, Elsevier/North-Holland, Amsterdam.
Martini, A.V., Martini, A. and Cardinali, G. (1993) *Antonie van Leeuwenhoek*, **63**, 145.
Maule, A.P. and Thomas, P.D. (1973) *Journal of the Institute of Brewing*, **79**, 137.
Meaden, P.G. (1990) *Journal of the Institute of Brewing*, **96**, 195.
Moll, M. (1981) *Journal of the Institute of Brewing*, **87**, 303.
Molzahn, S.W. (1993) *Proceedings of the 25th Congress of the European Brewery Convention, Oslo*, IRL Press, Oxford, p. 203.
Nykänen, L. and Suomalainen, H. (1983) *Aroma of Beer, Wine and Distilled Beverages*, Riedel, Dordrecht, New York and London.
Pedersen, M.B. (1983) *Proceedings of the 19th Congress of the European Brewery Convention, London*, IRL Press, Oxford, p. 457.
Pettipher, G.L., Mansel, R., McKinnon, C.H. and Cousins, C.M. (1980) *Applied and Environmental Microbiology*, **39**, 423.
Pfenninger, H.B. (1977) *Journal of the Institute of Brewing*, **83**, 109.
Put, H.M.C. and de Jong, J. (1980) In *Biology and Activities of Yeasts* (eds F.A. Skinner, S.M. Passmore and R.R. Davenport), Academic Press, London, p. 181.
Rainbow, C. (1981) In *Brewing Science*, Vol. 2 (ed. J.R.A. Pollock), Academic Press, London, p. 491.
Simpson, W.J., Fernandez, J.L. and Hammond, J.R.M. (1992) *Journal of the Institute of Brewing*, **98**, 33.
Speers, R.A., Tung, M.A., Durance, T.D. and Stewart, G.G. (1992) *Journal of the Institute of Brewing*, **98**, 525.
Sponholz, W.-R. (1993) In *Wine Microbiology and Biotechnology* (ed. G.H. Fleet), Harwood, Chur, p. 395.
Stewart, G.G. (1975) *Brewers' Digest*, **50** (March), 42.
Stratford, M. (1992) *Advances in Microbial Physiology*, **33**, 1.
Taylor, G.T. and Marsh, A.S. (1984) *Journal of the Institute of Brewing*, **90**, 134.

Thompson, A.N., Wright, D.M., Pawson, E.C. and Meaden, P.G. (1995) In *Proceedings of the Fourth Aviemore Conference on Malting, Brewing and Distilling 1994* (eds I. Campbell and F.G. Priest), Institute of Brewing, London, p. 213.
Van Oevelen, D., Spaepen, M., Timmermans, P. and Verachtert, H. (1977) *Journal of the Institute of Brewing*, **83,** 356.
Wickerham, L.J. (1951) *Technical Bulletin no. 1029*, US Department of Agriculture, Washington, DC.
Wiles, A.E. (1953) *Journal of the Institute of Brewing*, **59,** 265.

CHAPTER 8

Rapid detection of microbial spoilage

I. Russell and T.M. Dowhanick

8.1 INTRODUCTION

Minimization of the length of time required to detect the presence of spoilage microflora is a paramount concern in the food and beverage industry, for failure to keep such microorganisms to a minimum practicable level can lead to significant economic losses as a result of recalls, reduced shelf life and inconsistency in product quality. In the brewing industry, the last decade or more has witnessed the development of a plethora of novel or improved methods for the detection of microorganisms. The reasons for this include:

1. increased consumer awareness on product quality;
2. tightened government regulations;
3. increased competition among brewers due to declining consumption;
4. the increasing trend to avoid pasteurization of packaged beer;
5. technological advancements.

It has been fortuitous for the brewer that most microorganisms, particularly those that are pathogenic to humans, do not survive in a typical beer due to factors such as the presence of alcohol (usually in the range of 4–5.5% ethanol), subphysiological pH, anti-microbial hop components and an anaerobic environment. As a result, the range of microorganisms generally found in a brewery is relatively few in comparison to the diverse array of microbia found in the medical, environmental, cosmetic or food and dairy industries. Table 8.1 lists the genera of Gram-positive and Gram-negative bacteria, along with wild yeasts which have been detected in various breweries (Ingledew, 1979; Rainbow, 1981; Back, 1987; Dowhanick, 1990; Stenius *et al.*, 1991; see also Chapters 5, 6 and 7).

Brewing Microbiology, 2nd edn. Edited by F. G. Priest and I. Campbell.
Published in 1996 by Chapman & Hall, London. ISBN 0 412 59150 2

Table 8.1 Bacteria and yeast genera found in breweries

Gram-positive	*Gram-negative*	*Wild yeasts*
Bacillus	*Acetobacter*	*Brettanomyces*
Lactobacillus	*Acinetobacter*	*Candida*
Leuconostoc	*Alcaligenes*	*Cryptococcus*
Micrococcus	*Citrobacter*	*Debaryomyces*
Pediococcus	*Enterobacter*	*Endomyces*
Streptococcus	*Flavobacterium*	*Hansenula*
	Gluconobacter	*Kloeckera*
	Hafnia	*Pichia*
	Klebsiella	*Rhodotorula*
	Megasphaera	*Saccharomyces*
	Obesumbacterium	*Torulopsis*
	Pectinatus	*Zygosaccharomyces*
	Proteus	
	Pseudomonas	
	Zymomonas	

Detection techniques may or may not require microbial growth. For those requiring growth, selective conditions may be necessary to suppress the proliferation of desirable microorganisms such as brewer's yeast, in order to detect the presence of undesirable ones such as potential beer spoilage microorganisms. As described by Kyriakides and Thurston (1989), there are two broad groups of critical control points for microbiological sampling found in the brewing process. The first group encompasses locations where brewing yeasts are present, and where selective conditions (i.e. excluding the brewing yeast strain being employed) are required to detect contaminants. Examples of such critical control points include the yeast slurry, fermentation and ageing. The second group consists of locations where any microorganism should either be completely absent or present only in very low numbers. Such critical control points include raw materials, bright beer, finished product and strategic surfaces of process machinery such as filler heads. In the latter group, relatively non-selective conditions are employed to detect as broad a range of contaminating organisms as possible.

Detection implies establishing the presence or absence of particular microbes. Traditionally, microbial detection methods in a brewery most often begin with streaking or inoculating a sample to a specific medium (of a selective or non-selective nature), followed by incubation of the medium under defined conditions of temperature and environment (aerobic, anaerobic or CO_2 enriched) for a lengthy but specified period of time. While classic detection techniques usually result in accumulation of the required information, they are too slow to benefit the brewer in time to implement corrective measures. It is with this in mind that improvements in the speed of microbial detection and identification procedures would be very useful to the brewer. Yet, considerable caution must be exercised when the term 'rapid' is prefixed to any microbial detection method. In any detection scenario where a growth period is required in order for microbes to reach a particular titre, it must be remembered that microorganisms have their own specific growth rates. These rates may be very slow compared to commonly encountered clinical species such as *Escherichia coli*. While a rapid detection test for *E. coli*, with a mean generation time of 20–30 min, might require only a few hours incubation, the same rapid test for fastidious growers such as *Pediococcus damnosus* or *Lactobacillus lindneri*, with doubling times in excess of 5 h, could require a minimum period of several days (Barney and Kot, 1992).

With the relatively recent advances in molecular biology and genetic biotechnology, many techniques have been developed to characterize organisms at the molecular level and to study their genetic expression under a multitude of conditions. These techniques have been applied to the clinical, food and beverage areas, and to the environmental fields in order to achieve one or both of the following goals:

1. the detection of problematic organisms;
2. the identification and/or differentiation of organisms at the genus, species or subspecies level.

This chapter will offer an overview of several detection techniques. It will become apparent that some of the detection techniques are also identification techniques. For example, specific probes such as monoclonal antibodies or oligonucleotides not only detect the presence of a particular microorganism in, for example, a yeast slurry, but due to the high specificity of detection, these probes can also identify the particular microbe. Similarly, it can be argued that identification of that same contaminant in the yeast slurry signifies the detection of the contaminant. This type of methodology is in stark contrast to the use of identification kits such as API and others which require a pure isolate in order to perform the identification. Therefore, although this chapter is dedicated to detection techniques, some of the same techniques are also discussed in Chapter 9 but in the latter case the emphasis is on identification.

8.2 IMPEDIMETRIC TECHNIQUES (CONDUCTANCE, CAPACITANCE)

Impedimetric techniques are based on the observation that growth-related microbial activity can be measured in culture media over time by monitoring changes in the chemical and ionic composition in the medium (Harrison and Webb, 1979; Kilgour and Day, 1983; Firstenberg-Eden and Eden, 1984; Evans, 1985; Schaertel, Tsang and Firstenberg-Eden, 1987; Adams, 1988; Adams, Bryan and Thurston, 1989; Kyriakides and Thurston, 1989; Unkel, 1990; Barney and Kot, 1992; Fleet, 1992; Owens,

Konirova and Thomas, 1992; Pishawikar, Sinhal and Kulkarni, 1992; Foster, 1994). When a substance, such as a growth medium is subjected to an alternating electrical current, the impedance (resistance to the flow of the current through the medium) is affected by the conductance (ability of the medium to allow electricity to pass through it) and the capacitance (the ability to store an electrical charge). A pair of electrodes is submerged in a conducting culture medium to determine total impedance (the vectoral sum of conductance) which is associated with changes in the bulk ionic medium, and capacitance, which is associated with changes in close proximity to the electrodes (Schaertel, Tsang and Firstenberg-Eden, 1987). In the growth medium, large molecules such as carbohydrates lack electrical charge and consequently this increases the impedance of the medium while decreasing both the conductance and capacitance. As these uncharged macromolecules are metabolically broken down by microorganisms into smaller subunits such as bicarbonate, which carries a charge, the capacitance and conductance begin to increase while the impedance decreases. Thus, microbial growth can be monitored in the culture medium by measuring either decreases in total impedance, or increases in either conductance or capacitance.

When monitoring a culture using the appropriate medium, measurements of either impedance or conductance, when plotted against time, produce a curve similar, but not superimposable, to the respective growth curve. Impedimetric measurements require growth of microorganisms to a particular threshold value, which when reached can be accurately detected and measured by computerized accessories. For these instruments, threshold levels, or detection times (i.e. the time required to reach the threshold level in which a change in the electrical property occurs), usually correspond to microbial levels of about 10^5–10^6 colony forming units (CFU) ml^{-1}. Since detection times are calculated when threshold levels are reached, two factors significantly affect the detection times of healthy cells inoculated into a suitable medium. These are the quantity of cells initially inoculated and the natural growth rates or generation times of the respective cultures. The greater the quantity of viable microorganisms initially placed into the specialized sample cells, or the shorter their doubling times, the shorter the detection time. Figure 8.1 shows a hypothetical growth and conductance response of two microorganisms inoculated to the same initial concentration, but possessing different growth rates.

There are a number of advantages to using impedimetric technology. They include:

1. a time reduction of days in detecting even the most fastidious of microorganisms when compared to standard plating;
2. some selectivity for microbial growth depending on the medium employed (see Chapter 9);

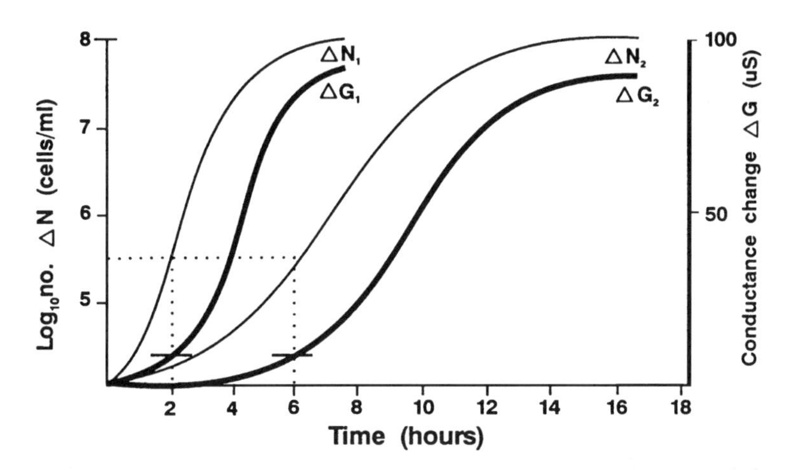

Fig. 8.1 Hypothetical growth and conductance response of two organisms (designated 1 and 2) inoculated to the same initial concentration but possessing different growth rates.

3. user-friendly automation for assessing multiple samples for threshold detection.

The disadvantages include:

1. the instruments are expensive to purchase and operate;
2. due to ionic interferences, not all types of media are amenable to electrometric analysis.

There are at least two good examples of how impedimetric methods may be used in the brewery situation. First, they have been used to replace the standard forcing test for bacterial spoilage by filtering a beer sample and incubating the filter in brewery wort in a Bactometer (Evans, 1982). Standard and impedimetric forcing tests were in complete agreement for 'spiked' beers but the latter were obtained in 2–4 days rather than up to 3 weeks for traditional forcing tests. A second brewery application is the detection of bacterial contamination in pitching yeasts (Kilgour and Day, 1983). Selectivity for bacterial growth was obtained by incubating pitching yeast samples in WLN broth containing cycloheximide to prevent yeast growth. Bacterial growth was detected in a Malthus instrument in less than 18 h depending on the inoculum size (10^6–10^7 cells ml^{-1} were detected immediately, but 10 cells ml^{-1} required the full 18 h). In practice, this allowed detection of 100 bacteria per million yeasts in 18 h, a sensitivity which compares favourably with traditional plating, but in about half the time.

8.3 MICROCALORIMETRY

Calorimetric or heat measurements can be performed by either measuring changes in quantities of heat by use of a thermometer, or by measuring thermoelectric effects produced in thermoelements as a result of heat passage. As microorganisms grow in a culture medium, they generate small but detectable amounts of heat, primarily as a result of catabolic biochemical reactions. These microfluctuations in heat (typically in the microwatt range) can be measured and recorded over time by instruments called microcalorimeters. Microcalorimeters designed for microbial measurements generally function by passing heat through thermoelectric couples, producing an electrical current which is proportional to the heat flow. The thermocouple produces a potential (in the millivolt range) which is amplified and read off a recorder, producing a thermogram. The beginning of a thermogram usually occurs during exponential growth of a microbial culture and increases as the culture continues to grow. As exponential growth is taken over by the onset of stationary phase, metabolic activity in the culture diminishes, as does the heat level in the culture (Pishawikar, Sinhal and Kulkarni, 1992).

Microcalorimeters are capable of detecting as few as 10^3 actively growing bacteria per ml and have been used for microbial detection in clinical applications (Russell *et al.*, 1975; Beezer *et al.*, 1978; Herman *et al.*, 1980) as well as in the food and dairy industries (Berridge, Cousins and Cliffe, 1974; Lampa *et al.*, 1974; Gram and Soggaro, 1985). While microcalorimetry has been mentioned in brewing literature (Hope and Tubb, 1985; Simpson, 1990), it does not appear that this technique has undergone much testing or implementation among brewers. However, with the recent increase in popularity of filter-sterilized beer, microcalorimetry may find application as an early indicator of microbial contamination in a manner similar to methods used with ATP bioluminescence.

8.4 TURBIDOMETRY

When microorganisms proliferate in a clear liquid broth, both the turbidity and absorbance of the broth increase. These changes to the broth can be detected and measured by a spectrophotometer or turbidometer. If the initial inoculum size is low and the broth is quite clear, the growth curve obtained will appear very similar in shape to that produced by impediometry (Firstenberg-Eden and Sharpe, 1991). With the use of an automated turbidometer (Bioscreen Analysis System, Labsystems, Helsinki, Finland), microbial growth has been assessed from a variety of food and dairy matrices (Jorgensen and Schultz, 1985; Mattila and Alivehmas, 1987; Mattila, 1987) and from beer (Haikara, Mattila-Sandholm and Manninen, 1990).

Haikara, Mattila-Sandholm and Manninen (1990) employed automated turbidometry in order to detect microbial contaminants in pitching yeast. These contaminants included *Lactobacillus casei*, *Enterobacter agglomerans*, *Pediococcus damnosus*, *Pectinatus cerevisiiphilus* and *Saccharomyces diastaticus*. As would be expected for microorganisms with naturally diverse growth rates, detection times varied significantly between the different species. In general, detection times using the automated turbidometer were approximately 1 day shorter than the time required to visually discern turbidity, and the detection limit for these studies was approximately 10^6 cells ml^{-1}. The time saved in using turbidometry, compared to standard plating, varied from species to species, but ranged anywhere from 2- to 8-fold, or 2–4 days.

8.5 FLOW CYTOMETRY

Flow cytometry combines both microscopic and biochemical analyses in a single technique to measure microorganisms in liquids. The cells are introduced into the centre of a rapidly moving fluid stream and are forced to flow single file out of a 50–100 µm orifice at a uniform speed (typically 1–10 m s^{-1}). As they do so, the cells are carried past a station where they are illuminated by a light source (laser) and measurements are made at rates of 10^4–10^6 cells min^{-1}. As the microorganisms in the flow stream pass through the light beam, the illuminating light is scattered by the cells and the intensity of light scattered at different angles yields information about cell size, shape, viability, density and surface morphology (Kruth, 1982; Salzman, 1982; Muirhead and Horan, 1984; Traganos, 1984).

There have been a number of applications of flow cytometry relevant to brewing described in the literature (Hutter, 1991; Donhauser *et al.*, 1992). Hutter (1992a) demonstrated that immunofluorescent staining could be used to mark the cells of a particular species of microorganism in order to facilitate identification and counting. Early work only allowed one species at a time to be counted, thus not proving very useful to the brewing industry. Newer developments in fluorescent dyes and dye–antibody conjugates, as well as the detection and spectral separation of fluorescent light in flow cytometers, now allow several species to be stained and counted in a single operation. Using species-specific antibodies marked with four different fluorochrome conjugates, Hutter (1992b) demonstrated that four species in a mixed culture (*Saccharomyces cerevisiae*, *Schizosaccharomyces pombe*, *Lactobacillus brevis* and *Pediococcus damnosus*) could be clearly distinguished and counted.

The major advantage to the method is sensitivity. The major disadvantages include slow processing of large volumes and small particulate debris interfering with the assay. There is also a high capital cost for multi-purpose cytometers and a high level of operator training is required.

8.6 ATP BIOLUMINESCENCE

An important cellular metabolite which can be used in biochemical assays in order to detect the presence of microorganisms is adenosine 5′-triphosphate (ATP). ATP is a high energy molecule that is found in all living organisms. ATP bioluminescence offers the opportunity to significantly reduce, from days to hours, the time required to identify the presence of living organisms. This technology is based on detection of the presence of ATP in samples using enzyme-driven light production, or bioluminescence. The ATP molecule can be assayed efficiently by a two-step reaction employing the luciferin–luciferase enzyme reaction, which is the basis of bioluminescence in fireflies:

Step 1: luciferin + luciferase + ATP + Mg^{2+} ⇒ (luciferin-luciferase-AMP) + pyrophosphate

Step 2: (luciferin-luciferase-AMP) + O_2 ⇒ oxyluciferin + luciferase + CO_2 + AMP + light

The amount of light that is produced correlates with the amount of ATP in the sample. When properly assayed, most of the ATP, but not necessarily all, is directly indicative of the presence of living organisms in the sample. There are commercially available reagent kits which can detect as low as 100 yeast cells per sample without enrichment. The required instruments (luminometers) have limits of detection as low as 1–2 yeast cells per sample. Thus, one yeast cell in a given sample, with a doubling time of 2–3 h, can be detected after approximately 20 h incubation. This results in detection of living microbes in considerably less time than that required for standard plating techniques. This method is also comparable in cost to standard plating when factors such as labour and use of disposables such as anaerobic system envelopes (when screening for microaerophilic or anaerobic microorganisms) are included.

The use of ATP-driven bioluminescence for the detection of yeast and bacteria in non-pasteurized beer or at certain locations along the brew path has become a useful and practical procedure in the brewing industry (Hysert, Kovecses and Morrison, 1976; Kilgour and Day, 1983; Dick *et al.*, 1986; Harrison, Theaker and Archibald, 1987; Krause and Barney, 1987; Avis and Smith, 1989; Miller and Galston, 1989; Simpson, 1989; Simpson *et al.*, 1989; Takamura *et al.*, 1989; Barney and Kot, 1992; Ogden, 1993; Dowhanick and Sobczak, 1994). In the last few years, improvements have been made both in development of more efficient reagents for the extraction of ATP from brewery microorganisms (Jago *et al.*, 1989; Simpson and Hammond, 1989) and in the development of more sensitive luminometers. A recent survey by Stanley (1992) reported that more than 90 luminometers from more than 60 companies are currently commercially available, and some of the more widely used reagents have undergone comparative evaluations (Griffiths and Phillips, 1989).

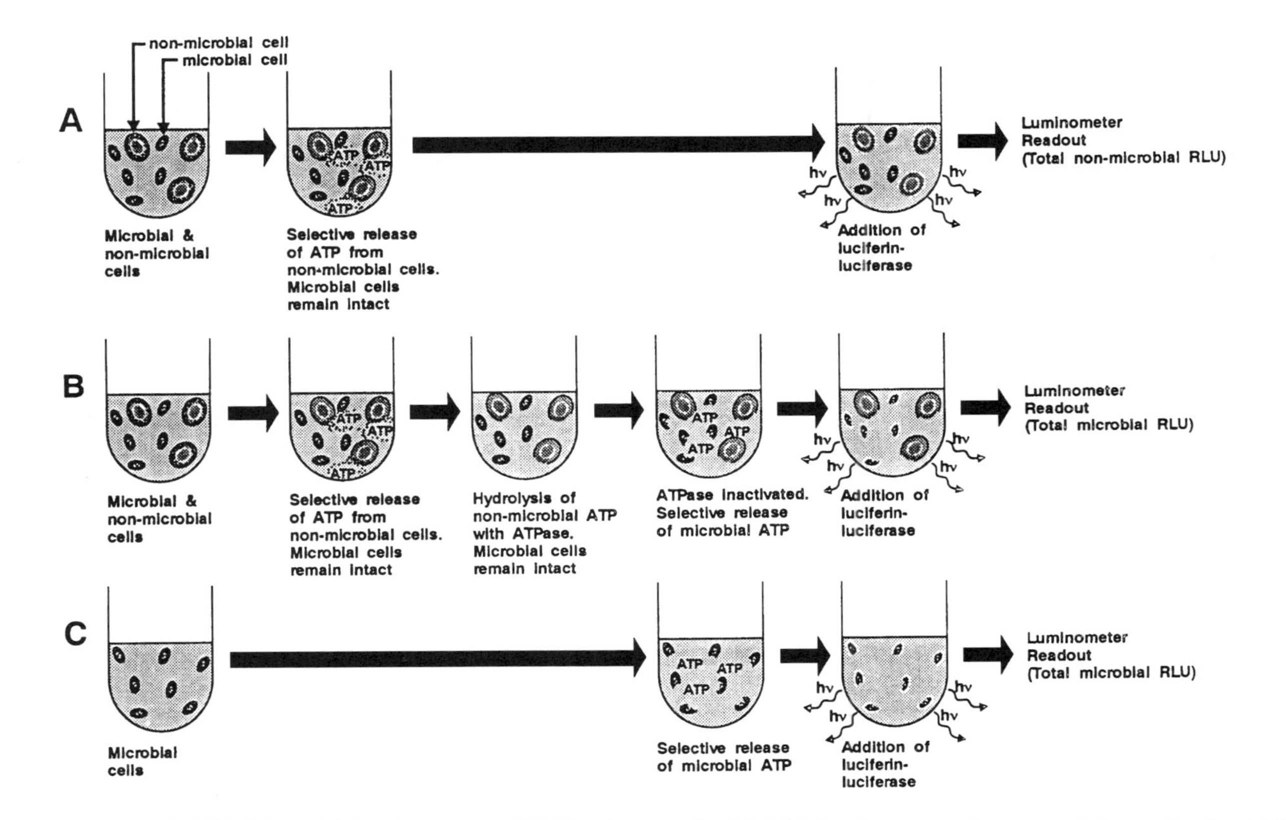

Fig. 8.2 Measurement of ATP-driven bioluminescence. (A) Total non-microbial bioluminescence from a mixture of microbial and non-microbial cells. (B) Total microbial bioluminescence from a mixture of microbial and non-microbial cells. (C) Total bioluminescence from microbial cells only. RLU, relative light units.

ATP bioluminescence as a quality control device for hygienic surface swabbing of process machinery, water analysis or beer analysis is currently in use in many breweries. As outlined in Fig. 8.2, the assay can be used to detect the presence of either microbial or non-microbial ATP. The result is qualitative, but not necessarily quantitative, as unknown samples may contain numerous species of microorganisms possessing vastly different ATP levels (e.g. yeasts versus bacteria). Quantitative levels of detection compared to numbers of organisms observed by plating techniques can be determined when pure cultures of organisms are assayed under reproducible conditions. This is because different organisms contain vastly different levels of ATP, at different phases of growth, on a per cell basis. For example, a typical yeast cell contains approximately 100 times the amount of ATP found in a bacterial cell (Hysert, Kovecses and Morrison, 1976). In addition, bacteria which vary significantly in size (such as an isolate of *Lactobacillus plantarum* compared to one of *Pediococcus damnosus*) will also vary in relative amounts of ATP per cell. Lastly, especially when conducting surface hygiene or water analyses, not all organisms extracted for ATP will necessarily develop into colonies when plated on the comparative solid media. Therefore, employment of bioluminescence for routine detection of either microbial or non-microbial ATP is useful as a qualitative, or semi-quantitative assay along the brew path.

8.7 MICROCOLONY METHOD

The microcolony method employs microscopy to detect growing cells which have not yet reached colony forming units visibly discernible to the naked eye. This method is essentially a rapid modification to membrane filtration of a sample followed by incubation of the membrane on a nutrient medium. In lieu of waiting several days for the growth of discernible colony forming units, microcolonies are selectively stained and then visualized microscopically within, at most, 24 h incubation. In the early stages of microcolony method development, stains such as ponceau S, janus green, methylene blue and safranin were used for the rapid detection of yeasts (Richards, 1970; Winter, York and El-Nakha, 1971; Barney and Helbert, 1975; Anon., 1975; Portno and Molzahn, 1975). After the microcolonies were stained, the membrane was dried, heated and cleared (usually with immersion oil), and then examined by transmitted light microscopy. An alternative microcolony method using stains was developed whereby the membrane filters and the cells were selectively stained rather than the cells alone (Tse and Lewis, 1984). Using this method, the filter membranes containing microcolonies were stained with bromocresol green and methyl red. This technique was originally developed to distinguish easily between microcolonies of bacteria from water samples. The bacterial colonies stained dark blue while the membrane filter stained pink. As with the direct staining of yeast cells, a notable disadvantage was that post-stain treatment of the membrane was lethal to the microcolonies, thereby eliminating the possibility of reincubating the filter membrane for further study.

Significant improvements to the microcolony technique were made by use of incident-light fluorescence microscopy, which allowed fluorescent dyes to be added directly to the nutrient medium. Such dyes, when incorporated by yeasts and other microorganisms, allowed direct fluorescence detection of the living cells without requiring further treatment of the membrane. While initial difficulties were encountered, as most fluorescent dyes were either not incorporated or growth inhibitory to many microorganisms, certain optical brighteners (i.e. fluorescent whitening agents found in most washing powders and soaps) were found to be readily taken up from nutrient media (Paton and Jones, 1971). Harrison and Webb (1979) utilized optical brighteners derived from 4,4′-diaminostilbene-disulphonic acid to develop a microcolony method for yeasts and moulds. Using the optical brightener Uvitex 2B (Ciba-Geigy), Harrison and Webb (1979) were able to not only detect fluorescing microcolonies against a black filter membrane, but were also able to differentiate between living and dead yeast cells, since the dead cells fluoresced uniformly while the living cells fluoresced primarily at sites of budding where chitin was most concentrated. In addition, these optical brighteners were non-inhibitory to cellular growth of both yeasts and moulds, and the filter membranes could be reincubated, allowing for continued examination of a sample over several days and/or further identification of the microorganisms if desired. The optical brighteners tested, however, were not suitable for staining bacteria.

The microcolony method was further refined, allowing for detection of both yeasts (brewing and wild) and brewery bacteria (including members of the genera *Lactobacillus*, *Pediococcus*, *Bacillus*, *Pectinatus*, *Hafnia* and *Escherichia*) using the optical brightener Tinopal CBS-X (Ciba-Geigy), a derivative of distryl-β-phenyl (Haikara and Boije-Backman, 1982). While some difficulties were observed in correlating aerobic bacteria microcolony counts with those of standard colonies, the correlation of yeasts and anaerobic bacteria using both methods was very good. While the microcolony technique is useful for the rapid detection of very low numbers of microorganisms, examination of the microbes is both laborious and tedious, resulting in operator fatigue (Haikara and Boije-Backman, 1982; Simpson, 1991; Barney and Kot, 1992). It is also noteworthy that microcolony methods, as with standard plating, do not give results which necessarily reflect the ability of proliferating microorganisms on a nutrient medium to actually grow in and spoil beer (Hope and Tubb, 1985).

8.8 DIRECT EPIFLUORESCENCE FILTER TECHNIQUE

The direct epifluorescent filter technique (DEFT) is a method which employs polycarbonate membrane filtration of samples (for concentration of microflora) combined with epifluorescent microscopy in order to visualize individual, fluorochrome-stained cells (Pettipher *et al.*, 1980; Kilgour and Day, 1983; Pettipher, 1985; Navarro *et al.*, 1987; Simpson, 1990; Barney and Kot, 1992; Fleet, 1992). This method, which is considerably more sensitive and rapid than the microcolony method, was initially developed as a quality control technique for counting bacteria in raw milk (Pettipher *et al.*, 1980). DEFT can be completed in approximately 30 min as no pre-growth stage is required, and it has the potential to detect single cells on a membrane filter. Polycarbonate membranes are used for filtration instead of standard cellulose acetate membranes as they have a much flatter, smoother surface which is better suited for microscopy (Pettipher, 1985).

Acridine orange has been the most commonly used fluorescent dye for DEFT because of its ability to bind to nucleic acids and theoretically differentiate metabolically active or living cells (which stain orange) from inactive or dead cells (which stain green). The reason for the observed difference in colour between living and dead cells is that acridine orange stains orange when it binds to single stranded DNA or to the relatively abundant levels of RNA which are found in metabolically active cells. When cells die, these single stranded nucleic acids are rapidly degraded by nucleases, while double stranded DNA molecules, which naturally fluoresce green, are not rapidly degraded in dead cells. However, in studies conducted on filtered brewery samples, differentiation of viable from non-viable cells was not successfully achieved, and nondescript stained debris was a particular problem (Kilgour and Day, 1983; Navarro *et al.*, 1987; Barney and Kot, 1992). Peladan and Leitz (1991) were able to improve results using acridine orange by adding a second stain, berberine sulphate, to the epifluorescent procedure, as the latter dye selectively attaches to the double stranded DNA of dead cells, giving them a green fluorescence while remaining impermeable to living cells. Others have reported that when using fluorescent dyes such as aniline blue, viablue or certain tetrazolium derivatives, viable and non-viable yeasts can easily and reliably be differentiated from each other (Koch, Bandler and Gibson, 1986; Hutcheson *et al.*, 1988; Betts, Bankes and Banks, 1989). In addition, some of the nondescript stained debris can be removed by prefiltering the sample through 12 μm pore size polycarbonate membranes (Navarro *et al.*, 1987).

When manually performed, DEFT (as well as the microcolony technique) is tedious, particularly when screening for low numbers of cells on large membrane filters. Such analyses can quickly lead to operator fatigue and, in effect, become an impractical procedure. To offset this difficulty, an automated version of DEFT can be performed by use of computer-assisted microscopic scanning of the filter membrane coupled to computer-enhanced image analysis (Kilgour and Day, 1983; Pettipher *et al.*, 1989; Simpson, 1990; Barney and Kot, 1992; Fleet, 1992; Niwa, 1993). Such automation is currently expensive and takes significant periods of time for computer processing, due to the large number of tasks performed by the computer (e.g. filter scanning, size separation and debris exclusion, colour recognition etc.). However, an automated counting system is marketed by Micro-Measurements Ltd (Saffron Walden, Essex, UK) which removes the problems associated with eye fatigue and reportedly gives good agreement with visual counts. Moreover, a completely automated DEFT system has been announced (Cobra 2024, Biocom, France) which features three computers controlling sample preparation, filtration, drying and image analysis. A throughput rate of 150 samples per hour is claimed with improved reproducibility and lower counting limits than manual systems.

8.9 PROTEIN FINGERPRINTING BY POLYACRYLAMIDE GEL ELECTROPHORESIS

Microorganisms generally possess two groups of proteins: the constitutively synthesized structural or regulatory 'housekeeping' proteins, and the differentially regulated polypeptides that are either induced or repressed as a result of environmental stimuli. The former category of proteins tends to be found under most conditions conducive to cell viability and growth, whereas the latter category is regulated by conditions that include the growth medium, growth temperature, nature of the gaseous environment and growth phase at the time of harvesting. Based on these observations, polyacrylamide gel electrophoresis (PAGE) of cellular proteins has been used in recent years as a means of classifying and identifying yeasts and other microorganisms (Kersters and De Ley, 1975; Van Vuuren *et al.*, 1981; Drawert and Bednar, 1983; Van Vuuren and Van Der Meer, 1987, 1988; Dowhanick *et al.*, 1990; Vancanneyt *et al.*, 1991; Newsom, O'Donnell and McIntosh, 1992; Chapter 9). In a series of steps, soluble proteins are extracted from pre-grown cells (quantities as low as single colony isolates) and subjected to electrophoretic separation. This separation can be made in one dimension based on differences among proteins in size using sodium dodecyl sulphate (SDS PAGE) or native PAGE, or based on differences in ionic charge using isoelectric focusing. Separations can also be performed in two dimensions by combining size and ionic charge separation in one gel. The resulting separation pattern or 'protein fingerprint' is specific to the gene expression of the isolate in question. As such, it can be analysed for relative differences or similarities to other strains, and based on this information, the microorganism can be categorized. Visual analysis can be used to compare the patterns, or densitometer/computer analysis can be used to numerically cluster closely related strains that display only minor but reproducible differences.

When determining whether or not two or more isolates are essentially the same, it is very important that the respective isolates are treated in the same manner, since the cellular inventory (and respective quantities) of gene products can be significantly altered by differences in growth conditions. Attempts to compare the protein profiles of one organism to another should include common pre-growth steps immediately prior to harvesting and preparation of cells. Once pre-growth of cells has been completed, extraction, electrophoretic separation and staining of proteins can be accomplished within the span of a few hours.

8.10 IMMUNOANALYSIS

Immunoanalysis has been considered a useful means of identifying contaminating microorganisms along the brew path because of its potential to detect or identify microbes in either a semi-quantitative or directly quantitative way (Campbell and Allan, 1964; Tsuchiya, Fukazawa and Kawakita, 1965; Dolezil and Kirsop, 1975; Hutter, 1978). Using polyclonal or monoclonal antibodies, different assays have been designed to differentiate or even speciate microorganisms from each other.

Rapid, semi-quantitative 'sandwich' immunoassays can be used to colorimetrically differentiate or identify relatively abundant ($> 10^4$) numbers of intact cells. Such assays employ secondary antibodies to which enzymes have been attached (e.g. alkaline phosphatase, horseradish peroxidase). Figure 8.3 is an example of a sandwich enzyme-linked immunosorbent assay (ELISA). Typical ELISAs employ a matrix such as a filter membrane, dipstick, test tube or microtitre plate coated with antibody specific to the desired antigen of interest. The sample is placed in contact with the immobilized antibody, and the antigen is allowed to bind to the matrix-bound antibody. The matrix is washed (leaving the bound antigen attached), and a second antigen-specific antibody–enzyme conjugate is allowed to bind to the matrix-bound antigen. The unbound secondary antibody is removed and a substrate is added to the matrix. This substrate is colorimetrically altered by the enzyme attached to the secondary antibody. This results in a semi-quantitative colorimetric change to the matrix if the specific antigen, such as a beer spoilage microorganism, is present. ELISA kits have been designed and are being used extensively for environmental testing (for pesticides and other compounds), in at-home pregnancy testing kits, workplace drug screening and for AIDS testing. In the food and beverage industry, the main application for these ELISA kits is in screening for organisms such as *Salmonella* and *Listeria*. To date such kits are not yet commercially available for beer spoilage microorganisms, but research has been undertaken (Umesh-Kumar and Nagarajan, 1991).

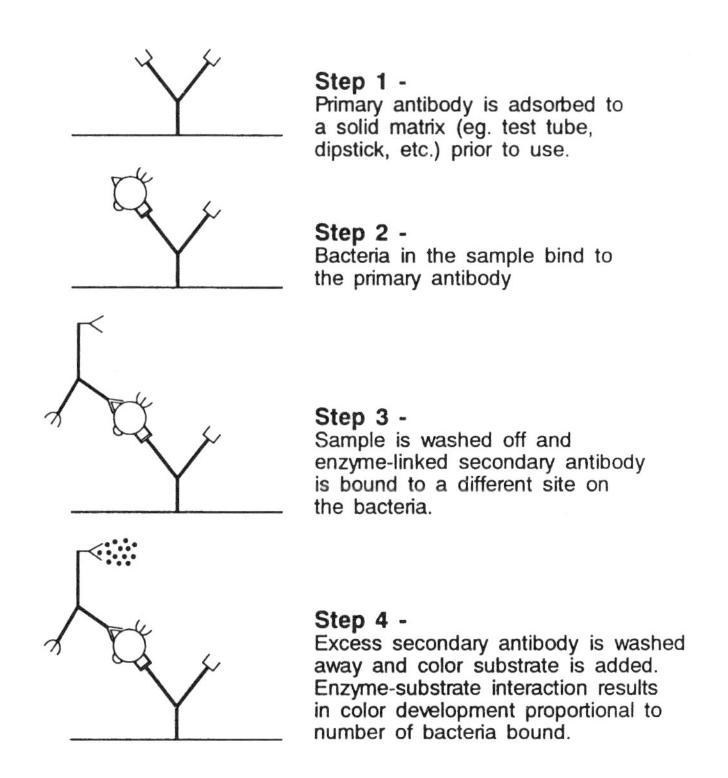

Fig. 8.3 Enzyme linked immunosorbent assay.

The major advantages these kits will offer is their low cost, speed and portability.

Antibodies have been designed and employed to identify brewery microorganisms (Legrand and Ramette, 1986; Phillips and Martin, 1988; Yasui, Taguchi and Kamiya, 1989, 1992; Hutter, 1992a,b; Vidgren *et al.*, 1992; Whiting *et al.*, 1992; Gares *et al.*, 1993; Laurent *et al.*, 1993). In these cases, single or very low numbers of cells (including microcolonies) have been visualized by immunofluorescent microscopy. Typically, specific antibodies are allowed to bind to contaminant microbes. After removal of unbound specific or primary antibodies, secondary 'indicator' antibodies (such as antibodies specific to the type of immunoglobulin from which the primary antibody is derived) to which fluorochromes (e.g. fluoroscein thiocyanate) have been attached are added. Unbound secondary antibodies are removed, and microscopic observation of samples under ultraviolet illumination is used to identify the presence of target microorganisms which are seen as brightly fluorescing cells.

As with many techniques requiring microscopy, eye fatigue is often a problem. It is also necessary to select antibodies that are specific to easily accessible (usually surface) cellular components which are abundant regardless of the growth conditions of the cells (e.g. antibodies raised against cytosolic glucose-repressed or stress-induced gene products would be of limited use). However, as with the use of any specific probe, once the potential problems of cross-reactivity of antibodies to other sources have been resolved, the technique offers fast, qualitative and quantitative analysis of brewery samples without requiring incubation periods for microbial growth to obtain threshold detection values.

8.11 HYBRIDIZATION USING DNA PROBES

Deoxyribonucleic acid (DNA) is the genetic material found in almost all living organisms, with the exception of certain viruses where the genetic material is ribonucleic acid (RNA). The biological properties which essentially define all organisms are manifest in sequences of nucleotides, which consist of four bases: adenine, cytosine, guanine and either thymine for DNA or uracil for RNA. The nucleotide sequences found in strains from different species are highly specific and remain, for the most part, constant from generation to generation. Given the magnitude of roughly millions of bases of DNA found in a typical yeast or bacterium, this detection/identification technique is capable of readily distinguishing specific sequences of anywhere from less than 100 nucleotides to several thousand nucleotides within total genomes.

Hybridization (reviewed extensively by Wetmur (1991)) is described as the formation of a double-stranded nucleic acid (either DNA to DNA or DNA to RNA) by base pairing between single-stranded nucleic acids derived (usually) from different sources. DNA–DNA hybridization and DNA–RNA hybridization techniques have been developed and utilized primarily by molecular biologists to isolate, characterize and study the expression of genes (DNA) and gene transcripts (RNA). Of particular interest when performing hybridizations is the ability to control the level of specificity of base pairing by altering conditions of the hybridization, such as temperature or salt concentrations. When conditions are used to maximize the stringency of the reaction, hybridization becomes a highly selective tool to either differentiate minute differences in DNA sequences that can be associated within and between species, or identify target sequences found specifically within the genetic information of certain organisms. The DNA probes employed are sensitive and can detect 10^4–10^7 organisms per test sample. Few problems with cross-reactivity are encountered once the system has been optimized. The advantage of the DNA–DNA hybridization assay, as compared to PAGE or immunoanalysis, is that detection of the target organism is not dependent on the products of gene expression, which as already discussed can vary with

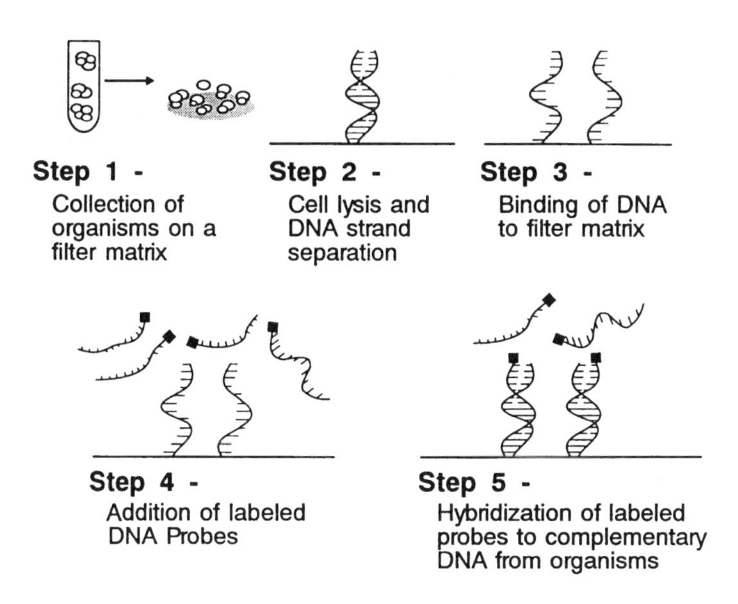

Fig. 8.4 DNA–DNA hybridization test.

growth conditions, physiological state of the organisms in the test sample etc.

Hybridization assays can be performed using single colony isolates, cells collected on membrane filters or purified nucleic acid digested with restriction endonucleases and size separated by gel electrophoresis (i.e. restriction fragment length polymorphisms (RFLPs)). Diagnostic assays based on DNA hybridization include the following:

1. propagation of the organism from the test sample to sufficient titre;
2. DNA release from the target organism;
3. a DNA probe specific for the organism;
4. a hybridization format;
5. labelling of DNA probes and detection of the resultant hybrids.

Figure 8.4 illustrates a typical assay which is described in more detail in Chapter 9.

A variety of DNA probes have been used to identify, differentiate or characterize yeasts and bacteria relevant to the brewing industry. These probes include genes derived from specific microorganisms (e.g. the *HIS4* or *LEU2* genes from *Saccharomyces cerevisiae* or the S-layer protein gene from *Lactobacillus brevis*), endogenous plasmids (such as those found in *Pediococcus damnosus*), transposable or ty elements, arbitrary repeated sequences such as poly-GT, or even viral DNA such as the single-stranded phage M13 (Decock and Iserentant, 1985; Martens, van den Berg and

Harteveld, 1985; Pedersen, 1983, 1985, 1986; Steffan *et al.*, 1989; Walmsley, Wilkinson and Kong, 1989; Delley, Mollet and Hottinger, 1990; Meaden, 1990; Colmin *et al.*, 1991; Lonvaud-Funel *et al.*, 1991; Lonvaud-Funel, Joyeux and Ledoux, 1991; Miteva, Abadjieva and Stefanova, 1992; Reyes-Gavilan, *et al.*, 1992; Vidgren *et al.*, 1992; Lonvaud-Funel, Guilloux and Joyeux, 1993). While this technology is not yet user friendly enough to be employed as a routine quality control test in the brewery quality control laboratory, the potential to produce hybridization kits does now exist. Using hybridization techniques and DNA probes, several commercial diagnostic kits are already available to the food industry. These kits are designed to detect microorganisms such as *Salmonella* or *Listeria* with high specificity.

8.12 KARYOTYPING (CHROMOSOME FINGERPRINTING)

Karyotyping is the determination of chromosomal size and number. Karyotyping has been used for many years to either characterize, differentiate or identify eukaryotic organisms. In higher eukaryotes such as animal and plant species, karyotyping can easily be performed by selectively staining DNA *in situ* (e.g. by use of the Feulgen reaction) and then viewing the clearly discernible chromosomes under a light microscope. The individual chromosomes can then be sorted by micromanipulation and identified. In lower eukaryotes such as yeast, the chromosomes, while still being considered as very large molecules, are too small in size to be readily karyotyped by selective staining and light microscopy. In yeast, karyotyping is achieved by the electrophoretic separation of whole chromosomes through an agarose gel. Conventional one-dimensional agarose gel electrophoresis, in which DNA fragments approximately 1/100 to 1/1000 the size of a yeast chromosome move through a uniform electric field, cannot resolve large DNA molecules such as yeast chromosomes, since such large molecules tend to migrate independently of their size. By applying two orientations of an electric field it is possible to karyotype yeast chromosomes, since smaller chromosomes have the ability to respond more quickly to changes in the electric field than larger ones (Bustamante, Gurrieri and Smith, 1993). Such changes in electric field to size separate large DNA molecules form the basis of pulsed-field gel electrophoresis (PFGE).

When chromosomes separated by PFGE are stained with ethidium bromide, they appear as discrete bands under ultraviolet light. Various methods for chromosome fingerprinting are available, with the primary difference being the orientation of the alternating fields during electrophoresis (Fig. 8.5). As with protein fingerprinting, computer-designed cluster analyses can be used to differentiate and/or speciate sample yeast isolates.

Karyotyping of *Saccharomyces* chromosomes has been well documented

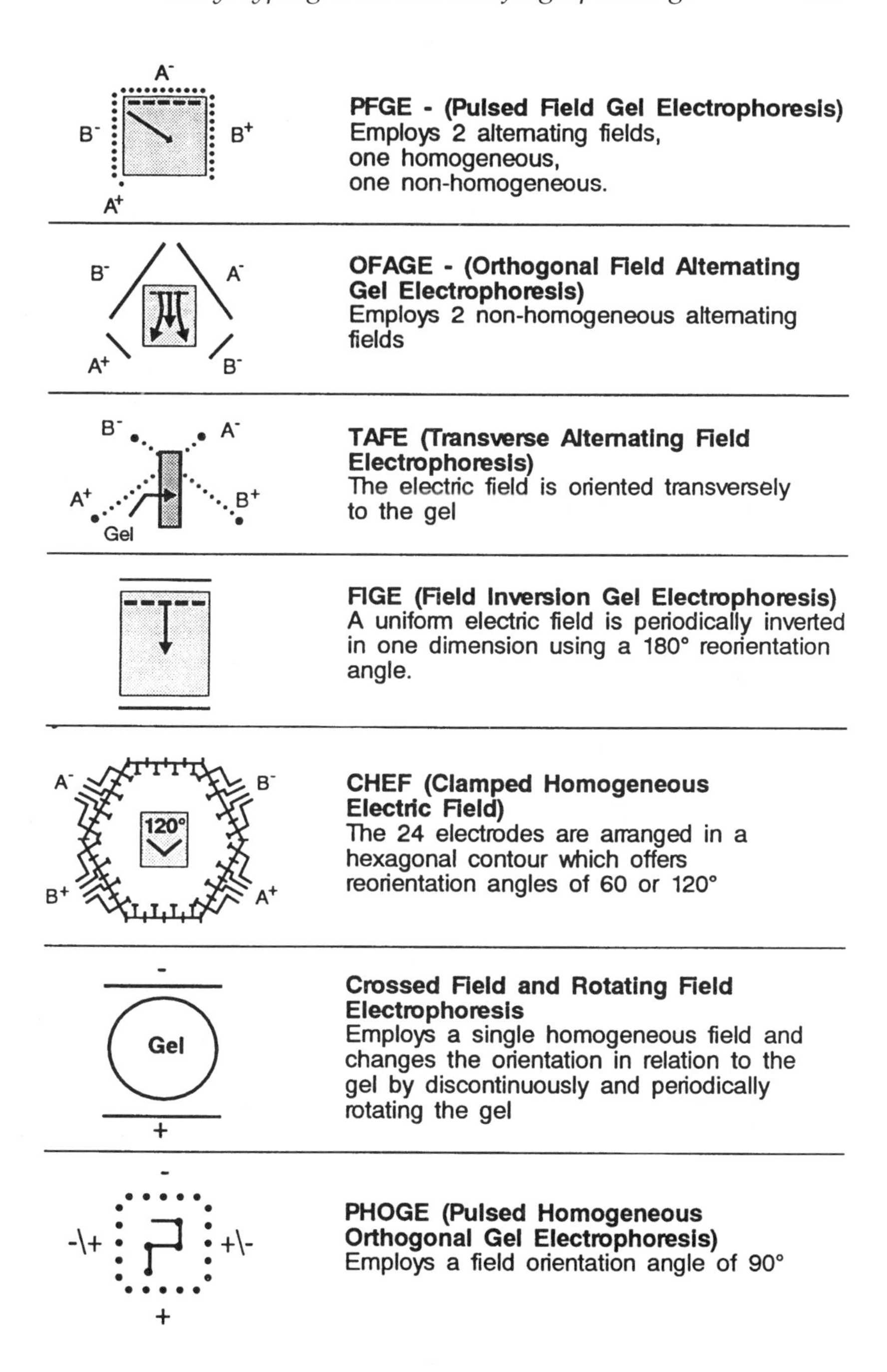

Fig. 8.5 Examples of common gel electrophoresis systems for karyotyping yeasts.

(De Jong *et al.*, 1986; Johnston and Mortimer, 1986; Casey, Xiao and Rank, 1988; Degre *et al.*, 1989; Casey, Pringle and Erdmann, 1990; Vezinhet, Blondin and Hallet, 1990; Querol, Barrio and Ramon, 1992), and has been utilized in the brewing industry not only as a research and development tool, but from a quality control standpoint as a means of differentiating or fingerprinting ale and lager production strains. This technique has been useful in differentiating closely related lager yeast strains through observed chromosomal polymorphisms, which by classical means of characterization (e.g. giant colony morphology, sporulation, growth characteristics etc.) have often been difficult or even impossible to differentiate (Casey, Pringle and Erdmann, 1990; Good *et al.*, 1993). Such chromosomal polymorphisms are the result of insertions, deletions and translocations of DNA fragments large enough (typically 10 kilobase pairs or greater) to be electrophoretically discerned. Pre-grown cells can be prepared and electrophoretically separated by PFGE within 48 h, although more than one run may be required to optimally separate chromosomes of similar size.

8.13 POLYMERASE CHAIN REACTION

The polymerase chain reaction (PCR) (Mullis *et al.*, 1986) is one of the newest and potentially most promising techniques to allow rapid detection of the presence of brewery microbial contaminants. PCR is a widely used *in vitro* method for amplifying very small amounts of selected nucleic acids (DNA or RNA) by several orders of magnitude over a short period of time (hours). PCR has revolutionized DNA technology by allowing any nucleic acid sequence to be simply, quickly and inexpensively generated *in vitro* in both great abundance and at a discrete length. Several publications describe the use of PCR as a rapid detection method for identification of pure cultures of foodborne pathogens (Bej *et al.*, 1990a,b; Bessesen *et al.*, 1990; Böddinghaus *et al.*, 1990). Problems have however been encountered when PCR is applied directly to various foods due to losses in sensitivity (Grant *et al.*, 1993).

At the molecular level, the PCR procedure consists of repetitive cycles of DNA denaturation, primer annealing, and extension by a highly thermostable DNA polymerase. This process is illustrated in Fig. 8.6. Two short DNA probes (also called primers) of typically 15–25 nucleotides flank the DNA segment to be amplified and are repeatedly heat denatured, hybridized to their complementary sequences, and extended with DNA polymerase. The two primers hybridize to opposite strands of the target sequence, such that synthesis proceeds across the region between the primers, replicating that DNA segment. The product of each PCR cycle is complementary to, and capable of binding, primers so the amount of DNA synthesized is doubled in each successive cycle. The original template DNA can be in a pure form and as a discrete molecule or it can be a very small part of a complex mixture of biological substances. After 20 replication

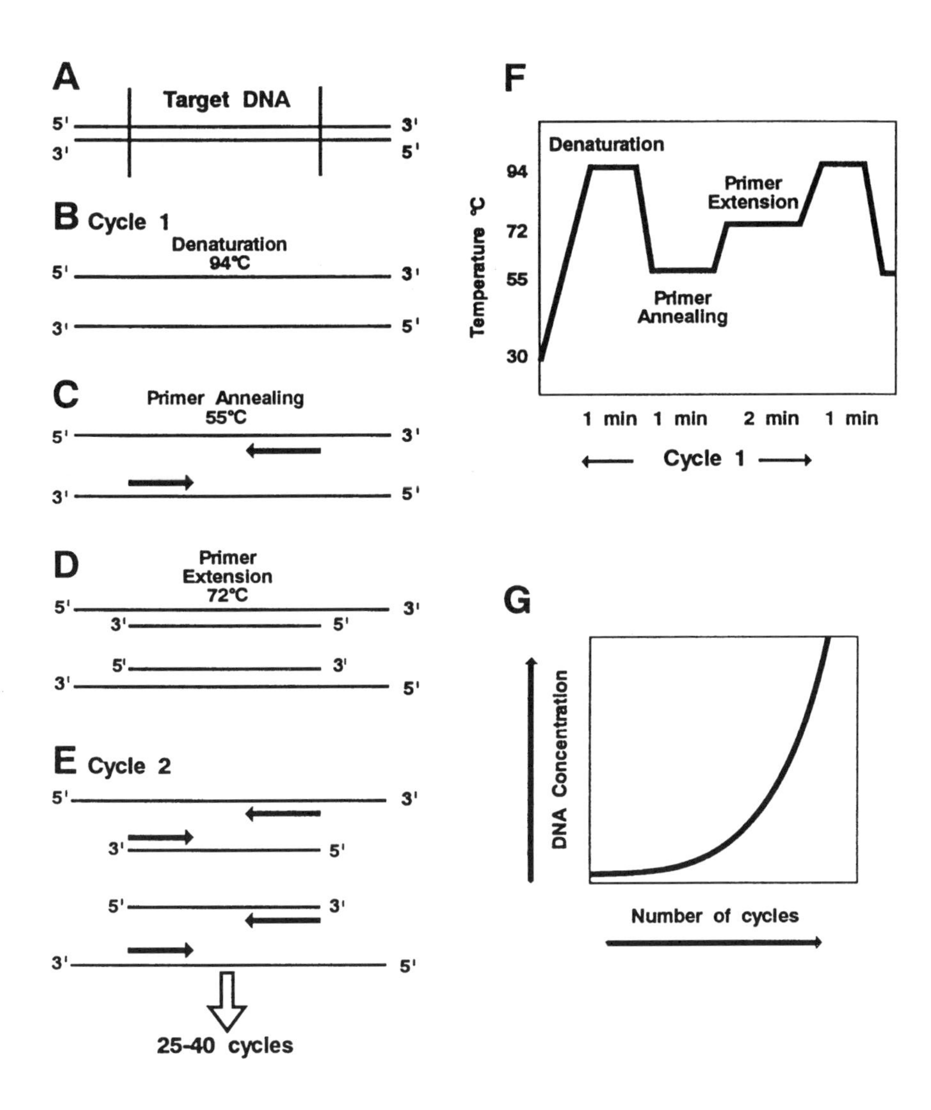

Fig. 8.6 Polymerase chain reaction. Target DNA (A) is heat denatured (B) at 94°C. Primers are annealed (C) at 55°C and then primer extension (D) proceeds at 72°C. The cycle (A–D) is then repeated (E) until 25–40 cycles have been completed. (F) Time–temperature representation of a typical PCR cycle. (G) Quantitation of amplified DNA product. Copies of amplified DNA increase exponentially as number of cycles increases.

cycles, the target DNA will (theoretically) have been amplified over a million-fold. Amplified samples can be size separated by one-dimensional agarose or polyacrylamide gel electrophoresis, stained with ethidium bromide and visualized under ultraviolet light.

To differentiate microorganisms such as yeasts or bacteria at the genus, species or subspecies level, primers are designed from comparative sequence analysis of variable or highly conserved regions of DNA, most notably ribosomal DNA (rDNA). Different probes can be designed to produce PCR products of various sizes, and the presence or absence of electrophoretically size-separated bands can be indicative of the presence or absence of specific genera or species in the test sample. DNA probes may also be chromogenically 'tagged' to produce PCR products of different colours for differentiation of subsets of microorganisms (Embury *et al.*, 1990; Kropp, Fucharoen and Embury, 1991). It should be possible to manipulate the specificity of PCR so that, if desired, nucleic acid differentiation could be accomplished for Gram-positive or Gram-negative bacteria in one scenario, or for *Pediococcus damnosus* or *Obesumbacterium proteus* in another. Such identification could be elucidated at the single cell level within a yeast slurry, a fermenter sample or virtually any sample vulnerable to microbial contamination. To date, although in its relative infancy in the brewing industry, PCR using specific DNA probes has been reported as effective in identifying different isolates of brewery microorganisms (Klijn, Weerkamp and de Vos, 1991; Tsuchiya *et al.*, 1992; Tsuchiya, Kano and Koshino, 1993a,b; DiMichele and Lewis, 1992; Prest, Hammond and Stewart, 1994).

8.14 RANDOM AMPLIFIED POLYMORPHIC DNA PCR

As already noted, highly specific probes which are carefully designed to amplify a specific target sequence typically range in size from 15 to 25 nucleotides. An alternative to the use of specific probes is the use of non-specific or randomly designed probes that range in size from 9 to 12 nucleotides. The use of these smaller, randomly designed or non-specific probes for PCR diagnostics is called random amplified polymorphic DNA or RAPD. RAPD PCR will typically produce several PCR products due to binding at a considerably greater number of different locations in a genome than would be observed with a specific probe. These products can then be electrophoretically size separated to give DNA fingerprints which can be genus or even species specific (Williams *et al.*, 1991; Waugh and Powell, 1992). RAPD fingerprints are dependent on the sequence of the primers, the reaction conditions for each cycle of amplification, and the source of the DNA. This technology has recently been employed to characterize and differentiate brewery microorganisms including members of *Pediococcus, Lactobacillus, Obesumbacterium* spp. as well as ale and lager yeasts (Savard and Dowhanick, 1993; Savard, Hutchinson and Dowhanick, 1993). In place of a mixture of oligonucleotide probes, RAPD PCR can be performed using highly repetitive, specific sequences such as poly-GT oligonucleotides. In the yeast genome, poly-GT sequences are located at the ends of chromosomes and are considered to be 'hot spots' for genetic recombination (Walmsley, Wilkinson and Kong, 1989).

8.15 SUMMARY

Significant advances have been and are being made in the field of rapid microbiological methods. Some of the rapid detection techniques discussed in this chapter still rely on microbial growth until a particular threshold value is attained. When using such techniques, it must be remembered that the concept of 'rapid' remains limited by the growth rate, growth conditions and initial inoculum size of the microbes being analysed. The primary benefit of using these techniques is that elucidation of the presence or absence of microbes can now be performed in only a fraction of the time previously required with classic plating techniques. In addition to detection alone, certain techniques now offer both rapid verification and identification of major beer spoilage microorganisms such as *Lactobacillus, Pediococcus, Obesumbacterium* and other potentially problematic microbes, to both the genus and species level. These methods truly constitute a major breakthrough for the brewing industry.

Rapid methods and automation in the brewing microbiology laboratory is an area of great potential and many exciting new developments can be expected in the coming years. Currently some of the techniques reviewed in this chapter remain better suited for a research and development laboratory rather than a quality control laboratory. However, as design of equipment and reagent kits coupled to advancements in methodology catch up to the needs of new and old customers alike, these advances in detection and identification will be able to quickly, efficiently and cost effectively assist the brewer in maintaining beer of the highest quality and consistency, through minimization of spoilage brought on by microbial contamination.

REFERENCES

Adams, M.R. (1988) *Analytical Proceedings*, **25,** 325.

Adams, M.R., Bryan, J.J. and Thurston, P.J. (1989) *Letters in Applied Microbiology*, **8,** 55.

Anonymous (1975) *Millipore Technical Bulletin AB807*.

Avis, J.W. and Smith, P. (1989) In *Rapid Microbiological Methods for Foods, Beverages and Pharmaceuticals* (eds C.J. Stannard, S.B. Petitt and F.A. Skinner), Blackwell Scientific Publications, Oxford, p. 1.

Back, W. (1987) *Brauwelt International*, **2,** 174.

Barney, M.C. and Helbert, J.R. (1975) *Master Brewers Association of America Technical Quarterly*, **12,** 23.

Barney, M. and Kot, E. (1992) *Master Brewers Association of America Technical Quarterly*, **29,** 91.

Beezer, A.E., Bettelheim, K.A., Al-Salihi, S. and Shaw, E.J. (1978) *Science Tools*, **25,** 6.
Bej, A.K., Mahbubani, M.H., Miller, R. *et al.* (1990a) *Molecular Cellular Probes*, **4,** 353.
Bej, A.K., Steffan, R.J., Dicesare, J.L., Haff, L. and Atlas, R.M. (1990b) *Applied and Environmental Microbiology*, **56,** 307.
Berridge, N.J., Cousins, C.M. and Cliffe, A.J. (1974) *Journal of Dairy Research*, **41,** 203.
Bessesen, M.T., Luo, Q., Rotbart, H.A. *et al.* (1990) *Applied and Environmental Microbiology*, **56,** 2930.
Betts, R.P., Bankes, P. and Banks, J.G. (1989) *Letters in Applied Microbiology*, **9,** 199.
Böddinghaus, B., Rogall, T., Flohr, T. *et al.* (1990) *Journal of Clinical Microbiology*, **28,** 1751.
Bustamante, C., Gurrieri, S. and Smith, S.V. (1993) *Trends in Biotechnology*, **11,** 23.
Campbell, I. and Allan, A.M. (1964) *Journal of the Institute of Brewing*, **70,** 316.
Casey, G.P., Pringle, A.T. and Erdmann, P.A. (1990) *Journal of the American Society of Brewing Chemists*, **48,** 100.
Casey, G.P., Xiao, W. and Rank, G.H. (1988) *Journal of the Institute of Brewing*, **94,** 239.
Colmin, C., Pebay, M., Simonet, J.M. and Decaris, B. (1991) *FEMS Microbiology Letters*, **81,** 123.
Decock, J.P. and Iserentant, D. (1985) *Proceedings of the 20th Congress of the European Brewery Convention, Helsinki*, IRL Press, Oxford, p. 195.
Degre, R., Thomas, D.Y., Ash, J. *et al.* (1989) *American Journal of Enology and Viticulture*, **40,** 309.
De Jong, P., De Jong, C.M., Meijers, R. *et al.* (1986) *Yeast*, **2,** 193.
Delley, M., Mollet, B. and Hottinger, H. (1990) *Applied and Environmental Microbiology*, **56,** 1967.
Dick, E., Wiedman, R., Lempart, K. and Hammes, W.P. (1986) *Chemie Mikrobiologie Technologie der Lebensmittel*, **10,** 37.
DiMichele, L.J. and Lewis, M.J. (1992) Presented at the Second Brewing Congress of the Americas, St Louis, 20–24 September.
Dolezil, L. and Kirsop, B.H. (1975) *Journal of the Institute of Brewing*, **81,** 281.
Donhauser, S., Eger, C., Hubl, T. *et al.* (1992) *Brauwelt*, **132,** 1301.
Dowhanick, T.M. (1990) *Brewers' Digest*, **65,** 34.
Dowhanick, T.M. and Sobczak, J. (1994) *Journal of the American Society of Brewing Chemists*, **52,** 19.
Dowhanick, T., Sobczak, J., Russell, I. and Stewart, G.G. (1990) *Journal of the American Society of Brewing Chemists*, **48,** 75.
Drawert, F. and Bednar, J. (1983) *Journal of Agriculture and Food Chemistry*, **31,** 848.
Embury, S.H., Kropp, G.L., Stanton, T.C. *et al.* (1990) *Blood*, **76,** 619.
Evans, H.A.V. (1982) *Journal of Applied Bacteriology*, **53,** 423.
Evans, H.A.V. (1985) *Food Microbiology*, **2,** 19.
Firstenberg-Eden, R. and Eden, G. (1984) *Impedance Microbiology*, Research Studies Press, John Wiley and Sons, Toronto.
Firstenberg-Eden, R. and Sharpe, A.N. (1991) In *Biotechnology and Food Ingredients* (eds I. Golgberg and R. Williams), Van Nostrand Reinhold, New York, p. 507.
Fleet, G. (1992) *Critical Reviews in Biotechnology*, **12,** 1.
Foster, A. (1994) Presented at the 60th Annual Meeting of the American Society of Brewing Chemists, Toronto, Canada.
Gares, S.L., Whiting, M.S., Ingledew, W.M. and Ziola, B. (1993) *Journal of the American Society of Brewing Chemists*, **51,** 158.
Good, L, Dowhanick, T.M., Ernandes, J.E. *et al.* (1993) *Journal of the American Society of Brewing Chemists*, **51,** 35.
Gram, L. and Søggaro, H. (1985) *Journal of Food Protection*, **48,** 431.
Grant, K.A., Dickinson, J.H., Payne, M.J. *et al.* (1993) *Journal of Applied Bacteriology*, **74,** 260.
Griffiths, M.W. and Phillips, J.D. (1989) In *Rapid Microbiological Methods for Foods,*

Beverages and Pharmaceuticals (eds C.J. Stannard, S.B. Petitt and F.A. Skinner), Blackwell Scientific Publications, Oxford, p. 13.
Haikara, A. and Boije-Backman, S. (1982) *Brauwissenschaft*, **35**, 113.
Haikara, A., Mattila-Sandholm, T. and Manninen, M. (1990) *Journal of the American Society of Brewing Chemists*, **48**, 92.
Harrison, J. and Webb, T.J.B. (1979) *Journal of the Institute of Brewing*, **85**, 231.
Harrison, M., Theaker, P.D. and Archibald, H.W. (1987) *Proceedings of the 21st Congress of the European Brewery Convention, Madrid*, IRL Press, Oxford, p. 168.
Herman, J.P.M., Jakubczak, E., Izard, D. and Leclerc, H. (1980) *Canadian Journal of Microbiology*, **26**, 413.
Hope, C.F.A. and Tubb, R.S. (1985) *Journal of the Institute of Brewing*, **91**, 12.
Hutcheson, T.C., McKay, T., Farr, L. and Seddon, B. (1988) *Letters in Applied Microbiology*, **6**, 85.
Hutter, K.-J. (1978) *Brauwissenschaft*, **31**, 278.
Hutter, K.-J. (1991) *Brauwissenschaft*, **44**, 216.
Hutter, K.-J. (1992a) *Brauwelt International*, **2**, 183.
Hutter, K.-J. (1992b) *Monatsschrift für Brauerei*, **45**, 280.
Hysert, D.W., Kovecses, F. and Morrison, N.M. (1976) *Journal of the American Society of Brewing Chemists*, **34**, 145.
Ingledew, W.M. (1979) *Journal of the American Society of Brewing Chemists*, **37**, 145.
Jago, P.H., Stanfield, G., Simpson, W.J. and Hammond, J.R.M. (1989) In *ATP Luminescence – Rapid Methods in Microbiology* (eds P.E. Stanley, B.J. McCarthy and R. Smither), Blackwell Scientific Publications, Oxford, p. 53.
Johnston, J.R. and Mortimer, R.K. (1986) *International Journal of Systematic Bacteriology*, **36**, 569.
Jorgensen, H.L. and Schultz, E. (1985) *International Journal of Food Microbiology*, **2**, 177.
Kersters, K. and De Ley, J. (1975) *Journal of General Microbiology*, **87**, 333.
Kilgour, W.J. and Day, A. (1983) *Proceedings of the 19th Congress of the European Brewery Convention, London*, IRL Press, Oxford, p. 177.
Klijn, N., Weerkamp, A.H. and de Vos, W.M. (1991) *Applied and Environmental Microbiology*, **57**, 3390.
Koch, H.A., Bandler, R. and Gibson, R.R. (1986) *Applied and Environmental Microbiology*, **52**, 599.
Krause, J.A. and Barney, M.C. (1987) Paper presented at the 53rd Annual Meeting of the American Society of Brewing Chemists, Cincinnati, Ohio.
Kropp, G.L., Fucharoen, S. and Embury, S.H. (1991) *Blood*, **78**, 26.
Kruth, H.S. (1982) *Analytical Biochemistry*, **125**, 225.
Kyriakides, A.L. and Thurston, P.A. (1989) In *Rapid Microbiological Methods for Foods, Beverages and Pharmaceuticals* (eds C.J. Stannard, S.B. Pettit and F.A. Skinner), Blackwell Scientific Publications, Oxford, p. 101.
Lampa, R.A., Mikelson, D.A., Rowley, D.B. *et al.* (1974) *Food Technology*, **28**, 52.
Laurent, F., Drocourt, J.L., Jakobsen, M. *et al.* (1993) *Proceedings of the 24th Congress of the European Brewery Convention, Oslo*, IRL Press, Oxford, p. 487.
Legrand, M. and Ramette, P. (1986) *Cerevisia*, **11**, 173.
Lonvaud-Funel, A., Joyeux, A. and Ledoux, O. (1991) *Journal of Applied Bacteriology*, **71**, 501.
Lonvaud-Funel, A., Guilloux, Y. and Joyeux, A. (1993) *Journal of Applied Bacteriology*, **74**, 41.
Lonvaud-Funel, A., Fremaux, C., Biteau, N. and Joyeux, A. (1991) *Food Microbiology*, **8**, 215.
Martens, F.B., van den Berg, R. and Harteveld, P.A. (1985) *Proceedings of the 20th Congress of the European Brewery Convention, Helsinki*, IRL Press, Oxford, p. 211.
Mattila, T. (1987) *Journal of Food Protection*, **50**, 640.
Mattila, T. and Alivehmas, T. (1987) *International Journal of Food Microbiology*, **4**, 157.

Meaden, P. (1990) *Journal of the Institute of Brewing*, **96,** 195.
Miller, R. and Galston, G. (1989) *Journal of the Institute of Brewing*, **95,** 317.
Miteva, V.I., Abadjieva, A.N. and Stefanova, Tz.T. (1992) *Journal of Applied Bacteriology*, **73,** 349.
Muirhead, K.A. and Horan, P.K. (1984) In *Advances in Cell Culture*, Vol. 3 (ed. K. Maramorosch), Academic Press, New York.
Mullis, K.B., Faloona, S.J. Scharf, S.J. *et al.* (1986) *Cold Spring Harbor Symposium on Quantitative Biology*, **51,** 263.
Navarro, A. Moreno, P. Carbonell, J.V. and Pinaga, F. (1987) *Proceedings of the 21st Congress of the European Brewery Convention, Madrid*, IRL Press, Oxford, p. 465.
Newsom, I.A., O'Donnell, D.C. and McIntosh, J. (1992) *Proceedings of the 22nd Convention of the Institute of Brewing (Australian and New Zealand Section), Melbourne*, Institute of Brewing, London, p. 189.
Niwa, M. (1993) *Brauwelt International*, **11,** 149.
Ogden, K. (1993) *Journal of the Institute of Brewing*, **99,** 389.
Owens, J.D., Konirova, L. and Thomas, D.S. (1992) *Journal of Applied Bacteriology*, **72,** 32.
Paton, A.M. and Jones, S.M. (1971) In *Methods in Microbiology*, Vol. 5A (eds J.R. Norris and D.W. Ribbons), Academic Press, London, p. 135.
Pedersen, M.B. (1983) *Carlsberg Research Communications*, **48,** 485.
Pedersen, M.B. (1985) *Carlsberg Research Communications*, **50,** 263.
Pedersen, M.B. (1986) *Carlsberg Research Communications*, **51,** 163.
Peladan, F. and Leitz, R. (1991) *Proceedings of the 23rd Congress of the European Brewery Convention, Lisbon*, IRL Press, Oxford, p. 481.
Pettipher, G.L. (1985) In *Rapid Methods and Automation in Microbiology and Immunology* (ed. K.-O. Habermehl), Springer-Verlag, New York, p. 629.
Pettipher, G.I., Mansell, R., McKinnon, C.H. and Cousins, C.M. (1980) *Applied and Environmental Microbiology*, **39,** 423.
Pettipher, G.L., Kroll, R.G., Farr, L.J. and Betts, R.P. (1989) In *Rapid Microbiological Methods for Foods, Beverages and Pharmaceuticals* (eds C.J. Stannard, S.B. Pettit and F.A. Skinner), Blackwell Scientific Publications, Oxford, p. 33.
Phillips, A.P. and Martin, K.L. (1985) In *Rapid Methods and Automation in Microbiology and Immunology* (ed. K.-O. Habermehl), Springer-Verlag, Berlin, p. 408.
Phillips, A.P. and Martin, K.L. (1988) *Journal of Applied Bacteriology*, **64,** 47.
Pishawikar, M.S., Sinhal, R.S. and Kulkarni, P.R. (1992) *Trends in Food Science and Technology*, **3,** 165.
Portno, A.D. and Molzahn, S.W. (1975) *Brewers' Digest*, **3,** 44.
Prest, A.G., Hammond, J.R.M. and Stewart, G.S.A.B. (1994) *Applied and Environmental Microbiology*, **60,** 1635.
Querol, A., Barrio, E. and Ramon, D. (1992) *Systematic and Applied Microbiology*, **15,** 439.
Rainbow, C. (1981) In *Brewing Science*, Vol. 2 (ed. J.R.A. Pollock), Academic Press, London, p. 491.
Reyes-Gavilan, C.G. de los, Limsowtin, K.Y., Tailliez, P. *et al.* (1992) *Applied and Environmental Microbiology*, **58,** 3429.
Richards, M. (1970) *Wallerstein Laboratory Communications*, **33,** 97.
Russell, W.J., Farling, S.R., Blanchard, G.C. and Boling, E.A. (1975) In *Microbiology* (ed. D. Schlessinger), American Society for Microbiology, Washington, DC, p. 22.
Salzman, G.C. (1982) In *Cell Analysis*, Vol. 1, (ed. N. Catsimpoolas), Plenum, New York, p. 111.
Savard, L. and Dowhanick, T.M. (1993) Presented at the 59th Annual Meeting of the American Society of Brewing Chemists, Tucson, 12–16 June.
Savard, L., Hutchinson, J. and Dowhanick, T.M. (1993) *Journal of the American Society of Brewing Chemists*, **52,** 62.

Schaertel, B.J., Tsang, N. and Firstenberg-Eden, R. (1987) *Food Microbiology*, **4**, 155.

Simpson, W.J. (1989) *Brewers' Guardian*, **118**, 20.

Simpson, W.J. (1990) In *Proceedings of the Third Aviemore Conference on Malting, Brewing and Distilling* (ed. I. Campbell), Institute of Brewing, London, p. 161.

Simpson, W.J. (1991) *Brewers' Guardian*, **120**, 30.

Simpson, W.J. and Hammond, J.R.M. (1989) In *ATP Luminescence – Rapid Methods in Microbiology* (eds P.E. Stanley, B.J. McCarthy and R. Smither), Blackwell Scientific Publications, Oxford, p. 45.

Simpson, W.J., Hammond, J.R.M., Thurston, P.A. and Kyriakides, A.L. (1989) *Proceedings of the 22nd Congress of the European Brewery Convention, Zurich*, IRL Press, Oxford, p. 663.

Stanley, P.E. (1992) *Journal of Bioluminescence and Chemiluminescence*, **7**, 77.

Steffan, R.J., Breen, A., Atlas, R.M. and Sayler, G.S. (1989) *Canadian Journal of Microbiology*, **35**, 681.

Stenius, V., Majamaa, E., Haikara, A. *et al.* (1991) *Proceedings of the 23rd Congress of the European Brewery Convention, Lisbon*, IRL Press, Oxford, p. 529.

Takamura, O., Nakatani, K., Taniguchu, T. and Murakami M. (1989) *Proceedings of the 22nd Congress of the European Brewery Convention, Zurich*, IRL Press, Oxford, p. 148.

Traganos, F. (1984) *Cancer Investigation*, **2**, 149.

Tse, K-M. and Lewis, C.M. (1984) *Applied and Environmental Microbiology*, **48**, 433.

Tsuchiya, T., Fukazawa, Y. and Kawakita, S. (1965) *Mycopathologia et Mycologia Applicata*, **26**, 1.

Tsuchiya, Y., Kano, Y. and Koshino, S. (1993a) *Journal of the American Society of Brewing Chemists*, **51**, 40.

Tsuchiya, Y., Kano, Y. and Koshino, S. (1993b) *Journal of the American Society of Brewing Chemists*, **52**, 95.

Tsuchiya, Y., Kaneda, H., Kano, Y. and Koshino, S. (1992) *Journal of the American Society of Brewing Chemists*, **50**, 64.

Umesh-Kumar, S. and Nagarajan, L. (1991) *Folia Microbiologica*, **36**, 305.

Unkel, M. (1990) *Brauwelt*, **4**, 100.

Vancanneyt, M., Pot, B., Hennebert, G. and Kersters, K. (1991) *Systematic and Applied Microbiology*, **14**, 23.

Van Vuuren, H.J.J. and Van Der Meer, L.J. (1987) *American Journal of Enology and Viticulture*, **38**, 49.

Van Vuuren, H.J.J. and Van Der Meer, L.J. (1988) *Journal of the Institute of Brewing*, **94**, 245.

Van Vuuren, H.J.J., Kersters, K., De Ley, J. and Toerien, D.F. (1981) *Journal of Applied Bacteriology*, **51**, 51.

Vezinhet, F., Blondin, B. and Hallet, J.N. (1990) *Applied Microbiology and Biotechnology*, **32**, 568.

Vidgren G., Palva, I., Pakkanen, R. *et al.* (1992) *Journal of Bacteriology*, **174**, 7419.

Walmsley, R.M., Wilkinson, B.M. and Kong, T.H. (1989) *Bio/Technology*, **7**, 1168.

Waugh, R. and Powell, W. (1992) *Trends in Biotechnology*, **10**, 186.

Wetmur, J.G. (1991) *Critical Reviews in Biochemistry and Molecular Biology*, **26**, 227.

Whiting, M., Crichlow, M., Ingledew, W.M. and Ziola, B. (1992) *Applied and Environmental Microbiology*, **58**, 713.

Williams, J.G.K., Kubelik, A.R., Livak, K.J. *et al.* (1991) *Nucleic Acid Research*, **18**, 6531.

Winter, F.H., York, G.K. and El-Nakha, H. (1971) *Applied Microbiology*, **22**, 89.

Yasui, T., Taguchi, H. and Kamiya, T. (1989) *Proceedings of the 22nd Congress of the European Brewery Convention, Zurich*, IRL Press, Oxford, p. 190.

Yasui, T., Taguchi, H. and Kamiya, T. (1992) *Proceedings of the 22nd Convention of the Institute of Brewing (Australian and New Zealand Section), Melbourne*, Institute of Brewing, London, p. 190.

CHAPTER 9

Methods for the rapid identification of microorganisms

C.S. Gutteridge and F.G. Priest

9.1 WHAT IS IDENTIFICATION?

Identification is correctly defined as an activity leading to the assignment of unidentified microorganisms to a particular class in a previously made classification (Norris, 1980). This definition notes the link between identification and classification, which is the ordering of microorganisms into groups. It is often emphasized that satisfactory schemes for identification can only be based on good classifications (Sneath, 1972), although in identification, speed is imperative and second only in importance to accuracy (Steel, 1962). Nevertheless, identification is understood by most microbiologists to mean the naming of an unknown culture against a developed taxonomic scheme. This activity provides a vital framework that allows microbiologists to recognize when they are working with the same microorganism and is thus fundamental to the pursuit of the science of microbiology.

Identification does need to be clearly distinguished from the process of constructing taxonomic groups. Once groups are known, individual characters can be weighted and used predictively. The production of keys and diagnostic tables and methods for discriminating between organisms that are easily confused is a traditional task for the diagnostic microbiologist.

Microbial identification is at its most intense in the clinical diagnostic laboratory where the daily workload of a medium-sized hospital laboratory may be between 170 and 350 samples (Bascomb, 1980). Only 10–30% of the specimens tested actually contain a pathogen and these are subsequently processed further. Testing for antibiotic sensitivity is the priority,

Brewing Microbiology, 2nd edn. Edited by F. G. Priest and I. Campbell.
Published in 1996 by Chapman & Hall, London. ISBN 0 412 59150 2

but pathogens are identified because this sometimes aids diagnosis and treatment, but more often provides epidemiological information to monitor the spread of disease. It is not surprising that the main developments in rapid microbiological identification in the past two decades have been targeted at the clinical laboratory and clinical problems.

Outside of the clinical laboratory, few microbiologists face the same intensity of identification problems. In industrial laboratories, unless they are specifically involved in screening for the isolation of organisms with novel properties, identification tasks are kept to a minimum. The food and beverage industry spends millions of pounds annually on microbiological testing, and although this is primarily enumeration, it does include the detection of specific organisms that are either potential pathogens or are involved in spoilage or biodeterioration problems.

Jarvis (1985) has commented that the food microbiologist would be prepared to carry out more identification if time could be saved on enumeration. There is much to be gained from a detailed understanding of the distribution of microorganisms in the food processing or brewing environment but surveys are rarely undertaken because of the resources they divert.

A further aspect of the identification problem in industrial laboratories is that it is often aimed at key properties and may not be designed along purely taxonomic lines. For example, in the study of the spoilage of chemically preserved beverages, the recognition that a yeast is preservative resistant is important. Preservative resistance is not necessarily exclusive to one genus or species. Thus, recognizing preservative resistance is a type of pragmatic identification that does not relate to a formally established taxonomy. For such tasks it is perhaps better to use the term characterization to describe the search for key properties in microorganisms. In this chapter methods for identifying and characterizing microorganisms are reviewed and techniques that offer rapid identification are described in more detail. The methods discussed are drawn from all areas of microbiology and examples are not restricted to the brewing industry. Future trends in the identification of microorganisms are considered.

9.2 LEVELS OF EXPRESSION OF THE MICROBIAL GENOME

A suitable scheme for relating the various techniques used to identify microorganisms was proposed by Norris (1980) and is shown in Table 9.1. Four levels of expression of the genetic information in microorganisms can be recognized. At Level 1, genetic information is expressed in terms of the DNA molecule and the genome of the cell. One of the earliest and simplest methods to characterize DNA is measurement of the base composition of the chromosome in terms of mol% guanine and cytosine (mol% G + C). This is not particularly useful since two organisms with the same

Table 9.1 Techniques for studying the genetic information in microbial cells at different levels of expression (adapted from Norris (1980))

Level		*Techniques*
1	The genome	DNA/DNA, DNA/RNA hybridization Gene probes %G + C Plasmids Phage
2	Proteins	Amino acid sequencing Electrophoresis Immunology
3	Cell components	Amino acid pools Light-scattering techniques Cell wall composition Chromatography Spectroscopy Bacteriocins Phage
4	Morphology and behaviour	Microscopy Motility Enzyme tests Physiology Nutrition

G + C content need not necessarily be related. For example cows and rats have about the same chromosomal G + C contents of around 44%. Nevertheless, if two organisms have very different base compositions, differing by more than 5%, they cannot be related. This negative implication of base composition can often be useful to exclude certain taxa when identifying difficult organisms. This is particularly the case if base composition can be detected rapidly and simply with minimum sample preparation, for example using differential scanning calorimetry (Miles, Mackey and Parsons, 1986) or high performance liquid chromatography (HPLC) (Tamaoka, 1994).

Complete nucleotide sequencing of the genetic material in a cell would provide an absolute identification but, although it might be feasible with modern genetic techniques, it is certainly not practical. Separation of the DNA molecule into single strands followed by hybridization with single strands from another organism can be used to produce a measure of the similarity of base sequences between DNAs of different microorganisms and hence offers a direct measure of relatedness. Indeed, DNA hybridization forms the basis of the generally accepted species definition in bacterial systematics; strains belonging to the same species should possess more than 70% DNA sequence homology (Grimont, 1988). Similar techniques can be used to compare DNA and RNA. These hybridization techniques have made an important contribution to microbial taxonomy over the last two decades. They have also spawned the development of gene probes (DNA probes) which are unique nucleotide sequences that will only hybridize with a specific gene(s) of interest. If the sequence is chosen judiciously its detection within the genome of a microorganism can be used as an accurate means of identification (Falkow, 1985; Schleifer, 1990).

At Level 2, the genetic information is expressed as proteins. Amino acid sequencing, like nucleotide sequencing, is expensive and time consuming, and is not practical for rapid identification. In contrast, electrophoresis, i.e. determining the ability of protein molecules to migrate under the influence of an electric field through a gel, the pores of which are of molecular dimensions, is a technique with wide applicability to taxonomic and identification problems. The electrophoresis of whole-protein extracts of microorganisms can be used as an identification method which, while it is not necessarily rapid, can be efficient in terms of sample throughput (Kersters *et al.*, 1994). Studies on the molecular structure of particular proteins are valuable for taxonomic studies but, again, are not practical for identification. Immunology is often directed at proteins and is an important diagnostic tool.

Level 3 is the cell composition level, which can be addressed by a host of analytical techniques applied to whole cells or particular cell extracts. For example, amino acid pools, cell wall structure and cellular and membrane lipids can be profiled using techniques such as gas chromatography (GC), HPLC and thin layer chromatography (TLC). Whole cells can be volatilized by pyrolysis and the products detected by pyrolysis gas chromatography (Py-GC) or pyrolysis mass spectrometry (Py-MS). Mass spectrometry *per se* can be used to identify precise structures for diagnostic purposes. In addition to MS, a range of physical techniques has been examined with a view to microbial characterization, including circular intensity differential scattering (CIDS), flow cytometry, impedance measurements, microcalorimetry, nuclear magnetic resonance and infrared (IR) and Raman spectroscopy (Nelson, 1991). Several of these require expensive and highly specialized equipment and we shall discuss later only the pyrolysis and IR techniques since these have most promise for strain recognition and typing. Immunology can also be targeted at non-protein cell components and sensitivity to bacteriophage and bacteriocins may also be related to cell structure.

The final level of expression, Level 4, has been called morphology and behaviour and is the level at which most conventional identification methods operate. Studies on the presence or absence of particular enzymes, either directly or by the detection of end-products, come into this category. It includes all the multi-test systems that are marketed to standardize the examination of microbial behaviour and to extend the range of tests available. Morphological studies, usually by microscopy, are also tests performed within Level 4.

The remainder of the chapter will be devoted to a more detailed examination of developments in identification techniques at each of these four levels. It is, however, important to recognize that the longer term aim of diagnostic microbiology must be to devise rapid methods that can be applied generally (i.e. to a wide range of microorganisms).

9.3 IDENTIFICATION AT THE GENOMIC LEVEL

We have noted above the ability of single-stranded DNA to hybridize with complementary DNA and how this can lead to the concept of the gene probe (DNA probe) which is surely one of the most powerful diagnostic tools to be introduced in the past decade. The attraction to the problem of microbial identification is that a probe consisting of > 20 nucleotides can be shown to be statistically unique and if it hybridizes with part of the microbial genome then this can be used to give an identification of unparalleled accuracy. The basic steps in the nucleic acid probe test are:

1. design of probe;
2. labelling of probe;
3. preparation of target DNA;
4. hybridization;
5. detection of hybrids.

Some of these steps will be addressed briefly here to give the reader an idea of the processes involved and to stress that they are becoming increasingly simple. Such hybridizations are carried out by undergraduate students in the International Centre for Brewing and Distilling and could be similarly performed in the brewery laboratory with commercial kits and other proprietary reagents. For a fuller discussion of the techniques, the reader should consult recent reviews such as those written by Schleifer (1990), Schleifer, Ludwig and Amann (1993) and Amann and Ludwig (1994).

9.3.1 The probe

There are two broad classes of probe, those directed to unknown parts of the chromosome and those targeted to specific genes. The first class is seldom used today, but the most typical example is the use of chromosomal DNA itself. For microorganisms conforming to the genomic definition of a species (i.e. strains showing more than 70% DNA sequence relatedness) chromosomal DNA can be used as the probe since it should possess less than 70% (and in practice usually only 30% or less) sequence homology with DNA from other species of microorganism. Thus the DNA should hybridize strongly with DNA from members of the same species and weakly with all other DNA samples. This is generally found to be the case and by inclusion of suitable control DNA samples, accurate identification by hybridization can be obtained (e.g. for the food poisoning bacterium *Campylobacter*; Chevrier *et al.*, 1989).

The specificity of chromosomal DNA is insufficient for accurate identification of closely related bacteria. Identification of such organisms requires specific gene fragments. One possibility is to clone random pieces of DNA and test them for specificity. There are many examples of this approach, and it has been used successfully for *Lactobacillus* (Delley, Mollet and Holtinger, 1990; Pilloud and Mollet, 1990) and *Streptococcus* (Schmidhuber, Ludwig and Schleifer, 1988).

Probes are most often directed at a particular locus in the target DNA. This could be the gene for a particular metabolic trait, a virulence factor (often used in clinical diagnostic microbiology) or a gene such as a ribosomal RNA (rRNA) gene. Probes aimed at protein coding genes may be derived from the cloned gene or, if this is not available but the sequence is published, an oligonucleotide can be synthesized which is complementary to part of that gene. One simple way to make a probe to any published gene sequence is to amplify the gene from chromosomal DNA using the polymerase chain reaction (PCR) and to include a labelled nucleotide in the reaction. In this way large amounts of labelled probe can be made in a single PCR. Typical probes to virulence factors include those for the haemolysin of *Listeria monocytogenes* (Datta *et al.*, 1988) and the enteropathogenic toxins of *Escherichia coli* (Sommerfelt *et al.*, 1988).

The 16S and 23S rRNA molecules are particularly suitable as targets for probes because they are present in high copy number (10^4–10^5 molecules per bacterial cell) thus greatly improving the sensitivity of the hybridization. Within these genes there are regions which are highly variable and differ significantly between species. These regions are valuable for species identification. Other areas in the genes are more conserved and sequences are identical within all species in a genus. These parts are therefore useful for generic probes. This combination of factors has led to identification through rRNA gene probes becoming very popular, and Schleifer, Ludwig and Amann (1993) list numerous applications of this technology.

9.3.2 Labelling of the probe and detection

One obstacle to the widespread introduction of DNA hybridization methods into the food industry was that the probes were initially labelled with radioisotopes for detection purposes. Although for some purposes radioisotopes are still preferred, giving greater sensitivity of detection, for most routine applications simple and robust non-radioactive labelling and detection systems are generally used. Two popular systems available as commercially produced kits are based on biotin and digoxigenin (a plant steroid) incorporation into the probe. After the hybridization, labelled probe is detected by enzyme (e.g. alkaline phosphatase) conjugated avidin or antibody reactions respectively. The bound antibody is then detected colorimetrically. These systems have improved markedly over the past decade and now offer extreme sensitivity, simplicity of operation and none of the hazards associated with radioactivity.

9.3.3 The target

The target DNA can be purified directly from colonies on an isolation or quality control Petri dish of microorganisms. In these 'colony hybridizations', the microorganisms are removed to a nitrocellulose or nylon membrane, lysed and the DNA partially purified by treatment of the membrane with protease and detergent. The DNA is then bound to the membrane by heat or UV treatment and hybridized to the labelled probe. After development, hybridizing colonies are revealed.

Colony hybridizations often suffer from high background signals due to cellular material binding the probe in a non-specific way. In these circumstances, it is better to purify DNA from microbial cultures using a routine procedure. The DNA is then denatured and spotted onto a membrane using a 'dot-blot' or 'slot-blot' manifold. These pieces of apparatus allow the immobilization of large numbers of DNA samples on a membrane in a specific array of dots (or slots). The membrane can then be hybridized with a specific probe DNA and positive hybridization reactions are revealed as shown in Fig. 9.1.

An innovative detection/identification procedure is the reverse dot-blot hybridization which has particular application in quality control procedures. A sample of beer could be collected and 'forced' to provide microbial biomass. Total DNA is then prepared and labelled with a non-radioactive label using a standard kit. The membrane is impregnated with dots of denatured DNA from reference organisms such as representative lactobacilli or spoilage yeasts. The labelled DNA from the sample is hybridized with the filter and any bound DNA is subsequently detected. Any hybrids such as the *Lactobacillus brevis* DNA showing up indicate that *L. brevis* was present in the original sample.

The biggest barrier to the adoption of these hybridization techniques in the quality control or research laboratory is the perceived difficulty of working with molecular techniques. We stress that much of the complication of the techniques has been removed with the introduction of labelling and detection kits by most of the major biochemical manufacturers. Once a potential specific target has been identified, if the DNA sequence is available a probe can be readily prepared by PCR and a highly specific, economically realistic detection/identification procedure can be developed.

At present only one line of products for use in food microbiology is marketed as a complete probe kit: the Gene-Trak Systems (Integrated Genetics, Framingham, MA, USA). These are based on an ingenious 'dipstick' technology targeted at regions of the 16S rRNA. The probe has attached to it a 'tail' carrying a fluorescein label. Dipsticks specific for *E. coli*,

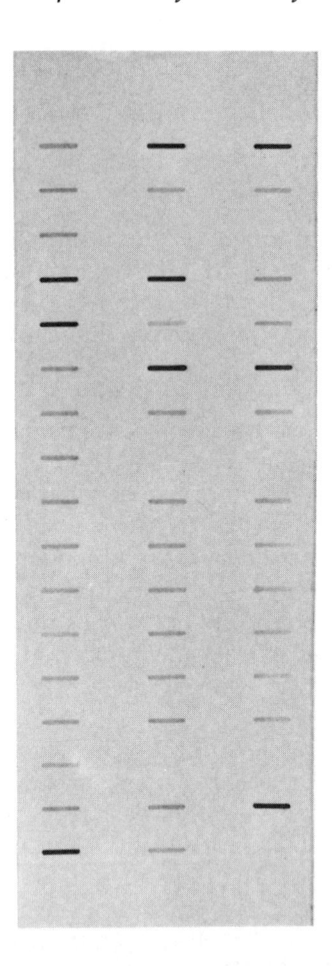

Fig. 9.1 Example of a slot-blot hybridization reaction. Chromosomal DNA samples were loaded onto a membrane using a slot-blot manifold and hybridized with a non-radioactively (digoxigenin) labelled DNA probe. Positive reactions are evident as dark bands, negative reactions as light (background) bands.

Salmonella and *Listeria* are currently marketed together with some clinically relevant bacteria such as *Yersinia* and *Staphylococcus*.

9.4 TECHNIQUES FOR EXAMINING PROTEINS

9.4.1 Immunology

Immunological techniques are not applied exclusively to proteins but this is a convenient time to consider them as many of the most successful immunological identification systems detect specific proteins. To chronicle fully all the exciting developments in immunology over the past decade and forecast their impact on diagnostic microbiology is beyond an article of this sort. The principal developments have been well publicized. Monoclonal antibody technology (Kohler and Milstein, 1975) provides the means of mass producing antibodies of defined specificity, and enzyme-linked immunoassays (ELISA) provide a convenient, non-radioisotopic means of detecting antigen–antibody reactions (for an introduction to monoclonal antibodies see Macario and Conway de Macario (1985), for ELISA techniques see Voller, Bidwell and Bartlett (1980) and Claussen (1988) and for general applications of serology in bacterial identification see Bowden (1993)).

As an example of the developing impact of immunology on microbial identification we can consider the problem of detecting the food poisoning organism, *Salmonella*. The fluorescent antibody (FA) technique was one of the first immunological techniques applied to the detection of salmonellae (Thomason, Cherry and Moody, 1957; Thomason, 1980). The principle of the FA is the application of conjugates of *Salmonella* antiserum with fluorescein isothiocyanate to smears of bacteria on microscope slides. When viewed against a dark background, *Salmonella* cells fluoresce orange-green (Schrade, 1984). Methodology based on FA was approved for use in the USA in 1976 following improvements in the quality of the reagents over a number of years. However, even with improved antisera, the FA technique has not become widely used. The main problem has been the high level of false positives (5–9%), exacerbated by the obvious difficulties in using a microscopical technique to examine an antigen–antibody reaction on hundreds of samples per day.

A different approach was reported by Sperber and Diebel (1969), who called their technique enrichment serology. Again, problems with the specificity of antisera gave an unacceptable level of false positives (13–15%) and the technique failed to detect some serotypes of *Salmonella*, including *S. agona* (Mohr, Trenk and Yeterian, 1974).

Krysinski and Heimsch (1977) developed the first enzyme immunoassay for salmonellae, but despite improvements by Minnich (1978) and Swaminathan and Ayres (1980) the antibody preparation still suffered from cross-reactions with other enteric bacteria.

The first enzyme immunoassays for salmonellae to employ a monoclonal antibody were reported by Robison, Pretzman and Mattingly (1983). The MOPC 467 protein was originally isolated by Potter (1970) and was later found to be specific for a flagella determinant found on several *Salmonella* strains (Smith, Miller and Whitehead, 1979). The MOPC 467 monoclonal could detect, specifically, salmonellae in mixed cultures with no cross-reactions with other enteric bacteria (Mattingly and Gehle, 1984). Unfortunately it could not detect all the known *Salmonella* serotypes. Mattingly (1984) developed a second monoclonal (6H4) by immunizing mice with strains of salmonellae that could not be detected by the MOPC 467 system. Mattingly *et al.* (1985) describe the ELISA system resulting from these developments. The monoclonal antibodies are attached to a solid-phase matrix (a ferrous metal bead) which can be transferred between the wells of a microtitre plate for exposure to the antigen, the enzyme-conjugated antibody and the substrate. The system uses a horseradish peroxidase enzyme producing a green colour if salmonellae are present.

The ELISA is used by first raising the *Salmonella* population to 10^5–10^6 cells ml^{-1} by the normal pre-enrichment/selective enrichment procedures, and then the assay can be applied directly, taking 2–4 h to perform. Conventional procedures take 24–48 h to isolate salmonellae and confirm suspect colonies by biochemical and serological techniques.

The development of a usable ELISA system for salmonellae is typical of the enormous potential for modern immunological methods. Monoclonal techniques simplify the search for specific antisera which are required for accurate typing. ELISA techniques simplify the detection of antigen–antibody reactions, replacing more difficult techniques such as radio-immunoassays. The main limitation on the use of immunological techniques for microbial identification is the effort, in terms of manpower, required for screening clones and developing assays. This will probably limit applications to those pathogens or organisms of industrial importance that are identified with a frequency which will support the commercialization of an ELISA kit. Indeed the plethora of commercial kits available today based on monoclonal antibodies and ELISA methodology testifies to the effectiveness of the technology (Sharpe, 1994).

9.4.2 Electrophoresis

Electrophoresis of whole cell protein extracts is another technique that has found its major application in taxonomic and characterization studies, but is becoming increasingly important in rapid identification. The expression of the microbial genome produces more than 2000 protein molecules within a microbial cell. Techniques that can address the characterization of these proteins obviously have enormous potential in diagnostic microbiology. Jackman (1985) has proposed a standard polyacrylamide gel electrophoresis (PAGE) technique which has been adopted by the National Collection of Type Cultures as the foundation of a system for the computer identification of medically important bacteria. A key feature of the technique is the standardization (as far as is practicable) of the conditions under which the organisms are first grown and then harvested and disrupted, releasing their proteins. Jackman carried out cell disruption in sodium dodecyl sulphate (SDS) buffer at 100°C, which has the added benefit of killing the cells, allowing pathogens to be handled safely. This is followed by electrophoresis in a homogeneous, discontinuous, SDS-containing alkaline buffer. A similar system has been developed in the laboratory of

K. Kersters (Vauterin, Swings and Kersters, 1993; Pot, Vandamme and Kersters, 1994). Both systems result in gels which are scanned by a densitometer following standardization by reference to marker proteins. The results from both approaches are complex banding patterns in which each band represents several protein species since the resolution of one-dimensional electrophoresis is insufficient to separate all proteins in the cell. The quantitative data from the densitometer are amenable to analysis by computer taxonomic techniques to provide cluster or species fingerprints which can be stored in a computer. For identification, unknown organisms can be compared with a library of fingerprints.

The taxonomic studies to which electrophoresis has been applied have been reviewed by Kersters *et al.* (1994). The most common findings are that electrophoretic patterns are most discriminatory at the species, subspecies or biotype levels. The main advantages of the electrophoretic approach are the efficiency with which large numbers of strains can be compared and, provided a cell disruption method can be found, its applicability to different types of microorganism. The main disadvantage is that it requires a large number of cells for the preparation of the initial extract.

Within the brewery, protein electrophoresis is particularly suited to identification of lactic acid bacteria. It has proven to be useful for identification of the *L. acidophilus* (Pot *et al.*, 1993) and *L. casei* groups (Hertl *et al.*, 1993) as well as leuconostocs (Dicks, Van Vuuren and Dellaglio, 1990).

A novel electrophoretic technique, introduced by Robert Silman of St Bartholomew's Hospital, London, was commercialized as the AMB-ID system (V.A. Howe, London). In this technique, single colonies of the organism to be identified are incubated from 30 min to 2 h (depending on the type of organism) in a medium containing ^{35}S-methionine. The methionine is taken up during protein synthesis and radiolabels the proteins. Growth and protein synthesis is halted by the addition of a buffer and the cells are disrupted by heating at 100°C for 2 min. The extract is then electrophoresed for 1 h in a gradient polyacrylamide gel system. The protein bands are detected and recorded by a specially developed beta scanner producing a fingerprint which can be matched against a computer library.

9.5 METHODS THAT EXAMINE ASPECTS OF CELL COMPOSITION

A host of analytical techniques can be applied to the study of the cellular composition of microorganisms. Although originally these techniques were applied largely to problems of taxonomy and characterization they are increasingly being used for rapid identification. Many of them have great potential to become more important in diagnostic microbiology and are perhaps held back by the lack of analytical expertise and equipment available to most microbiologists.

9.5.1 Physical techniques

One of the most intriguing techniques to be suggested as a rapid identification method is circular intensity differential scattering (CIDS) (Gregg *et al.*, 1985). CIDS involves the differential scattering of left and right circularly polarized light, generated by a laser and allowed to impinge on a microbial suspension in a cuvette. A detector measures the intensity of light at different scattering angles producing a spectrum. Spectra can be accumulated at different wavelengths. The physical basis of the CIDS spectra is still poorly understood but has been proposed to be the three-dimensional 'packaging' of helical molecules, principally those in the microbial genome.

Gregg *et al.* (1985) envisage future instruments, linked to a flow cytometer, that will separate cells so that organisms can be examined one at a time and a CIDS spectrum obtained. There is little published work to justify the optimism for CIDS as a general identification technique. Salzman, Griffith and Gregg (1982) have shown that different CIDS spectra were obtained from five different crude influenza virus preparations, two strains of *Salmonella typhimurium* and three strains of *Escherichia coli*.

Flow cytometry (for reviews see Muirhead, Horan and Poste (1985) and Nelson (1991)) is another physical technique with promise for microbial identification. A flow cytometer is a device in which cells are introduced into the centre of a fast-moving fluid stream and are forced to flow in a single file out of a 50–100 μm orifice at a uniform speed within a layer of fluid. The cells pass a measurement station where they are illuminated by a light source (usually a tunable laser). Measurements are made at rates of 104–106 cells min^{-1}. When a particle in the flow stream passes through the light beam, the illuminating light is elastically scattered by the cell and the intensity of scatter at different angles can yield information about cell size, shape, viability, density and surface morphology. In most applications of flow cytometry, fluorochromes are used to label the cellular component(s) of interest. The fluorescence emitted by these molecules, when excited by the illuminating laser beam, can yield information on the expression of target molecules within single cells. Flow sorters are then able to separate physically the cells of interest from the heterogeneous population.

Phillips and Martin (1985) provide an example of the potential use of flow cytometry in diagnostic microbiology. Using a fluorescein-conjugated antibacterial IgG, they were able to separate *Bacillus anthracis* spores above a background of Gram-negative bacteria.

CIDS and flow cytometry are expensive techniques in terms of equipment costs and have had little or no impact on diagnostic microbiology. However, they are techniques that can be applied to a heterogeneous population of cells with the potential to separate and type each individual member of the population. Thus they are rapid techniques that might be applied to mixed cultures, whereas the vast majority of identification techniques will only work with pure cultures.

9.5.2 Analytical techniques

Goodfellow and Minnikin (1985) have pointed out that the nature of the layer binding the bacterial cell has held a special place in microbial systematics since the invention of the Gram stain over 100 years ago. Salton (1952) was the first to introduce methods for isolating and purifying cell walls. Since then the walls of many different bacteria have been analysed and several procedures developed to detect specific components. Examples of these are shown in Table 9.2. O'Donnell, Minnikin and Goodfellow (1985) have shown how many of these analytical procedures can be integrated to generate a chemo-systematic profile of actinomycetes. However, most of the analytical procedures that have been applied to microbial characterization are slow because they require the production of a small quantity of biomass for extraction and the methods are generally not automated. Consequently, they have been largely applied to classical taxonomic studies.

Gas chromatographic techniques, however, have been used extensively for microbiological characterization. Good reviews have been published by Odham, Larsson and Mardh (1984), Jantzen and Bryn (1994) and Embley and Wait (1994). Gas chromatography is a general-purpose analytical technique for separating covalent compounds of moderate molecular weight. In simple terms, a sample is introduced into a stream of heated carrier gas (the mobile phase) where its components are volatilized and swept through a chromatography column containing a stationary phase. In the column the components are selectively retarded according to their interactions with the stationary phase. The separated components enter a detector and are recorded and quantitated as chromatographic peaks. Gas chromatographs can be readily connected to mass spectrometers (GC-MS) to effect the identification of the separated components.

As discussed, GC techniques can be used to generate information about many constituents of microbial cells. For the identification of microorganisms they have, until recently, found particular application in anaerobic bacteriology where strict anaerobes can be identified by the determination of their fermentation end-products. The VPI *Anaerobe Laboratory Manual* (Holdeman, Cato and Moore, 1977) describes many of the essential procedures. In some situations GC can be applied to extracts of infected materials to make a direct diagnosis of the contaminating organism. Examples include the presence of short-chain fatty acids in pus specimens containing anaerobes (Phillips, Taylor and Eykyn, 1980) and in amniotic fluid (Gravett *et al.*, 1982).

The major development in GC for identification purposes, however, is the introduction of complete microbial identification systems based on this technology. Originally introduced by Hewlett Packard and now marketed as the MIDI system (Newark, Delaware, US), the basis of the method is the extraction of whole cell fatty acids and GC profiling of their methyl esters

Table 9.2 Chemotaxonomic information from bacterial cell envelopes (adapted from Goodfellow and Minnikin (1985))

Site	*Component*	*Techniques*	*Reference*
Plasma membrane	Isoprenoid quinones	HPLC, MS, RPTLC	Collins (1985)
	Lipid-soluble pigments		
	Lipoteichoic acids		
	Polar lipids		
	Fatty acids	GC	Moss (1985)
	Isoprenoid ethers	HPLC, MS, RPTLC	Jantzen and Bryn (1985)
	Other long-chain components		
	Proteins		
Walls	Peptidoglycans and analogues	GC	O'Donnell, Minnikin and Goodfellow (1985)
	Polysaccharide sugar patterns		
	Teichoic acids and analogues		
Outer membrane Gram-negative bacteria	Lipopolysaccharides (K and O antigens)		
	Polar lipids	TLC	Dobson *et al.* (1985)
Outer membrane Gram-positive bacteria	Bound lipids–mycolic acids	TLC	Dobson *et al.* (1985)
	Free lipids		
	Glycolipids		
	Sulphoglycolipids		
	Waxes		

HPLC, high-performance liquid chromatography; MS, mass spectrometry; RPTLC, reverse-phase thin layer chromatography; GC, gas chromatography; TLC, thin layer chromatography.

(FAME profiles). The methodology is based on the work of C.W. Moss at the Centers for Disease Control, Atlanta, USA (Moss, 1981; Moss and Dees, 1976, 1978; Moss, Dees and Guerrant, 1980) and uses standardized cultivation of the microorganisms and preparation of the FAME samples. Bacteria are grown for 24 h as spread plates and the cells are saponified in a strong base. After saponification the solution is acidified and the liberated fatty acids are methylated to increase their volatility for GC. The methyl esters are extracted and 1 μl of a washed extract injected on to a 50 m methylsilicone fused silica capillary column. The system is completed by a computer for recording data and matching FAME profiles against a reference library.

The library covers a number of Gram-negative and Gram-positive genera but the technique can be applied to any microorganism and individual reference libraries constructed. The production of FAME profiles takes approximately 60–90 min (60 min sample preparation time) although samples are prepared and analysed in batches and the actual preparation time per sample is quoted at less than 4 min. Examples of FAME profiles for two Gram-negative bacteria are shown in Fig. 9.2 (Miller, 1984). The technique of FAME profiling has wide-ranging potential as an identification tool as evidenced by the introduction and success of the MIDI system. Possible disadvantages are, firstly, that running any GC system requires considerable dedication and analytical expertise, and secondly, it is still not clear whether FAME profiles vary sufficiently to allow typing of strains as well as genera and species.

Some analytical techniques can be applied directly to whole cells with no requirement to make an extract. One example is infrared (IR) spectrometry which has received increasing attention as a simple procedure to obtain fingerprints of microorganisms (for reviews see Giesbrecht *et al.*, 1985; Nichols *et al.*, 1985; Magee, 1993; Naumann, Helm and Schultz, 1994). Attempts to use IR spectroscopy to characterize microorganisms were made in the 1950s and 1960s (e.g. Norris, 1959), but no successful routine technique could be established because of the time-consuming preparation procedures, the low scanning speeds of the conventional spectrometers and the difficulties in discriminating the complex and relatively similar spectra. Many of these problems have been overcome with the modern Fourier transform (FT) IR instruments.

The IR spectrum of any compound is known to express a unique 'fingerprint' and it is this characteristic that allows IR spectroscopy to be used for identification using spectral data libraries. The recent microbiological applications rely on FT-IR spectra of whole cells which comprise the vibrational features of all cell components, that is DNA/RNA, proteins, membrane and cell wall constituents (Helm *et al.*, 1991). FT-IR therefore probes the total composition of a given organism in a single experiment and allows selectivity to a very high level, for example at the strain or serotype level. Moreover, since whole cells are being tested complicated

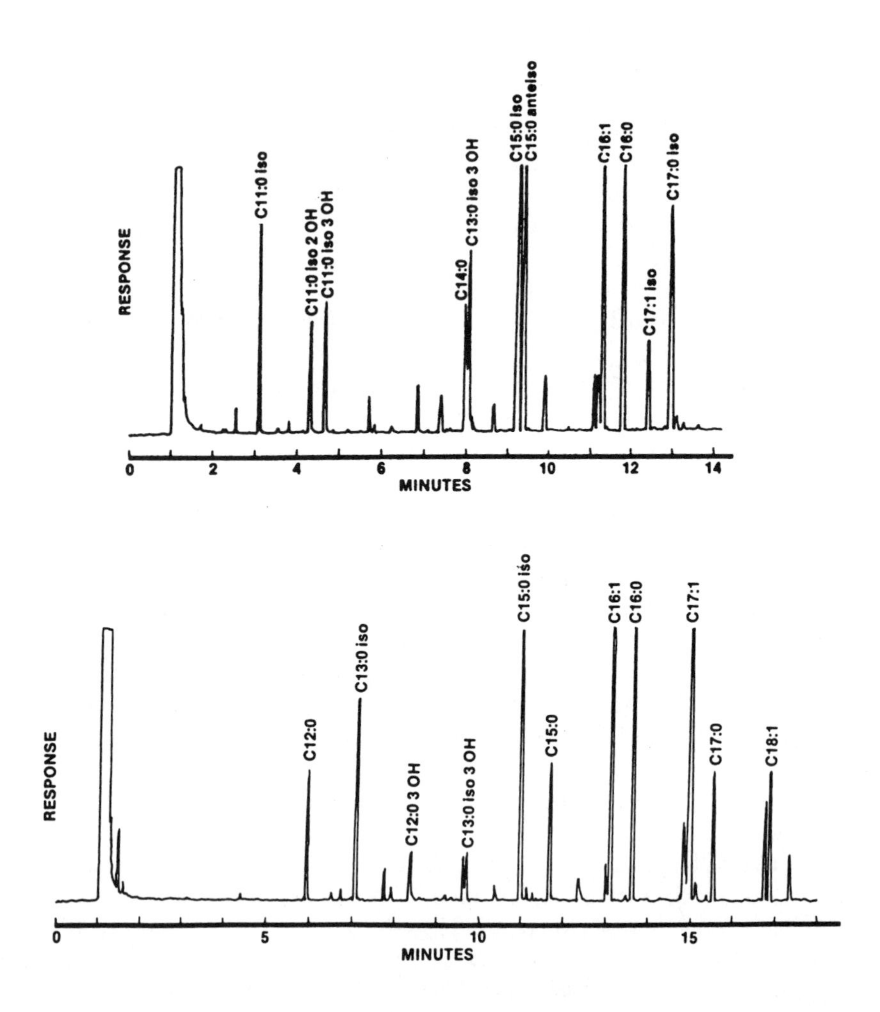

Fig. 9.2 Fatty acid methyl ester profiles for *Xanthomonas maltophilia* (top trace) and *Pseudomonas putrefaciens* (bottom trace). (From Miller (1984), reproduced by kind permission of Hewlett Packard.)

and time-consuming preparation of cell constituents is avoided. The procedure has been used to distinguish intrageneric classification and provide identification of staphylococci (Helm, Labischinski and Naumann, 1991) and in principle the method would be of value for lactic acid and yeasts. Some typical examples of FT-IR spectra are shown in Fig. 9.3.

Ultraviolet resonance Raman spectrometry (UVRRS) applied to bacterial identification has been reviewed by Nelson and Sperry (1991) and Magee (1993). This method depends on spectra produced by UV excitation of bacteria at various wavelengths yielding information on nucleic acids,

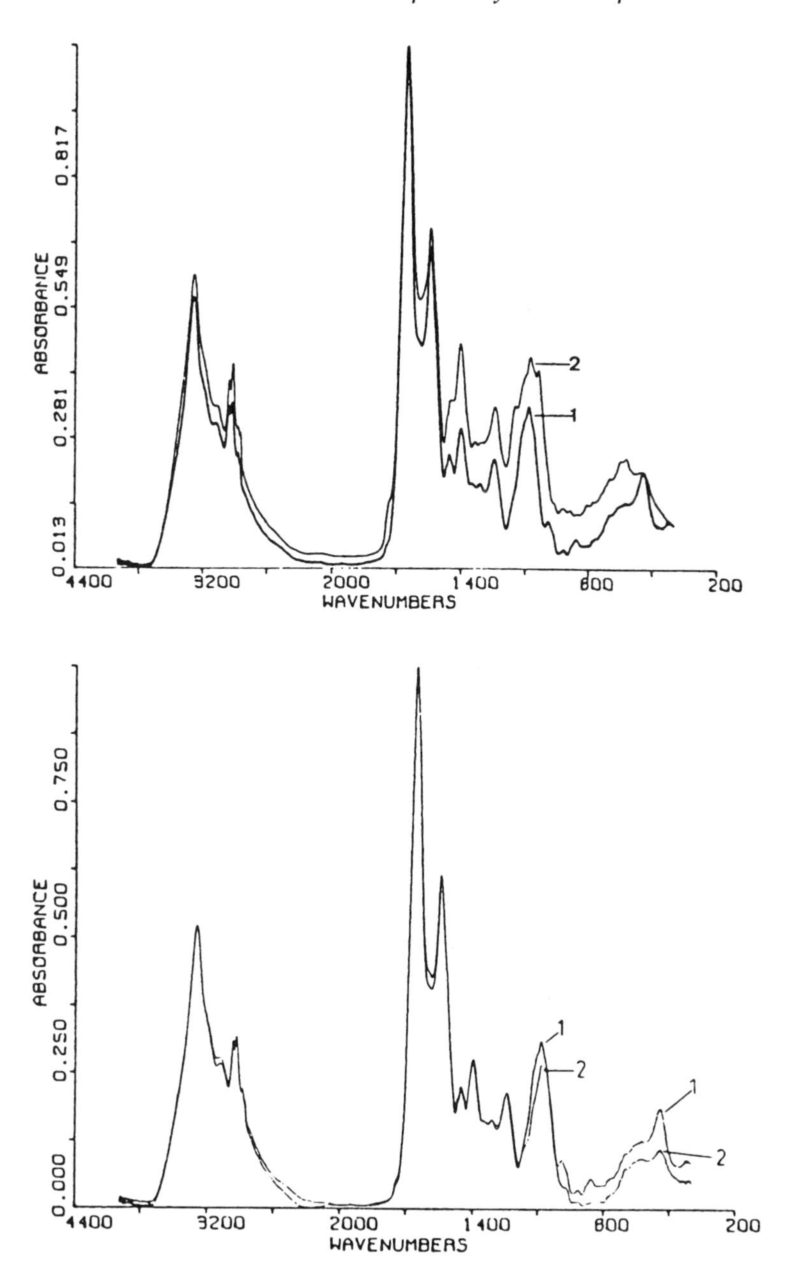

Fig. 9.3 Fourier transform infrared spectra of: (top trace) 1, *Streptococcus pyogenes*; 2, *Klebsiella pneumoniae*; (bottom trace) 1, *Streptococcus pyogenes* Group A; 2, *Streptococcus agalactiae* Group B. (From Giesbrecht *et al.* (1985), reproduced by kind permission of Springer-Verlag.)

proteins and cell wall constituents. This powerful method is capable of distinguishing bacteria at the species and strain level but it is in its infancy and routine methods for its application have yet to be described.

Pyrolysis, which is the controlled thermal degradation of a compound, in this case microbial biomass, in an inert atmosphere produces a complex mixture of low molecular weight volatile compounds. This mixture can be separated using GC and the constituents detected to produce a 'fingerprint' known as a pyrogram. This technique of pyrolysis gas chromatography (Py-GC) has been applied by several groups to the characterization of microorganisms (Gutteridge and Norris, 1979; Gutteridge, Mackey and Norris, 1980; Magee, Hindmarch and Meecham, 1983; Magee, 1993; Goodfellow *et al.*, 1994). The main difficulties with Py-GC are the slow speed per sample (modern capillary columns will separate over 100 components and an analysis can take up to 2 h) but most important of all is the lack of quantitative reproducibility as columns age and need to be replaced. To overcome these problems, Meuzelaar and Kistemaker (1973) combined pyrolysis with mass spectrometry (Py-MS) producing an instrument with fast analysis times (3 min/sample) and no column to degrade. Various instrumental developments have taken place (Meuzelaar *et al.*, 1976; Shute *et al.*, 1984; Aries, Gutteridge and Ottley, 1986) culminating in the design of dedicated, low-cost, commercial systems with automated sample input such as the PYMS 200X and RAPyD 400 machines marketed by Horizon Instruments (Heathfield, Sussex, UK).

The attraction of Py-MS as a characterization technique, apart from the fast analysis times, is that it can be applied to a single microbial colony taken directly from the surface of a culture plate and transferred to the pyrolysis filament. Thus, preparation steps are kept to an absolute minimum. In addition, it is capable of discriminating organisms at taxonomic levels below species and providing fingerprints of individual strains. An excellent example of this is provided by Wieten, Meuzelaar and Haverkamp (1984) who showed the differentiation of *E. coli* strains with and without the K1 polysaccharide antigen. A spectrum of the purified antigen (colominic acid) accounted for the key differences between the spectra of the whole bacteria. Other similar examples, particularly for epidemiological typing of pathogens, have been described by Freeman *et al.* (1990) and reviewed by Magee (1993) and Goodfellow *et al.* (1994).

One of the earlier disadvantages of Py-MS was the requirement for complex statistical techniques to handle the large amounts of quantitative data produced. A typical pyrolysis mass spectrum is shown in Fig. 9.4. The intensities of all masses above m/z 12 and below m/z 300 are recorded, although, in practice, microorganisms only give signals in the range m/z 30–160 depending on the type of mass spectrometry employed (Tas *et al.*, 1985). Observations on a wide variety of microorganisms suggest that spectra of different genera or species are qualitatively similar and discrimination has to be based on complex combinations of small, but reproducible

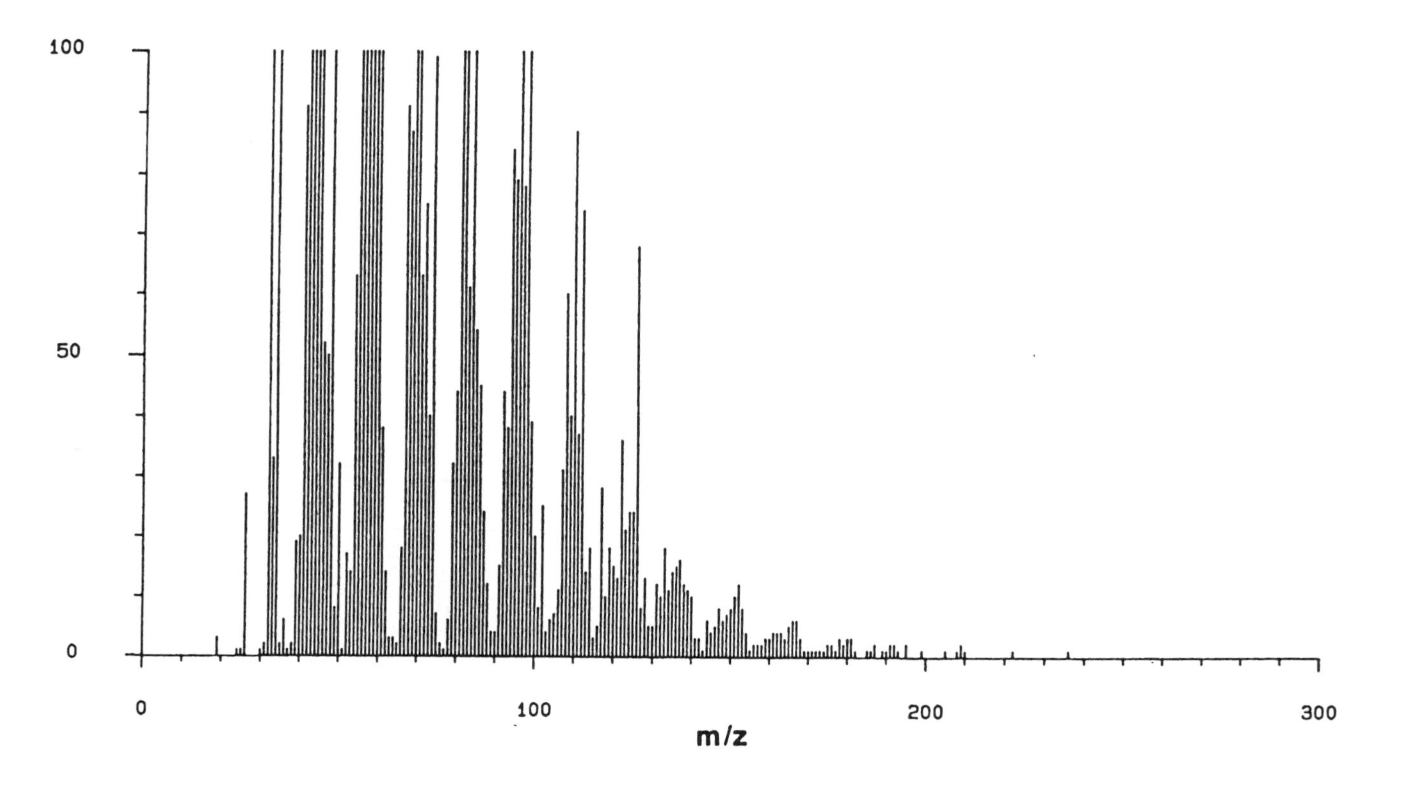

Fig. 9.4 Pyrolysis mass spectrum of *Brettanomyces naardenensis*.

variations in mass intensities. This requires the sorting of matrices containing tens of thousands of data points which can only be realistically accomplished by using multivariate statistical techniques such as principal components analysis, canonical variates analysis and cluster analysis. These methods and their application to Py-MS data have been discussed by Gutteridge, Vallis and MacFie (1985) and Magee (1993). Fortunately these computational problems have now been resolved with the introduction of dedicated Py-MS instruments as described above. Furthermore, developments in the use of artificial neural networks offer a powerful system for the discrimination of samples based on Py-MS data (Goodacre, Kell and Bianchi, 1992).

Further concerns about the use of Py-MS for identification of microorganisms relate to the long-term stability of the mass spectrometers used and the problem this may pose for the construction of reference libraries of key spectra. These problems have yet to be investigated fully but can be overcome by always including suitable reference strains.

Most microbiological applications of Py-MS involve the fingerprinting of batches of isolates to examine short-term problems. The rapid screening of batches of new isolates to gain an appreciation of the number of types of microorganism present is an important application of Py-MS. Figure 9.5

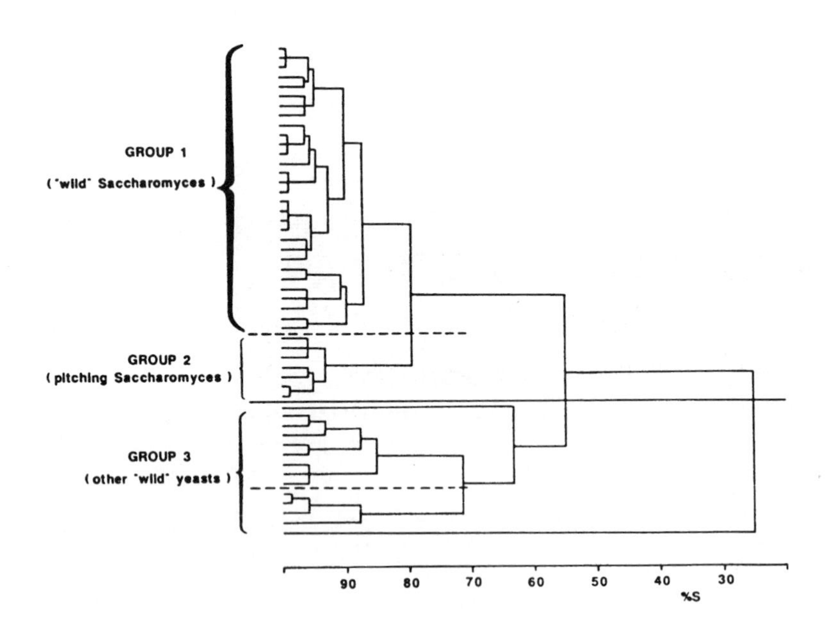

Fig. 9.5 Dendrogram of the relationships between 51 yeast isolates based on the analysis of pyrolysis mass spectrometry data.

shows a dendrogram obtained after the computer analysis of spectra for 51 yeast strains isolated from a home-brew product. Data were collected and analysed in 48 h producing a functional classification that divided the yeasts into four groups. Subsequently, conventional tests were applied to the strains and it was found that Py-MS had successfully discriminated wild *Saccharomyces* strains (Group 1) from the pitching yeasts (Group 2) and other wild yeasts (Group 3). The strains in Group 3 could be subdivided by Py-MS but the taxonomic basis of this was not determined. This example indicates the potential for Py-MS in screening and epidemiological investigations. We use Py-MS in our laboratory to get a 'first look' at the relationships between isolates when we have sampled the microflora of an environment for the first time. This allows us to make a selection of strains for further study and typing, saving a great deal of time in the process. This application of Py-MS is convenient because it avoids the thorny problems of data bases and long-term stability. There is no doubt that Py-MS can be used for the identification of microorganisms but whether it can be made sufficiently routine to appeal to the ordinary diagnostic laboratory remains in doubt.

One final technique that can be applied to whole cells, even single cells, is the laser microprobe mass analyser (LAMMA). LAMMA (Leybold-Heraeus, Koln, Germany) is a specialized laser pyrolysis MS apparatus. A specimen on a grid is located by light microscopy and a laser beam is focused along the same path. When the laser is activated a small section of the grid is pyrolysed and the pyrolysis products are detected and quantified by a time-of-flight mass spectrometer. LAMMA has been used by Bohm, Kapr and Schmitt (1985) to differentiate *Bacillus anthracis, B. cereus* and *B. thuringiensis* based on the analysis of single vegetative cells. One interesting feature of LAMMA is that it can also detect some inorganic ions and can be used to study, for example, Na, K and Ca metabolism. Nevertheless, because of the extreme capital costs, LAMMA remains a curiosity in terms of microbial identification despite its interesting ability to analyse single cells.

9.6 DEVELOPMENTS IN TECHNIQUES FOR STUDYING MORPHOLOGY AND BEHAVIOUR

9.6.1 Selective media, DEFT and impedance

Studies on the morphology and behaviour of microorganisms are several stages removed from the microbial genome. However, this is the level at which most conventional identification systems operate and important developments continue to be made which, by and large, can be implemented by most microbiologists as they do not require specialized analytical equipment. For example, the ability of microbiologists to construct selective media for the isolation of specific groups or types of microorganism continues unabated. The copper sulphate medium for the detection of *Saccharomyces* and non-*Saccharomyces* wild yeasts in the presence of brewing yeasts (Taylor and Marsh, 1984) is a fine example of characterization for pragmatic purposes. Raka-Ray agar (Saha, Sondag and Middlekauff, 1974), which is used for the selective isolation of lactobacilli from brewing processes, is another (Chapter 6).

The use of microscopy will continue to be the first line of attack for a microbiologist faced with an identification problem. The direct epifluorescent filter technique (DEFT), which is described in detail by Pettipher (1983, 1989), was developed as a rapid counting procedure for dairy samples. DEFT involves the recovery of organisms from a sample by filtration, staining *in situ* with a fluorescent dye, and examination under an epifluorescent microscope. The use of fluorescence staining under the carefully controlled conditions of the DEFT can improve the objectivity of microscopical examination and permit tentative identification of types of organisms in situations where there is previous experience, such as the examination of bovine materials for mastitis-causing organisms. The application to brewing microbiology has been fully described in Chapter 8.

Another technique devised for rapid enumeration but with potential for recognition of specific groups of microorganisms is electrical detection, otherwise known as conductance or impedance detection (Firstenberg-Eden and Eden, 1984; Easter and Gibson, 1989). In general terms, the growth of microorganisms in a medium will alter the electrical resistance and the changes can be monitored continuously by passing a small current through the medium between electrodes, and comparing it with an uninoculated control or an electronic reference. Three electrical signals can be distinguished: conductance, capacitance and impedance.

The major application of electrical detection is as a replacement for plate counts as a means of detecting growth (see Chapter 8). The development of selective media that produce appropriate electrical responses can allow this technology to be used for the detection of specific classes of microorganism. For example, Easter and Gibson (1985) developed an electrical detection system for salmonellae. Following pre-enrichment in buffered peptone water modified by the addition of dulcitol and trimethylamine oxide and selective enrichment in selenite-cystine broth with similar modifications, salmonellae can be detected by their ability to give a fast ($100\ \mu S\ h^{-1}$) and large ($> 600\ \mu S$) conductance change. Other enteric bacteria produce a small change or no change at all (Fig. 9.6). The principle of the test is the reduction of trimethylamine *N*-oxide to trimethylamine, a reaction which occurs during post-mortem bacterial spoilage of marine fish muscle and is accompanied by a significant change in conductance. This test can reduce the time taken for the detection of salmonellae by 24 h although its specificity can be improved by running it in parallel with other media and by using immunology as a confirmatory tool. Nevertheless,

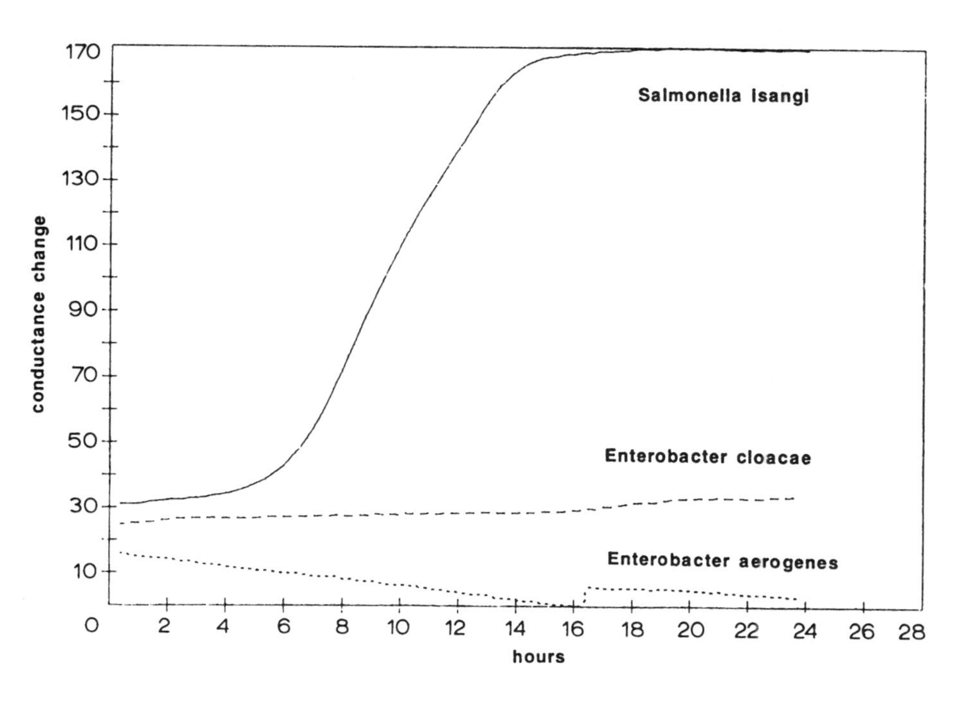

Fig. 9.6 Change in the conductance of a medium containing trimethylamine oxide inoculated with single strains of *Salmonella isangi*, *Enterobacter cloacae* and *Enterobacter aerogenes*.

electrical detection systems have the potential to be useful for the recognition of many broad classes of microorganism provided some key aspect of physiology can be adapted to produce an electric response. This limitation will perhaps not permit the use of electrical systems for the delineation of specific genera or species.

9.6.2 Miniaturized methods and commercial systems

The major development in rapid identification over the past 10–15 years has been the miniaturization of conventional biochemical and nutritional tests into commercially manufactured kits. A vast range of commercial systems is now available mainly, but not exclusively targeted at organisms of clinical significance. For the identification of members of the Enterobacteriaceae and other Gram-negative bacteria there are several systems available, including the API20E (bioMérieux Basingstoke, UK), Micro-ID (General Diagnostics, Morris Plains, USA) and Enterotube (Roche, Geneva, Switzerland), with results available in 4–24 h depending on the system used. The main advantage comes from standardization of reagent manufacture which produces reproducible identifications. Sneath (1974) estimated the probability of errors occurring with conventional identification tests within a laboratory to be 2–4% and between laboratories 6–10%. Within-laboratory studies of commercial identification systems yielded average probabilities of errors comparable with conventional tests. Definitive data on between-laboratory reproducibility for commercial identification systems are not readily available but Holmes, Dowling and Lapage (1979) expect it to be superior to conventional methods. Commercial identification systems are backed up with computerized collections of reference profiles generated from thousands of strains which can be used to produce an assignment of an unknown based on a probability score.

Recent microbiological developments of miniaturized systems have been in two opposing directions. Firstly, kits have been produced for more specialized groups of organisms, including several for clinical yeasts, e.g. Mycotube (Roche), Auxodisk (Sobioda, New York, USA), Uni-Yeast-Tek System (Corning Medical, Halstead, UK), Micro-drop Assimilation Test System (Clinical Science, Washington, USA) and API20C (bioMérieux Ltd). In the other direction, systems such as the API50CH allow the profiling of the carbohydrate assimilation pattern of any organism for which a suitable inoculation and incubation protocol can be established. Logan and Berkeley (1981) used API20E, API50E and APIZYM test kits to produce a taxonomic scheme for the Gram-positive, aerobic, spore-forming genus *Bacillus* which was in accord with schemes produced by conventional biochemical and morphological tests. As a result, identification of *Bacillus* spp. can be achieved by the current API50CH system backed by a computer-based pattern-matching program. In my own laboratory we have adapted the API50CH kit for the identification of yeasts involved in

Table 9.3 Examples of API50CH profiles for three *Brettanomyces* strains

Carbohydrate	B. anomalus *NCYC 615*	B. intermedius *CBS 73*	B. naardenensis *NCYC 813*
Glycerol	+	+	–
Erythritol	–	–	–
D-Arabinose	–	–	–
L-Arabinose	–	–	–
Ribose	–	–	–
D-Xylose	–	–	+
L-Xylose	–	–	–
Adonitol	–	–	–
β-Methylxyloside	–	–	–
Galactose	+	+	+
D-Glucose	+	+	+
D-Fructose	+	+	+
D-Mannose	+	+	+
L-Sorbose	–	–	–
Rhamnose	–	–	–
Dulcitol	–	–	–
Inositol	–	–	–
Mannitol	–	–	+
Sorbitol	–	–	+
α-Methyl-D-mannoside	–	–	–
α-Methyl-D-glucoside	+	+	–
N-Acetylglucosamine	+	+	+
Amygdalin	+	+	+
Arbutin	+	+	–
Aesculin	+	+	+
Salicin	+	+	–
Cellobiose	+	+	+
Maltose	+	+	+
Lactose	+	–	–
Melibiose	–	–	–
Sucrose	+	+	–
Trehalose	+	+	+
Inulin	–	–	–
Melezitose	+	+	+
D-Raffinose	–	+	–
Starch	+	–	–
Glycogen	–	–	–
Xylitol	–	–	+
β-Gentiobiose	+	+	+
D-Turanose	+	+	–
D-Lyxose	+	–	–
D-Tagatose	–	–	–
D-Fucose	–	–	–
L-Fucose	–	–	–
D-Arabitol	+	+	–
L-Arabitol	–	–	–
Gluconate	–	–	–
2-Ketogluconate	–	–	–
3-Ketogluconate	–	–	–

+, growth; –, no growth.

the spoilage of soft drinks. Used in conjunction with a few fermentation tests and some morphological observations, it can produce a profile that can be compared with an existing yeast data base, such as that published by Barnett, Payne and Yarrow (1983) which is now also available on computer. The adaptation of this system for yeast profiling is not a simple matter. Care has to be taken to avoid carry-over of substrates. Long incubation periods, compared with bacteria, may be necessary, increasing the risk of contamination. Nevertheless, perseverence can produce a profiling system geared to the identification needs of your own laboratory. Examples of API50CH profiles for three type cultures of *Brettanomyces* are shown in Table 9.3.

9.6.3 Automated systems

Technically, miniaturized systems are also proceeding in two directions: automation to provide rapid and reproducible sample handling, and a switch to enzyme- or metabolism-based tests to avoid the delays involved in traditional growth-based tests. Miniaturization of biochemical tests, accompanied by photo-optical detection of growth, are the principles behind the VITEK System marketed in the UK by bioMérieux. The VITEK is an approach to automation of screening, identification and antimicrobial assay in the laboratory. Plastic transparent test cards are factory-filled with chemical substrates. The cards are rehydrated with a suspension of the unknown organism and are incubated in stacks. Detection of growth by monitoring changes in optical density is automatic and the system uses a computer to produce a probabilistic identification score against a library of standard profiles. Three standard test cards have proved popular, those for Gram-negative identification, Gram-positive identification and yeast biochemical profiles. Emphasis was on organisms of clinical significance and the yeast card covers mainly *Candida* and *Cryptococcus* species. However, bioMérieux have been working to extend the application of VITEK into the industrial laboratory. Sister instrumentation for automated blood cultures (VITAL), automated immunoanalysis (VIDAS) and automated fluorescence microscopy (COBRA) has also been introduced.

Biolog (California, USA) have introduced methodology for identification based on the indication of carbon substrate utilization by the inclusion of tetrazolium-based indicator of respiration with the substrate. Carbon sources specific for Gram-negative or Gram-positive bacteria or yeasts and tetrazolium salt are dispensed in microtitre dishes which are purchased complete. Standard inoculation of the 96 wells with a microbial suspension and incubation for 4 and 16 h provides a biochemical fingerprint of the microorganism for the 95 substrates (one control well). These patterns can be read by eye or with an automated reader and compared with a database using standard computer algorithms. Although Biolog did not provide particularly high accuracy for identification of enterobacteria in a recent comparative survey (Stager and Davis, 1992), Biolog have one of the most extensive databases for microbial identification and recent introduction of trays for lactobacilli and yeasts make this system attractive to the brewery microbiologist.

Other automated systems such as those from Sensititre and Baxter Diagnostics are described below since they are based on enzyme tests.

9.6.4 Enzyme-based systems

A different and more rapid approach to the identification of clinical isolates is based on enzyme tests for characterizing the microorganisms (Bascomb, 1980, 1985; Manafi, Kneifel and Bascomb, 1991). Conventional methods rely on microbial growth and, with the dilution of the bacterial population that occurs during inoculation, at least 10–17 generation times are required to increase numbers to levels where enough metabolites are produced to trigger a detection system, such as a drop in pH indicated by a colour change. In addition, many conventional tests detect products or effects of enzymes that have to be induced or are the result of multienzyme pathways. Tests targeted at the detection of constitutive enzymes or their direct products are faster and more specific. Further advantages of enzyme activity tests are:

1. amplification of the signal is more easily achieved by optimizing conditions for enzyme activity than by increasing the concentration of a product;
2. the shorter incubation times required for enzyme activity testing minimize concerns over chance contamination;
3. enzyme activity tests are more sensitive and can be completed with lower numbers of bacteria than conventional tests;
4. equipment for monitoring enzyme activities is available.

Three categories of enzyme test can be recognized:

1. qualitative tests, where the presence of an enzyme is detected subjectively by observation of a change in the appearance, colour or fluorescence of an enzyme/substrate mixture;
2. studies of isoenzymes using electrophoresis;
3. quantitative tests in which activities are measured and expressed in terms of the quantity or rate of reaction product formation.

Enzymes whose activities have been used for identification have been mainly catabolic, including esterases, glycosidases, lipases, phosphatases, peptidases and proteases (Manafi, Kneifel and Bascomb, 1991). Detection has been made easier by the introduction of a variety of synthetic chromogenic and fluorogenic substrates where the chromogen or fluorogen is attached to a simple moiety such as an amino acid, monosaccharide, fatty acid, phosphate or sulphate ion. Quantitative measurements of activity require some degree of automation if large numbers of determinations are needed. Bascomb (1985) has chronicled developments in the use of continuous flow analysers, discrete analysers, fast centrifugal analysers, specialized analysers and multi-well plates.

Bascomb (1984) used a fluorimetric system in multi-well plates. Pre-prepared plates using dried substrates can be stored at 5–8°C. Plates were inoculated manually and incubated for 2 h after which reagents were added and the plates were incubated for a further 15 min to allow the reaction to occur. The fluorescence of each well was recorded automatically and fed into a computer running a discriminant analysis procedure for identification. Unknowns could be typed within 2–5 h of receiving the primary isolation plate. A comparison between identifications based on rapid fluorescence enzyme tests and conventional methods for 1543 cultures, representing 24 genera, showed an overall agreement of 88% (Bascomb, 1985). In general, the enzyme activity approach to identification is a good one that can give genuinely rapid results and it could be specifically adapted to the identification of brewery microorganisms.

Complete microbial identification systems based on enzyme activities are produced by Sensititre using fluorescently labelled substrates in 96-well microtitre plates for the identification of Gram-negative bacteria. Test plates are incubated off line and may be read after 5 or 18 h using a dedicated plate reader. Walkaway and autoScan are both manufactured by Baxter Diagnostics. These computer-controlled machines enable incubation of multi-well test plates and automatic interpretation of test results. Fluorogenic panels of enzyme substrates specific for Gram-negative rods, Gram-positive cocci and yeasts are available. In comparative studies, the VITEK, Sensititre and Baxter Diagnostics products all gave high and accurate identifications (Stager and Davis, 1992).

9.7 FUTURE TRENDS IN RAPID IDENTIFICATION

In order to make some sensible predictions about the future of rapid identification some comments on the costs, speed and applicability of the techniques discussed have been collated in Tables 9.4 and 9.5. Table 9.4 shows the capital cost of investing in the necessary equipment to use a technique and gives a comparative estimate of the cost of developing an identification system and of the running costs per sample. Obviously development costs depend on the application. The use of VITEK to identify Enterobacteriaceae from a novel source can be achieved with the existing test card(s) and is comparatively cheap. In contrast, the use of VITEK to identify non-clinical yeasts probably requires the development of a new test card and would consequently be more expensive. Development costs must also be judged alongside the benefits that might accrue from having a better identification method. ELISA and gene probe assays may be

Table 9.4 Comparison of costs for various identification techniques

Technique	*Capital cost (£ thousands)*	*Development cost*	*Cost per test**
Level 1: genetics			
Gene probes	None	High	High
Level 2: proteins			
ELISA	<5	High	High
Electrophoresis	10–15	Moderate	Low
AMB-ID system	25–45	Moderate	Moderate
Level 3: cell composition			
CIDS	>100	High	Low
Flow cytometry	50–100	High	Low
HP FAME GC system	20–30	Moderate	Moderate
FT-IR	30–40	High	Low
Py-MS	50–60	High	Low
LAMMA	>100	High	Moderate
Level 4: morphology and behaviour			
Selective media	None	Moderate	Low
DEFT	10–15	Low	Low
Impedance	20–40	Moderate	Moderate
Commercial kits	None	Low	Moderate/high
VITEK	>30	Moderate	Moderate/high
Enzyme tests	<5	High	Low

* Costs per test: low <£1, moderate £1–2, high >£2.

expensive to develop (£5000 to £50 000) but if successful they might provide accurate diagnosis with minimum sample preparation and save on running costs. The costs per test in Table 9.4 are rated low (< £1), moderate (£1–£2) or high (> £2). ELISAs and gene probes are costly on reagents for individual tests but these costs are continually declining in real terms with improved technology and the cost savings of high volume and the benefits of producing accurate identifications will often justify the higher cost. With respect to capital costs, it is clear from Table 9.4 that the analytical techniques that address cell composition all have high equipment costs. Microbiology is geared to the use of high volumes of disposables, in contrast to chemistry, and the need to invest in expensive analytical equipment is beyond the experience and influence of most microbiologists. Such entrenched attitudes are a real obstacle to the enhanced use of analytical techniques for microbial identification.

In Table 9.5, speed per sample is a judgement based on the amount of preparation time and the time taken by the analysis. Applicability is taken to mean whether the technique can be applied to different types of microorganism (assuming that the user will construct his/her own reference

Table 9.5 Comparison of speed and applicability for various identification techniques

Technique	*Speed per sample*	*Volume throughput*	*Applicability*
Level 1: genetics			
Gene probes	Moderate	High	Wide
Level 2: proteins			
ELISA	Moderate	High	Restricted
Electrophoresis	Moderate	High	Wide
AMB-ID system	Moderate	High	Wide
Level 3: cell composition			
CIDS	Fast	Low	Not known
Flow cytometry	Fast	High	Not known
HP FAME GC system	Moderate	High	Wide
FT-IR	Fast	Moderate	Not known
Py-MS	Fast	High	Wide
LAMMA	Fast	Moderate	Not known
Level 4: morphology and behaviour			
Selective media	Slow	Low	Restricted
DEFT	Fast	Low	Restricted
Impedance	Moderate	High	Restricted
Commercial kits	Moderate/slow	Low	Wide
VITEK	Fast/moderate	High	Wide
Enzyme tests	Fast	High	Wide

libraries). Thus, an individual ELISA or gene probe assay is restricted in the sense that it only has one 'target' per development budget. In contrast, investment in FAME profiling by GC gives the user a versatile tool that can be turned to a wide range of identification problems.

In conclusion, it seems obvious that, in the short term, the demand for accurate and rapid detection of organisms of public health significance will fuel the development of an increasing range of gene probes and ELISA assays. The cost of these developments, however, will restrict the commercial opportunities to those where large numbers of identifications per year can be guaranteed. In laboratories where identification and characterization of microorganisms is carried out less intensively and covers more unusual groups, the technologies of enzyme tests, FAME profiling, Py-MS and electrophoresis (including the AMB-ID system) will prosper because the users can establish, rapidly, their own database. Interestingly, all four of these approaches require the use of sophisticated computing techniques to handle the data generated. The computer power to allow this to happen on-line is now available at a reasonable cost. In the longer term, techniques with the ability to identify, directly, members of a mixed population must gain in prominence if diagnostic microbiology is to advance.

ACKNOWLEDGEMENTS

We would like to thank Mike Arnott and Renny Ison for permission to refer to some of their unpublished data.

REFERENCES

Amann, R. and Ludwig, W. (1994) In *Bacterial Diversity and Systematics* (eds F.G. Priest., A. Ramos-Cormenzana and B. Tindall), Plenum Press, New York, p. 115.
Aries, R.E., Gutteridge, C.S. and Ottley, T.W. (1986) *Journal of Analytical and Applied Pyrolysis*, **9**, 81.
Barnett, J.A., Payne, R.W. and Yarrow, D. (1983) *Yeasts: Characteristics and Identification*, Cambridge University Press, Cambridge.
Bascomb, S. (1980) In *Microbiological Classification and Identification* (eds M. Goodfellow and R.G. Board), Academic Press, London, p. 359.
Bascomb, S. (1984) In *New Horizons in Microbiology* (eds A. Sanna and G. Morace), Elsevier, Amsterdam, p. 241.
Bascomb, S. (1985) In *Rapid Methods and Automation in Microbiology and Immunology* (ed. K.-O. Habermehl), Springer, Berlin, p. 367.
Bohm, R., Kapr, T. and Schmitt, H.U. (1985) *Journal of Analytical and Applied Pyrolysis*, **8**, 449.
Bowden, G.H.W. (1993) *Handbook of New Bacterial Systematics* (eds M. Goodfellow and A.G. O'Donnell), Academic Press, London, p. 429.
Chevrier, D., Larzul, D., Megrand, F. and Guesden, J.L. (1989) *Journal of Clinical Microbiology*, **27**, 321.
Claussen, J. (1988) In *Laboratory Techniques in Biochemistry and Molecular Biology*, Vol. 1 (eds R.H. Burdon and P.H. van Knippenberg), Elsevier, New York.
Collins, M.D. (1985) In *Chemical Methods in Bacterial Systematics* (eds M. Goodfellow and D.E. Minnikin), Academic Press, London, p. 237.
Datta, A.R., Wentz, B.A., Shook, D. and Trucksess, M.W. (1988) *Applied and Environmental Microbiology*, **54**, 2933.
Delley, M., Mollet, B. and Holtinger, H. (1990) *Applied and Environmental Microbiology*, **56**, 169).
Dicks, L.M.T., Van Vuuren, H.J.J. and Dellaglio, F. (1990) *International Journal of Systematic Bacteriology*, **40**, 83.
Dobson, G., Minnikin, D.E., Minnikin, S.M. *et al.* (1985) In *Chemical Methods in Bacterial Systematics* (eds M. Goodfellow and D.E. Minnikin), Academic Press, London, p. 237.
Easter, M.C. and Gibson, D.M. (1985) *Journal of Hygiene, Cambridge*, **94**, 245.
Easter, M.C. and Gibson, D.M. (1989) *Progress in Industrial Microbiology*, **26**, 57–100.
Embley, T.M. and Wait, R. (1994) In *Chemical Methods in Prokaryotic Systematics* (eds M. Goodfellow and A.G. O'Donnell), John Wiley, Chichester.
Falkow, S. (1985) In *Rapid Methods and Automation in Microbiology and Immunology* (ed. K.-O. Habermehl), Springer, Berlin, p. 30.
Firstenberg-Eden, R. and Eden, G. (1984) *Impedance Microbiology*, Research Studies Press, Letchworth.

Freeman, R., Goodfellow, M., Gould, F.K. *et al.* (1990) *Journal of Medical Microbiology* **32,** 283.

Giesbrecht, P., Naumann, D., Labischinski, H. and Barnickel, G. (1985) In *Rapid Methods and Automation in Microbiology and Immunology* (ed. K.-O. Habermehl), Springer, Berlin, p. 198.

Goodacre, R., Kell, D.B. and Bianchi, G. (1992) *Nature,* **359,** 594.

Goodfellow, M. and Minnikin, D.E. (1985) In *Chemical Methods in Bacterial Systematics* (eds M. Goodfellow and D.E. Minnikin), Academic Press, London, p. 1.

Goodfellow, M., Chun, J., Atalan, E. and Sanglier, J.-J. (1994) In *Bacterial Diversity and Systematics* (eds F.G. Priest., A. Ramos-Cormenzana and B. Tindall), Plenum Press, New York, p. 87.

Gravett, M.G., Eschenbach, D.A., Spiegel-Brown, C.A. and Holmes, K.K. (1982) *New England Journal of Medicine,* **306,** 725.

Gregg, C.T., McGregor, D.M., Grace, W.K. and Salzman, G.C. (1985) In *Rapid Methods and Automation in Microbiology and Immunology* (ed. K.-O. Habermehl), Springer, Berlin, p. 184.

Grimont, P.A.D. (1988) *Canadian Journal of Microbiology,* **34,** 541.

Gutteridge, C.S. and Norris, J.R. (1979) *Journal of Applied Bacteriology,* **47,** 5.

Gutteridge, C.S., Mackey, B.M. and Norris, J.R. (1980) *Journal of Applied Bacteriology,* **49,** 165.

Gutteridge, C.S., Vallis, L. and MacFie, H.J.H. (1985) In *Computer-assisted Bacterial Systematics* (eds M. Goodfellow, D. Jones and F.G. Priest), Academic Press, London, p. 369.

Helm, D., Labischinski, H. and Naumann, D. (1991) *Journal of Microbiological Methods* **14,** 127.

Helm, D., Labischinski, H., Schallehn, G. and Naumann, D. (1991) *Journal of General Microbiology,* **137,** 69.

Hertel, C., Ludwig, W., Pot, B. *et al.* (1993) *Systematic and Applied Microbiology,* **16,** 463.

Holdeman, L.V., Cato, E.P. and Moore, W.E.C. (1977) *Anaerobe Laboratory Manual,* 4th edn, VPI, Blacksburg.

Holmes, B., Dowling, J. and Lapage, S.P. (1979) *Journal of Clinical Pathology,* **32,** 78.

Jackman, P.J.H. (1985) In *Chemical Methods in Bacterial Systematics* (eds M. Goodfellow and D.E. Minnikin), Academic Press, London, p. 115.

Jantzen, E. and Bryn, K. (1985) In *Chemical Methods in Bacterial Systematics* (eds M. Goodfellow and D.E. Minnikin), Academic Press, London, p. 145.

Jantzen, E. and Bryn, K. (1994) In *Chemical Methods in Prokaryotic Systematics* (eds M. Goodfellow and A.G. O'Donnell), John Wiley and Sons Ltd., Chichester, p. 21.

Jarvis, B. (1985) In *Rapid Methods and Automation in Microbiology and Immunology* (ed. K.-O. Habermehl), Springer, Berlin, p. 593.

Kersters, K., Pot, B., Dewettinck, D., Torck, V., Vancanneyt, M., Vauterin, L. and Vandamme, P. (1994) In *Bacterial Diversity and Systematics* (eds F.G. Priest., A. Ramos-Cormenzana and B. Tindall), Plenum Press, New York, p. 51.

Kohler, G. and Milstein C. (1975) *Nature,* **256,** 495.

Krysinski, E.P. and Heimsch, R.C. (1977) *Applied and Environmental Microbiology,* **33,** 947.

Logan, N.A. and Berkeley, R.C.W. (1981) In *The Aerobic Endosporeforming Bacteria: Classification and Identification* (eds R.C.W. Berkeley and M. Goodfellow), Academic Press, London, p. 105.

Macario, A.J.L. and Conway de Macario, E. (1985) *Monoclonal Antibodies Against Bacteria,* Vols. 1 and 2, Academic Press, New York.

Magee, J. (1993) In *Handbook of New Bacterial Systematics* (eds M. Goodfellow and A.G. O'Donnell), Academic Press, London, p. 383.

Magee, J.T., Hindmarch, J.M. and Meecham, D.F. (1983) *Journal of Medical Microbiology*, **16,** 483.
Manafi, M., Kneifel, W. and Bascomb, S. (1991) *Microbiological Reviews*, **55,** 335–348.
Mattingly, J.A. (1984) *Journal of Immunological Methods*, **73,** 147.
Mattingly, J.A. and Gehle, W.D. (1984) *Journal of Food Science*, **49,** 807.
Mattingly, J.A., Robison, B.J., Boehm, A. and Gehle, W.A. (1985) *Food Technology*, **39,** 90.
Meuzelaar, H.L.C. and Kistemaker, P.G. (1973) *Analytical Chemistry*, **45,** 587.
Meuzelaar, H.L.C., Kistemaker, P.G., Eshuis, W. and Engel, H.W.B. (1976) In *Rapid Methods and Automation in Microbiology* (eds S.W.B. Newsom and H.H. Johnston), Learned Information, Oxford, p. 225.
Miles, C.A., Mackey, B.M. and Parsons, S.E. (1986) *Journal of General Microbiology*, **132,** 939.
Miller, L. (1984) Hewlett Packard Gas Chromatography Application Note 228–37.
Minnich, S.A. (1978) MS Thesis, University of Idaho.
Mohr, H.K., Trenk, H.L. and Yeterian, M. (1974) *Applied and Environmental Microbiology*, **27,** 324.
Moss, C.W. (1981) *Journal of Chromatography*, **203,** 337.
Moss, C.W. (1985) In *Rapid Methods and Automation in Microbiology and Immunology* (ed. K.-O. Habermehl), Springer, Berlin, p. 232.
Moss, C.W. and Dees, S.B. (1976) *Journal of Clinical Microbiology*, **4,** 492.
Moss, C.W. and Dees, S.B. (1978) *Journal of Clinical Microbiology*, **8,** 772.
Moss, C.W., Dees, S.B. and Guerrant, G.O. (1980) *Journal of Clinical Microbiology*, **12,** 127.
Muirhead, K.A., Horan, P.K. and Poste, G. (1985) *Bio/Technology*, **3,** 337.
Naumann, D., Helm, D. and Schultz, C. (1994) In *Bacterial Diversity and Systematics* (eds F.G. Priest., A. Ramos-Cormenzana and B. Tindall), Plenum Press, New York, p. 67.
Nelson, W.H. (1991) *Modern Techniques for Rapid Microbiological Analysis*, VCH Publishers, New York.
Nelson, W.H. and Sperry, J.F. (1991) In *Modern Techniques for Rapid Microbiological Analysis* (ed. W.H. Nelson), VCH Publishers, New York, p. 7.
Nichols, P.D., Henson, J.M., Guckert, J.B. *et al.* (1985) *Journal of Microbiological Methods*, **4,** 79.
Norris, J.R. (1980) In *Microbiological Classification and Identification* (eds M. Goodfellow and R.G. Board), Academic Press, London, p. 1.
Norris, K.P. (1959) *Journal of Hygiene*, **57,** 326.
Odham, G., Larsson, L. and Mardh, P.-A. (1984) *Gas Chromatography/Mass Spectrometry Applications in Microbiology*, Plenum Press, London.
O'Donnell, A.G., Minnikin, D.E. and Goodfellow, M. (1985) In *Chemical Methods in Bacterial Systematics* (eds M. Goodfellow and D.E. Minnikin), Academic Press, London, p. 131.
Pettipher, G.L. (1983) *The Direct Epifluorescent Filter Technique for the Rapid Enumeration of Microorganisms*, Research Studies Press, Letchworth.
Pettipher, G.L. (1989) *Progress in Industrial Microbiology*, **26,** 19–56.
Phillips, A.P. and Martin, K.L. (1985) In *Rapid Methods and Automation in Microbiology and Immunology* (ed. K.-O. Habermehl), Springer, Berlin, p. 408.
Phillips, T., Taylor, E. and Eykyn, S. (1980) *Infection*, **8,** S155.
Pilloud, N. and Mollet, B. (1990) *Systematic and Applied Microbiology* **13,** 345.
Pot, B., Vandamme, P. and Kersters, K. (1994) In *Chemical Methods in Prokaryotic Systematics* (eds M. Goodfellow and A.G. O'Donnell), John Wiley & Sons Ltd, Chichester, p. 493.
Pot, B., Hartel, C., Ludwig, W. *et al.* (1993) *Journal of General Microbiology*, **139,** 513.
Potter, M. (1970) *Federation Proceedings*, **29,** 85.

Robison, B.J., Pretzman, C.I. and Mattingly, J.A. (1983) *Applied and Environmental Microbiology*, **45,** 1816.
Saha, R.B., Sondag, R.J. and Middlekauff, J.E. (1974) *Proceedings of the 9th Congress of the American Society of Brewing Chemists*, American Society of Brewing Chemists, St Paul, MN, **5,** 1.
Salton, M.R.J. (1952) *Biochimica et Biophysica Acta*, **8,** 510.
Salzman, G.C., Griffith, J.K. and Gregg, C.T. (1982) *Applied and Environmental Microbiology*, **44,** 1081.
Schleifer, K.H. (1990) *Food Biotechnology*, **4,** 585.
Schleifer, K.H., Ludwig, W. and Amann, R. (1993) In *Handbook of New Bacterial Systematics* (eds M. Goodfellow and A.G. O'Donnell), Academic Press, London, p. 464.
Schmidhuber, S., Ludwig, W. and Schleifer, K.-H. (1988) *Journal of Clinical Microbiology*, **26,** 1042.
Schrade, J.P. (1984) In *FDA Bacteriological Analytical Manual*, 6th edn, FDA, Washington, DC, Chapter 8.
Sharpe, A.N. (1994) *Food Research International*, **27,** 237.
Shute, L.A., Gutteridge, C.S., Norris, J.R. and Berkeley, R.C.W. (1984) *Journal of General Microbiology*, **130,** 343.
Smith, A.M., Miller, J.S. and Whitehead, D.S. (1979) *Journal of Immunology*, **123,** 1715.
Sneath, P.H.A. (1972) In *Methods in Microbiology*, Vol. 7A (eds J.R. Norris and D.W. Ribbons), Academic Press, London, p. 29.
Sneath, P.H.A. (1974) *International Journal of Systematic Bacteriology*, **24,** 508.
Sommerfelt, H., Kalland, K.H., Raj, P. *et al.* (1988) *Journal of Clinical Microbiology*, **26,** 2275.
Sperber, W.H. and Diebel, R.H. (1969) *Applied and Environmental Microbiology*, **17,** 533.
Stager, C.E. and Davis, J.R. (1992) *Clinical Microbiology Reviews*, **5,** 302–327.
Steel, K.J. (1962) In *Microbial Classification* (eds G.C. Ainsworth and P.H.A. Sneath), Cambridge University Press, Cambridge, p. 20.
Swaminathan, B. and Ayres, J.C. (1980) *Journal of Food Science*, **45,** 352.
Tamaoka, J. (1994) In *Chemical Methods in Prokaryote Systematics* (eds M. Goodfellow and A.G. O'Donnell), John Wiley, Chichester, p. 463.
Tas, A.C., van der Greef, J., de Waart, J. *et al.* (1985) *Journal of Analytical and Applied Pyrolysis*, **7,** 249.
Taylor, G.T. and Marsh, A.S. (1984) *Journal of the Institute of Brewing*, **90,** 134.
Thomason, B.M. (1980) *Journal of Food Science*, **44,** 381.
Thomason, B.M., Cherry, W.B. and Moody, M.D. (1957) *Journal of Bacteriology*, **74,** 525.
Vauterin, L., Swings, J. and Kersters, K. (1993) In *Handbook of New Bacterial Systematics* (eds M. Goodfellow and A.G. O'Donnell), Academic Press, London, p. 251.
Voller, A.D., Bidwell, D. and Bartlett, A. (1980) In *Manual of Clinical Immunology*, 2nd edn (eds N.R. Rose and H. Friedman), American Society for Microbiology, Washingon, p. 54.
Wieten, G., Meuzelaar, H.L.C. and Haverkamp, J. (1984) In *Gas Chromatography/ Mass Spectrometry Applications in Microbiology* (eds G. Odham, L. Larsson and P.-A. Mardh), Plenum Press, London, p. 335.

CHAPTER 10

Cleaning and disinfection in the brewing industry

M. Singh and J. Fisher

10.1 INTRODUCTION

'Cleaning and disinfection in the brewing industry' encompasses such a broad area that it would be impracticable to concentrate on any particular aspect in detail. Detergents based on caustic, acids and enzymes have all been used; active ingredients of disinfectant compositions from most of the oxidizing and non-oxidizing types have been considered at some stage. All systems have some disadvantages, and technology has yet to reach a stage where a universally acceptable product is available. The choice of detergent and disinfectant must be a carefully considered compromise.

The objectives of this chapter are to provide a broad view of cleaning and disinfection in terms of methods of application and the types of chemicals used, and it is not intended to be a detailed academic portrayal of detergent and disinfectant theory.

10.2 DEFINITIONS

It is important to define the terminology used in this section. Some are standard British Standard definitions (Anon., 1976), whereas other terms have evolved within the industry and are occasionally misused. The reasons why something is cleaned and the standard of cleaning required are obviously very closely interlinked. The standard of cleaning can be broadly divided into three classifications, the first three of the definitions given below.

1. Physical cleanliness: visually clean to a satisfactory standard.

Brewing Microbiology, 2nd edn. Edited by F. G. Priest and I. Campbell.
Published in 1996 by Chapman & Hall, London. ISBN 0 412 59150 2

2. Chemical cleanliness: clean to a standard where anything in contact (e.g. product) with the cleaned surface suffers no contamination. This may also be defined as 'water break free': clean water will completely wet the surface and drain as a continuous film without forming rivulets or droplets.
3. Microbiological cleanliness: cleaned to a level such that no physical or microbiological contamination is present.
4. Detergent: a cleaning agent. By a combination of physical and chemical processes a detergent removes soil from a substrate.
5. Soil: any substance which is in the wrong place. For example, a glass of stout is a fortifying beverage. However, spilt on one's clothes, it becomes soil, quite likely to leave a dark, difficult-to-remove stain.
6. Disinfection: the destruction of microorganisms, but not usually bacterial spores; it does not necessarily imply killing all microorganisms, but reducing them to a level which is harmful neither to health nor to the quality of perishable goods. The term is applicable to the treatment of inanimate objects and materials and may also be applied to the treatment of the skin and other body membranes and cavities (Anon., 1976). Chemicals used for disinfection are termed disinfectants.
7. Chemical sterilizing agent (sterilant): misnomers used as synonyms for disinfectant.
8. Sanitization: a term used mainly in the food and catering industry: a process of both cleaning and disinfecting utensils and equipment (sanitizer: a chemical agent used for sanitization).
9. CIP: cleaning in place without the need to dismantle.

10.3 STANDARDS REQUIRED WITHIN A BREWERY

The concept of why anything should be cleaned, with the results required, can be applied specifically to the brewing process. Even simplification of the discrete stages within the overall process leads to over a dozen stages, each with its own plant, soil type and the level of cleanliness desired. The main processes in a brewery are summarized in Table 10.1.

As Table 10.1 indicates, cleaning within the brewery is indeed a complex operation involving many types of plant and soil. The actual result required varies from one area to another, but in general, CIP methods are used to achieve microbiologically clean plant. Cleaning water of good and consistent standard greatly facilitates the operations and provides a basis for high-quality results.

In order to make recommendations for the type of detergent, apart from knowledge of the process, soil types and method of application, it is necessary to have information on water quality. The following is suggested as the minimum information necessary:

Table 10.1 Levels of cleanliness required and cleaning methods for various brewing operations

Process	*Plant*	*Soil type*	*Result required*	*How obtained*
Malting	Floors Screens	Particulate	Physically clean	Manually
Milling	Mills Rollers	Particulate	Physically clean	Manually
Mashing	Mash tun	Particulate Starch Sugar Protein Scale Tannin	Chemically clean	Manually or CIP
Straining	Mash tun Lauter tun Mash filter Strainmaster	Particulate Starch Sugar Protein	Chemically clean	Manually or CIP
Boiling	Copper (Hop kettle)	Particulate Starch Sugar Protein Tannin Scale	Chemically clean	Manually or CIP
Separation	Hop strainer Sedimentation vessel Whirlpool	Hops Trub Tannin	Microbiologically clean	CIP with detergent/ disinfectant or sanitizer
Cooling	Heat exchanger	Protein Scale Rust Particulate	Microbiologically clean	CIP with detergent/ disinfectant or sanitizer
Fermentation	Open or closed fermentors	Yeast Protein Oxidation products Tannin Sugar Scale	Microbiologically clean	CIP with detergent/ disinfectant or sanitizer or Manually, followed by disinfecting rinse
Separation	Yeast press Centrifuge	Yeast Protein	Microbiologically clean	CIP with detergent/ disinfectant or sanitizer or Manually, followed by disinfecting rinse

Table 10.1 *continued*

Process	*Plant*	*Soil type*	*Result required*	*How obtained*
Conditioning	Closed vessels	Yeast Protein Scale	Microbiologically clean	CIP with detergent/ disinfectant or Manually, followed by disinfecting rinse
Clarification	Filters Centrifuge	Yeast Protein	Microbiologically clean	CIP with detergent/ disinfectant or sanitizer
Storage Transport	Bright-beer tanks Road tankers	Yeast Protein Scale	Microbiologically clean	CIP with detergent/ disinfectant or sanitizer
Pasteurization	Heat exchanger	Protein Scale	Microbiologically clean	CIP with detergent/ disinfectant or sanitizer
Packaging (Kegs)	Kegs Casks Fillers	Protein Scale Particulate deposits	Microbiologically clean	CIP with detergent/ disinfectant or sanitizer
Packaging (Bottles)	Bottles Filler	Paper Glue Scale Mould	Microbiologically clean	In machine by temperature and causticity CIP detergent/ disinfectant or sanitizer
In-bottle pasteurization		Algae Scale	Microbiologically clean	Dosing of biocide
Can-warming		Particulate deposits		Scale control

1. water hardness (temporary and total);
2. total dissolved solids;
3. pH;
4. microbiological quality.

Table 10.2, showing data provided by the South Staffordshire Water Authority, gives the outline water analysis for a bore hole and the town supply in the Burton-on-Trent (UK) area. Untreated bore-hole water would give rise to scale formation if used with alkaline or caustic detergents, or if heated. The high level of dissolved solids could give rising problems.

Table 10.2 Analysis of bore-hole water from a Burton-on-Trent brewery and the town water supply

	Bore-hole supply	*Town supply*
Total hardness (as ppm $CaCO_3$)	791	232
Ca^{2+} (as ppm $CaCO_3$)	473	160
Mg^{2+} (as ppm $CaCO_3$)	318	72
Total dissolved solids (ppm)	1060	375
Total alkalinity (as ppm $CaCO_3$)	516	103
pH	8.3	7.2

Obviously bore-hole water would not be the liquor of choice for cleaning, even though it may be relatively inexpensive and readily available. The town supply is relatively hard and although it may be satisfactory for cleaning, a detergent of lower specification could be used if this liquor was softened. It is usually more cost effective to soften water by ion exchange than chemically by sequestering agents.

Many of the problems attributable to water hardness are eliminated by the use of acid detergents and sanitizers. However, these types of product have certain other limitations compared with caustic detergents, and also their use is restricted mainly to light- and medium-duty cleaning.

10.4 CLEANING METHODS AVAILABLE

A variety of cleaning techniques are used in the brewery. Examples of the following would be found in most operations:

1. manual cleaning (bucket and brush);
2. immersion cleaning (soak baths, tackle baths);
3. high-pressure spray cleaning (general environmental cleaning, e.g. walls and floors);
4. foam cleaning, gel cleaning (fillers, small open fermenters);
5. CIP (of vessels and pipework) by recirculation using either high or low pressure devices.

The brewing process involves the movement of a liquid from one area to another. At each stage the composition of this liquid changes as it undergoes complex reactions and is heated, cooled and filtered.

The composition of the soils deposited by these processes also alters as previously described (Table 10.1). The greater part of the plant is large, enclosed, inaccessible and not amenable to manual cleaning. It is not surprising therefore that CIP techniques have evolved to cope with the

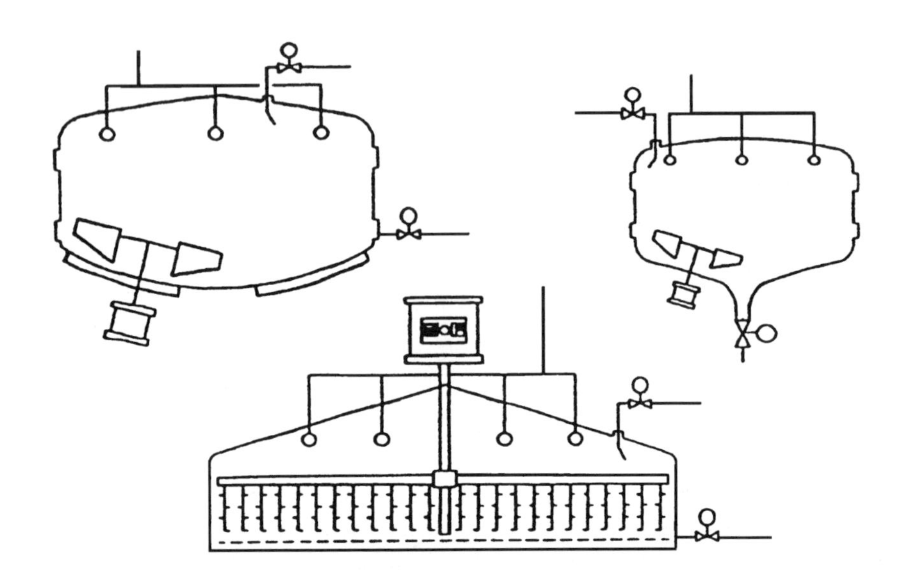

Fig. 10.1 Sprayball configuration in brewhouse vessels.

demands of towering cylindro-conical fermenters and the kilometres of pipework associated with the modern brewery.

It is beyond the scope of this chapter to detail the engineering aspects of modern cleaning systems. However, a basic understanding of the principles of CIP and the types of system available is necessary, as this chapter is largely concerned with cleaning and disinfection by CIP.

For cleaning vessels, there are two principal methods of applying detergents:

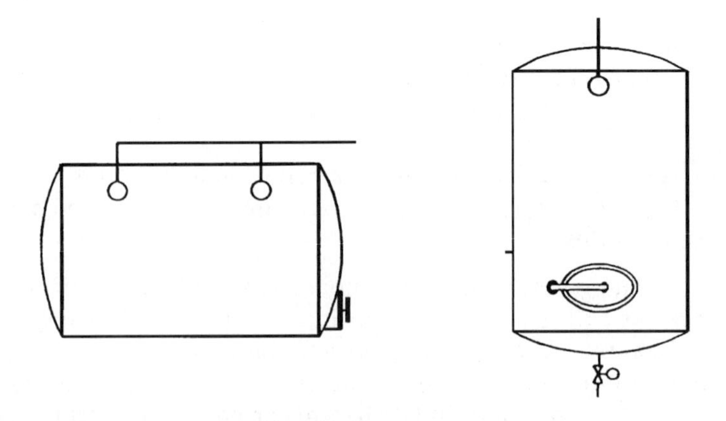

Fig. 10.2 Sprayball configuration in conditioning tanks.

Fig. 10.3 Rotating spray head in fermenting vessels.

1. through low-pressure devices such as spray balls placed strategically to give the desired coverage (Figs 10.1 and 10.2) and
2. through rotating high-pressure sprays (Fig. 10.3).

The cleaning of pipework is carried out by recirculating detergent solution through the pipes. Cleaning velocities and associated flow rates are presented in Table 10.3, which shows that there is an optimum flow range to achieve the best results.

Numerous types of cleaning systems (CIP sets) are available to match closely the equipment to cleaning requirements. There are three basic types:

1. total loss system;
2. partial recovery system;
3. full recovery system.

Diagrams of these three systems are shown as Figs 10.4, 10.5 and 10.6. The major difference between these systems is that the total loss system neither

Table 10.3 Cleaning velocities and associated flow rates ($l\ min^{-1}$)

Pipe outside diameter (inches)	*Cleaning velocity ($m\ s^{-1}$)*							
	0.9	*1.2*	*1.5*	*1.8*	*2.1*	*2.4*	*2.7*	*3.0*
1.0	27.8	37.1	46.3	55.6	64.8	74.1	83.4	92.6
1.5	62.5	83.4	104.2	125.1	145.9	166.8	187.6	208.4
2.0	111.2	148.2	185.3	222.3	259.4	296.4	333.5	370.6
2.5	173.7	231.6	289.5	347.4	405.3	463.2	521.7	579.0
3.0	250.2	333.5	416.9	500.3	583.6	667.0	750.4	833.8
4.0	444.7	592.9	741.1	889.3	1038	1185	1334	1482
6.0	1001	1334	1668	2001	2335	2668	3002	3335
8.0	1779	2372	2964	3557	4150	4731	5336	5929
12.0	4002	5336	6670	8004	9338	10672	12006	13340
	A		B			C		

Range A: poor cleaning; B: efficient cleaning; C: no improvement in cleaning, increasing risk of water hammer and damage to pipework and fittings.

recovers dilute detergent for reuse nor recovers water of pre- or post-rinses. Recovery systems not only save dilute detergent to be topped up and reused, they also save post-rinses for subsequent pre-rinse application.

In all cases, the cleaning cycles involve a pre-rinse to remove gross loose soil, and detergent recirculation to clean the vessel and pipework. After discarding or recovering the dilute detergent, the vessel is rinsed with clean water. A disinfectant may be added to this rinse water. Sometimes a pulsed rinse cycle ('burst rinsing') is used to ensure more efficient use of rinse liquor.

Total loss systems are more likely to be found in applications where

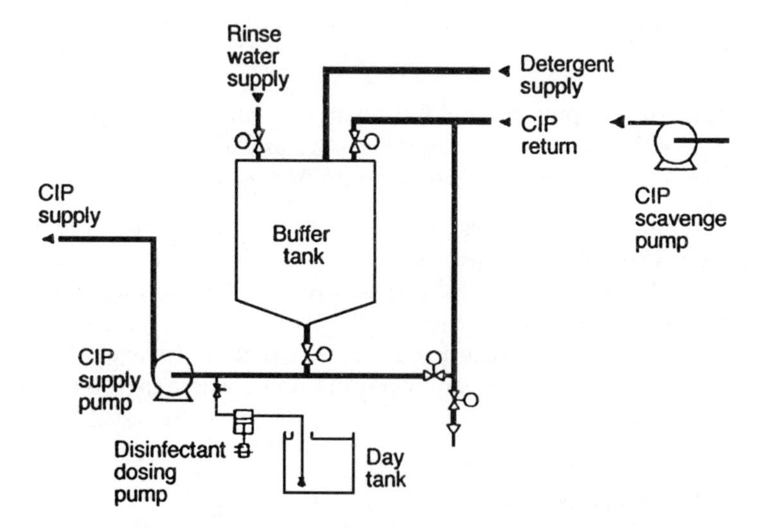

Fig. 10.4 Single use (total loss) system.

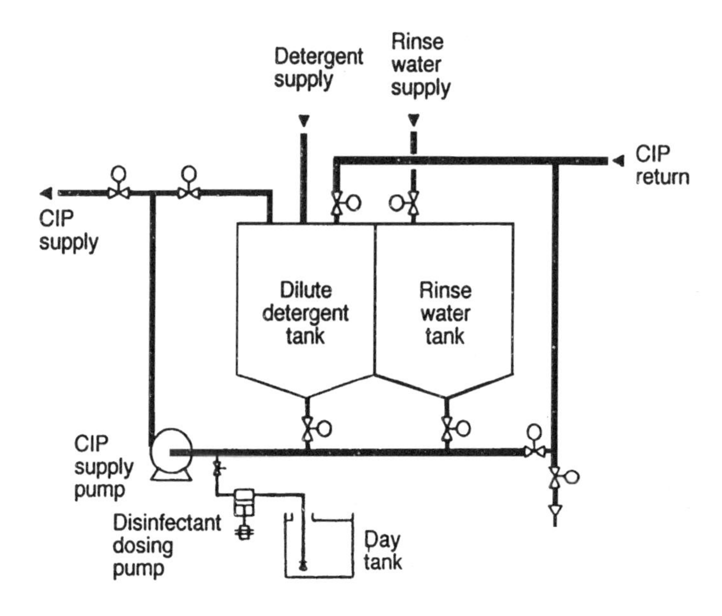

Fig. 10.5 Partial recovery system.

gross soiling is present and a recovered detergent may suffer heavy contamination, e.g. yeast recovery plants. In areas of light soiling, e.g. bright beer tanks, buffer tanks and road tankers, a recovery system may well be the best choice.

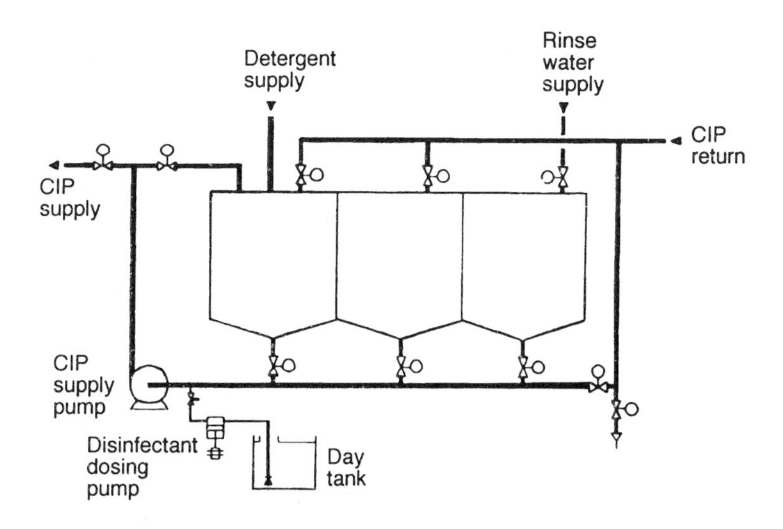

Fig. 10.6 Full recovery system.

The detergent process is a complex mixture of chemical and physical effects. There is a basic energy requirement to achieve satisfactory cleaning:

total cleaning energy required = chemical energy + mechanical energy + thermal energy

As a 'rule of thumb', plant used for hot processes (e.g. mashing, pasteurization) must be cleaned hot, and plant used for cold processes may be cleaned cold. In all cases, when a satisfactory cleaning regime has been established, it is vital to take into account the energy balance equation and to ensure that the total cleaning energy is available.

10.5 SOIL COMPOSITION

It is important to recognize how soil changes as we progress from the hot brewhouse through fermentation to cold conditioning: denatured proteins are gradually replaced by tannins and minerals. The physical volume of soil changes from one process to another and the effects of time and temperature further modify soils. The detergent chosen for the cleaning task must be reactive against at least a proportion of the soil if an acceptable cleaning standard is to be achieved.

10.6 PROCESS OF DETERGENCY

At least four distinct stages are involved.

1. The wetting of the surface to allow intimate contact between detergent and soil.
2. Chemical action on the soil. Here, at least three separate interactions are possible:
   (a) action of acids on mineral scales to dissolve the inorganic scale deposits;
   (b) hydrolysis reactions to solubilize proteins;
   (c) the saponification by caustic reaction on 'oily' components of soils.
3. The dispersion of large to finely divided soil particles.
4. The suspension in solution of any removed soil.

Cleaning of brewery plant by various soak, spray and recirculation techniques involves to a greater or lesser extent all of the above physical and chemical processes. The type of detergent and method of application determine how effectively the soil is removed. The following sections deal with the chemistry of detergents with reference to the removal of brewing soils.

10.7 CHEMISTRY OF DETERGENTS

Nearly all cleaning processes in a brewery employ either strongly acidic or strongly alkaline conditions (Fig. 10.7). The reason is simple: soil is most readily removed by chemical reaction and the two principal reactions, solution of inorganic soil and organic residues, occur in conditions of excess acidity (H^+) or alkalinity (OH^-).

The reactions are solubilization of mineral scale by acids:

$$CaCO_3 + 2HCl \rightarrow CaCl_2 + H_2O + CO_2\uparrow$$

solubilization of proteins:

$$\text{[complex protein]} \ \frac{(H^+) \ \rightarrow}{(OH^-) \ \rightarrow} \ -NH\,CH_2\,CO\,NH\,CH_2\,CO\,NH- \quad \text{'peptide chain'}$$

and solubilization of oils and fatty acids:

$$R\text{-}COOH + NaOH \rightarrow R\text{-}COONa + H_2O$$

(R = linear saturated or unsaturated alkyl group $\geq C_8$).

The use of near-neutral detergents is highly desirable because of handling, storage and effluent considerations. Currently the use of this type of product is limited mainly to manual operations where chemical energy is supplemented by physical action. For most CIP operations the use of detergents based on caustic soda prevails. However, sophisticated acidic detergents and sanitizers are becoming more widely adopted.

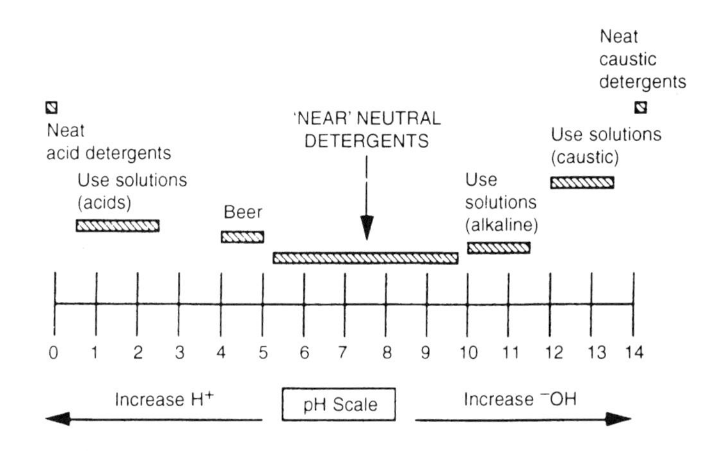

Fig. 10.7 pH scale showing relative positions of caustic, alkaline, 'neutral' and acid detergents.

10.8 CAUSTIC AND ALKALINE DETERGENTS

Caustic soda is a very reactive compound. It is the basis of most heavy-duty detergents used in the brewery. Undoubtedly, caustic soda has a number of advantages over other chemicals which may be considered in detergent compositions. The anti-microbial effect of NaOH solutions was demonstrated by Hobbs and Wilson (1942): Table 10.4 shows the lethal effect on *Bacillus subtilis* spores.

In washing beer bottles, temperatures around 70°C and causticities of 2–4% w/w are frequently employed. Under such conditions, a very pronounced anti-microbial effect is achieved and, provided the rinse water is of potable quality, microbiologically clean bottles may be obtained.

The principal disadvantages of NaOH as a detergent component for brewery applications are its reactions with water hardness salts and CO_2 atmospheres. Precipitation of water hardness salts could be one of:

1. $Ca(HCO_3)_2 + 2NaOH \rightarrow CaCO_3 \downarrow + Na_2CO_3 + H_2O$
2. $MgSO_4 + 2NaOH \rightarrow Mg(OH)_2 \downarrow + Na_2SO_4$
3. $CaSO_4 + Na_2CO_3 \rightarrow CaCO_3 \downarrow + Na_2SO_4$

The reaction with CO_2 is

$$2NaOH + CO_2 \rightarrow Na_2CO_3 + H_2O$$

It is chemically impossible to prevent these reactions. However, the deleterious effects of water hardness precipitation can be largely obviated by the formulation of carefully chosen additives into the caustic soda. Sequestrants are the principal additives used to prevent the precipitation of water hardness salts. Sequestrant technology has advanced significantly over recent years and the next section deals with the role of this group of compounds in brewery formulations.

Table 10.4 Percentage concentration of NaOH (w/v) required to destroy 25% of a population of *Bacillus subtilis* spores at various temperatures

Temperature		*Time (min)*				
°F	°C	*1*	*2*	*4*	*8*	*16*
120	49.0	2.44	1.64	1.10	0.74	0.50
130	54.5	2.15	1.44	0.97	0.65	0.44
140	60.0	1.90	1.28	0.86	0.58	0.39
150	65.5	1.66	1.12	0.75	0.51	0.34
160	71.0	1.46	0.89	0.66	0.45	0.30

10.9 SEQUESTRANTS

The sequestrants currently in use in detergent formulations can be subdivided into two main categories, based principally on their mode of action.

10.9.1 Stoichiometric sequestrants

This type functions by chelation, a process in which metal ions are held in a stable structure by the sequestrant, usually in a fixed ratio of metal ions to sequestrant. Important members of the group for brewery cleaning operations are:

1. amino carboxylic acids, nitrilotriacetic acid (NTA) and ethylenediamine tetra-acetic acid (EDTA);
2. hydroxycarboxylic acids (gluconic acid and its derivatives);
3. polyphosphates (sodium tripolyphosphate, sodium hexametaphosphate).

(a) Amino carboxylic acids

As their tri- and tetra-sodium salts, NTA and EDTA are widely used in detergent compositions. Both are useful additives if soil contains a high proportion of inorganic scale, since these sequestrants are able to form stable chelates with metal ions. This chelating action helps to break down the mineral content of soils, facilitating removal. The higher stability constants (Table 10.5) of NTA and EDTA chelates with metal ions indicate species of greater thermodynamic stability (Anon., 1972). The stability constant of the metal ion/sequestrant complex MZ is given by the expression:

Table 10.5 Logarithms of the stability constants of nitrilotriacetic acid and ethylenediamine tetra-acetic acid

Metal ion	*NTA*	*EDTA*
Mg^{2+}	5.4	8.7
Ca^{2+}	6.4	10.6
Mn^{2+}	7.4	13.8
Al^{3+}	–	16.1
Fe^{2+}	8.8	14.3
Zn^{2+}	10.7	16.5
Pb^{2+}	11.4	18.0
Ni^{2+}	11.5	18.6
Cu^{2+}	13.0	18.8
Fe^{3+}	15.9	25.1

$$K_{MZ} = \frac{[M^+][Z^-]}{[MZ]}$$

As Table 10.5 shows, the log K_{MZ} values of Ca with NTA and EDTA are 6.4 and 10.6, respectively. Therefore the ratios of the concentration of undissociated complex [MZ] to the dissociated complex $[M^+][Z^-]$ are $10^{6.4}$:1 and $10^{10.6}$:1 for NTA and EDTA, respectively. Clearly EDTA forms the more stable complex, and on this evidence would be the preferred choice for removing mineral scale containing a high proportion of Ca salts.

Table 10.6 shows the complexing efficiency of two commercial grades of Na_3NTA and Na_4EDTA within the pH ranges specified. It is evident that Na_3NTA is the more efficient agent in terms of actual amount of metal ion complexed per gram of active sequestrant (Anon., 1972). Brewery detergents containing these types of sequestrant are normally optimized mixtures, their composition depending on their final applications. For example, simply to soften water by removing Ca^{2+} and Mg^{2+} in the form of soluble chelates, NTA would be more efficient than EDTA. However, for removing tenacious scale as found in heat exchangers, EDTA would be more efficient.

(b) Hydroxycarboxylic acids

Gluconic acid is the most commonly used of this type of sequestrant. It is very soluble in caustic soda and has good long-term and high-temperature stability. To be fully effective as a chelating agent, gluconic acid must be used in the presence of free caustic alkali. At pH 11, 1 g sodium gluconate

Table 10.6 Complexing efficiency of 1 g of sodium salts of nitrilotriacetic acid and ethylenediamine tetra-acetic acid at 25°C

Metal ion	*Na3NTA*	*Na4EDTA*
Ca^{2+}	140–156 mg (pH 9.0–12.0)	110 mg (pH 8.5–11.5)
Mg^{2+}	78 mg (pH ?–10.0)	64 mg (pH 8.0–10.5)
Cu^{2+}	247 mg (pH 3.0–12.0)	167 mg (pH 1.5–13.0)
Fe^{3+}	217 mg (pH 1.5–3.0)	147 mg (pH 1.0–6.0)

will chelate only about 25 mg $CaCO_3$, whereas in 3% NaOH solutions (pH 14), 1 g will chelate 325 mg $CaCO_3$. Sodium gluconate may be formulated into high levels (> 40% w/w) NaOH, principally as liquid mixtures. This type of product finds widespread use in brewhouse cleaning operations and may be used at up to 5% w/w NaOH under 'boil out' conditions.

Sodium gluconate is also a very effective sequestrant for Fe^{3+} and Al^{3+}. This feature is extremely valuable in formulating detergents for washing beer bottles, since removal and suspension of aluminium is necessary for dealing with neck foils and labels. Additionally, removal of rust rings from bottle necks is facilitated by sodium gluconate.

(c) Polyphosphates

Polyphosphates are a very versatile sequestrant group but because of limited solubility and stability in caustic solutions, are more commonly incorporated in powdered detergents. Polyphosphates not only behave as stoichiometric sequestrants at high concentrations, they also have a threshold effect. Furthermore, they help to build up the detergency of formulations by improving dispersion and rinsing properties.

10.9.2 Threshold sequestrants

This type of sequestrant functions at a sub-stoichiometric level. A relatively small amount of a threshold sequestrant can prevent water hardness salts from producing a scale deposit. Unlike stoichiometric sequestrants which form soluble chelates, threshold sequestrants act by modifying the physical characteristics of a precipitate. In this respect they are said to behave as crystal growth modifiers in that they can distort crystal lattices, thereby preventing scale formation. They do not actually prevent the precipitation of insoluble salts. Threshold sequestrants find application in CIP and bottle-washing detergents, usually in combination with stoichiometric sequestrants.

Polyphosphates can exhibit threshold effects at sub-stoichiometric levels but the most notable threshold sequestrants are probably phosphonic acid derivatives such as amino-tris-(methylenephosphonic acid) (ATMP) and 1-hydroxyethane diphosphonic acid (HEDP). Polyacrylic acid derivatives can behave both as threshold agents and as dispersants and are often included in caustic and alkaline CIP detergents.

It is clear that sequestrants play an important role in alkaline and caustic detergent formulations. Amongst other factors, water conditions and soil composition in breweries determine the choice of product. A well-formulated detergent will allow some margin of variation in these parameters by combining the properties of both stoichiometric and threshold sequestrants.

10.10 ACIDS

Acids have been on the brewery detergent scene for many years. Originally used on an *'ad hoc'* basis mainly for descaling, their use has spread more and more to CIP cleaning of vessels and pipework from fermentation through to racking. One of the major reasons for the introduction of acids is compatibility with CO_2. The use of CO_2-compatible detergents offers savings in time, detergent and CO_2. Elmore (1980) discussed experimental results of CO_2 losses with a caustic detergent routine (Table 10.7): the effect of CO_2 on caustic solutions is substantial, even where the level of CO_2 is only 15%. The removal of CO_2 from a vessel is highly desirable before cleaning with a caustic-based detergent. However, where time is at a premium, this may not be practicable. It takes approximately 5 h to evacuate CO_2 from a vessel through the only exit available, a standard manway door, to achieve a level of CO_2 which is not detrimental to the caustic solution (Table 10.8).

The most commonly used acid for brewery applications is phosphoric acid. Blended with suitable surfactants it can form the basis of a good CIP detergent; also, it is often used in keg-washing products. When blended with nitric acid, the mixture is more aggressive but offers bacteristatic properties if enough nitric acid is present. Also, HNO_3 is cheaper than H_3PO_4 and has a passivating influence on stainless steel, but is totally destructive to any copper, brass or phosphor-bronze in the system.

In its raw form, sulphuric acid is rather corrosive to stainless steel, and should be formulated with a corrosion inhibitor if regular use is envisaged. H_2SO_4 is the cheapest source of acidity but its inherent detergency is not as good as H_3PO_4. It is used in some sanitizer formulations where very low pH is a prerequisite for most effective use of the biocide.

HCl can be very effectively inhibited from attacking mild steel with the use of cationic corrosion inhibitors. When formulated in this way, the

Table 10.7 Losses of CO_2 and NaOH during a caustic detergent routine with 2% NaOH

	CO_2					
Initial CO_2 (%)	*After pre-rinse (%)*	*After detergent cycle (%)*	*After final rinse (%)*	*Loss in complete cycle (%)*	*Fall in NaOH (%)*	*Increase in Na_2CO_3 (%)*
97	85	66	57	40	1.00	0.90
50	34	15	13	37	1.15	0.95
40	32	18	16	24	0.75	0.60
32	25	9	7	25	0.75	0.55
23	13	4	3	20	0.45	0.45
15	12	5	5	10	0.70	0.60
0	–	–	–	–	0.15	0.09

Table 10.8 Rate of evacuation of CO_2 from a brewery vessel

Time (h)	*CO_2 content (%)*
0	98
0.5	45
1	25
1.5	20
2	13
2.5	9
3	7
3.5	6
4	5
4.5	4
5	3
5.5	3

products are used mainly for removing water hardness scale from bottle-washing machines.

Sulphamic acid is available as a powder and has good descaling properties. It can be formulated into liquids, but tends to hydrolyse. Since it is quite expensive and does not offer any significant advantages over H_3PO_4, it has limited application.

For any acid, corrosion characteristics need careful consideration. The mineral acids commonly used in brewery detergents are highly corrosive to mild steel and must not be used without effective corrosion inhibitors in the formulation. Figures 10.8 and 10.9 show the corrosion profiles of HCl, HNO_3, H_3PO_4 and H_2SO_4 against stainless steel types 304 and 316. Clearly, nitric and phosphoric acids demonstrate far better compatibility with these metals and may be used without problems (Elmore, 1980). However, the level of Cl^- in dilution water for detergents should always be checked when considering acid detergents, because of the increased risk of pitting corrosion.

Organic acids such as citric, gluconic or tartaric are less powerful descalers than the mineral acids and much more expensive. They are not widely used in brewery detergents.

10.11 SURFACE ACTIVE AGENTS

Surface active agents, 'surfactants', are used in brewery detergents only in certain highly specialized areas, e.g. a type of surfactant is often used in bottle washing as a defoaming agent and in acidic CIP detergent formulations to improve detergency. Surfactants have four main functions in brewery detergent formulations.

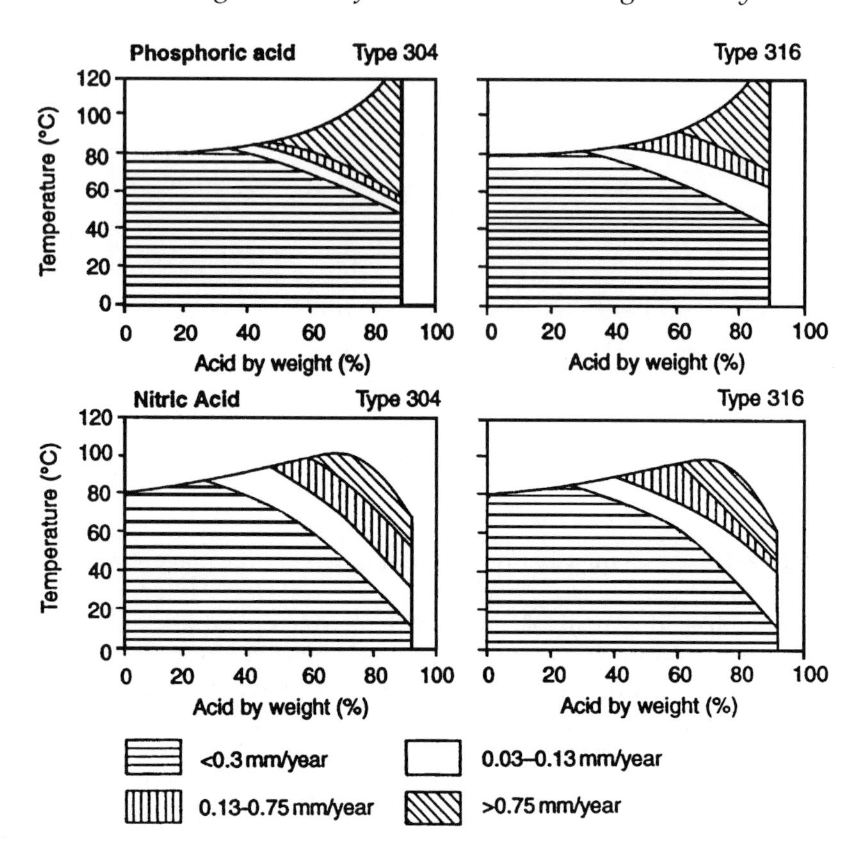

Fig. 10.8 Corrosion profile of phosphoric and nitric acids.

1. To reduce the surface tension of the liquid. This will help the wetting process.
2. A surfactant may produce foam, as a consequence of a reduction in surface tension and entrainment of air by physical action. The stability of this foam is then influenced by other surface effects. Foam generation is undesirable in CIP but vital for 'foam-cleaning' operations.
3. Some surfactants which are water insoluble can be used to defoam detergent solutions and application is found for these compounds in bottle-washing products.
4. Apart from helping in the physical processes of detergency by reducing surface tension, surfactants may also prevent readhesion of soil which has previously been removed by dispersion and micelle-forming action.

All surfactant molecules have a hydrophilic and hydrophobic structure. When added to water, the molecules concentrate at the surface in an attempt to have as much as possible of the hydrophobic portion out of the

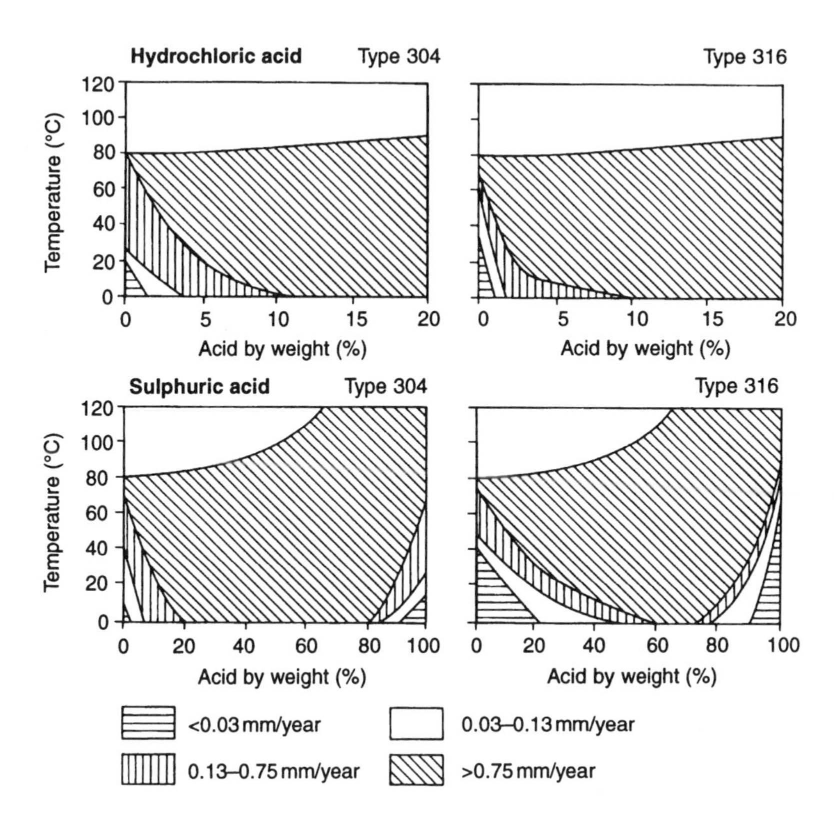

Fig. 10.9 Corrosion profile of hydrochloric and sulphuric acids.

water. Above a certain concentration, the molecules which cannot find space at the surface form micelles in the bulk solution, capable of holding soils in suspension and even of solubilizing them.

Surfactants are classified according to the charge on the dissociated species, the anionic being negatively charged and the cationic positively charged molecules. There is no net charge on non-ionic surfactants, and the amphoteric group may be of either charge, depending on pH.

In the anionic group are soaps which are good detergents and lubricants, and alkyl sulphonates and sulphates which are used in high-foaming 'neutral' detergents or 'washing-up liquids'. Soaps are often used in brewery conveyor lubricants.

Cationic surfactants include fatty amine salts which are used as corrosion inhibitors, especially in water treatment applications. Quaternary ammonium compounds (QACs) are used mainly for their biocidal properties. Most but not all QACs have an undesirable effect on beer head retention and have limited brewery applications.

For brewery cleaning, non-ionic surfactants play an important role. They have good hard-surface detergency and are used as both track lubricants and CIP products. Certain non-ionic surfactants also find application as defoamers.

The amphoteric compounds are a versatile, if somewhat expensive, group of compounds. Certain amphoterics are biocidal, whilst others are used for their 'mildness' in shampoos and bubble bath, but are not of great significance as brewery detergents.

It is envisaged that the use of specialized surfactants in brewery detergents will increase in the future. This may well be driven by a requirement to reduce caustic or acid strength of heavy-duty products for environmental reasons, or to tackle the cleaning problems associated with pH-sensitive parts of the plant where high or low pH would not be acceptable.

10.12 DISINFECTANTS AND SANITIZERS USED IN BREWERIES

Detergents are used in CIP, soak, spray and manual applications. Disinfectants can also be used in these areas, but there are special types used in the treatment of recirculating water such as cooling towers and pasteurizers. Water treatment biocides will be considered in section 10.15.

Table 10.1 showed the required levels of cleanliness at various stages of the brewing process. The next sections of this chapter will be concerned with the stages, from the brewhouse to the final packaged product, where disinfectants are required.

The disinfection stage is used as an extra guarantee of cleanliness, e.g. in CIP, and in some cases to disinfect and to prevent recontamination, e.g. soak baths. It does not compensate for a bad detergent stage or badly designed cleaning processes or equipment. Disinfection may be defined as 'the reduction of microbial numbers to an acceptable level', a level usually determined by the brewery for a particular area by sampling the final rinse of CIP or by swabbing surfaces of cleaned equipment. Usually in breweries the product is examined at a number of stages for the presence of bacteria and wild yeasts.

A disinfectant used in breweries should comply with the following factors:

1. compatible with other chemicals;
2. compatible with plant materials;
3. tolerant of hard water;
4. tolerant of soil;
5. low or non-foaming;
6. safe in use, non-irritant to skin;
7. non-tainting;
8. no effect on head retention;

9. broad anti-microbial activity at low concentration and low temperature;
10. economical in use;
11. of low environmental impact.

Unfortunately, no disinfectant meets all of these requirements. The chemistry and application advantages and disadvantages of the disinfectants most commonly used in the brewing industry will be discussed. Disinfectants can be considered as oxidizing and non-oxidizing types.

10.13 OXIDIZING DISINFECTANTS

10.13.1 Halogens

Usually chlorine or iodine disinfectants are used, but recently bromine has found an application.

(a) Chlorine

Chlorine gas was first used for fumigation in hospitals in 1791 (Sykes, 1972) but this usage has an important disadvantage: the gas is highly toxic to humans.

Active chlorine is supplied in two forms:

1. inorganic compounds containing hypochlorite ions either as a liquid, e.g. NaOCl, or as a powder, e.g. chlorinated trisodium phosphate $(Na_3PO_4.11H_2O)_4.NaOCl.NaCl$;
2. powdered organic chlorine-release agents, e.g. trichloroisocyanurate:

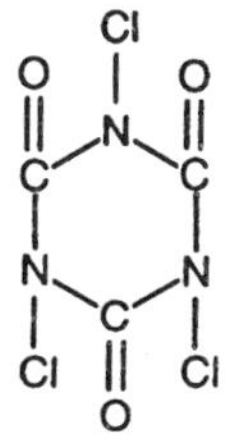

In solution, all of these compounds hydrolyse to produce hypochlorous acid and/or hypochlorite ions, depending on pH.

Acid → Alkaline

$$\underset{\text{chlorine gas}}{Cl_2} \rightleftharpoons \underset{\text{hypochlorous acid}}{HOCl} \rightleftharpoons \underset{\text{hypochlorite ion}}{OCl^-}$$

In breweries, NaOCl is normally used. It is most stable in a slightly alkaline solution, and for this reason the concentrate is supplied stabilized with NaOH at a pH of up to 12. In an in-use solution of between 50 and 300 ppm the pH will be between 7 and 9. At the optimum pH for disinfection, pH 5.0, the solution is less stable. Below pH 5.0 chlorine gas will be produced. NaOCl hydrolyses proteins making them more soluble and more easily removed.

NaOCl has many advantages: it is non-foaming, has some hard-water tolerance, does not leave an active residue and has a wide anti-microbial spectrum which includes activity against viruses and bacterial spores. It is fast acting and economical to use. The disadvantages are that it can be corrosive at high concentrations, irritating to skin, the in-use solution is unstable, it is affected by organic soil and may give taint problems due to formation of chlorophenol and chloramines.

A recent application of a chlorine compound is the use of chlorine dioxide for disinfection of process waters in breweries. The disinfecting power of ClO_2, an unstable, potentially explosive and toxic gas, has been known since the 1920s. It is soluble in water but unstable in solution. ClO_2 can be manufactured thus:

$$2NaClO_2 + 4HCl \rightarrow 4ClO_2 + 5NaCl + 2H_2O$$

Sodium chlorite itself is explosive: it is supplied in solution and mixed with HCl when the chlorine dioxide is required. Because of the hazardous nature of the reactants and product, expensive equipment is required to manufacture chlorine dioxide at the point of use. In most applications 0.1–0.5 ppm is sufficient to disinfect liquor but higher levels are required for a positive disinfection action on surfaces.

ClO_2 at use dilution overcomes some of the disadvantages of NaOCl in that it is non-tainting, non-corrosive and non-toxic. The reaction products of ClO_2 with organic substances are far less toxic than those of NaOCl.

(b) Iodine

Iodine itself is not very soluble in water and the vapour is irritating to the eyes which makes it difficult to handle. Iodine is a very reactive element, but it is this reactivity which makes it a good disinfectant. Compounds used in the brewing industry contain iodine complexed with surface-active agents which act as carriers or solubilizers. These complexes, known as iodophors, were first introduced in 1949 (Kelley, 1961). The complexes are water soluble and overcome the handling difficulties of I_2 but retain the disinfecting power. It is the free I_2 which is the disinfecting agent. I_2 is released by the iodophor on dilution. The optimum pH for anti-microbial activity is around 5.

Acid				Alkaline
I_2	⇌ HOI	⇌ OI^-	⇌ IO_3^-	
greatest activity	some activity	inactive	inactive	

The surface-active agents provide better wetting and organic soil penetration, so iodophors are less affected by soil than hypochlorite.

Iodophors have a broad anti-microbial spectrum which is similar to hypochlorite, although they tend to be less active against bacterial spores. Like hypochlorite they are fast acting but tend to be more expensive. They are used in soak baths and spray applications at up to 10 ppm available I_2 but only low-foam iodophors can be used in CIP. In solution, iodophors are yellow-brown in colour. The colour can be an advantage, e.g. in a soak bath the colour indicates the presence of active iodine. In-use solutions are unstable; as the iodine dissipates the solution will go colourless. The colour can be a disadvantage by causing staining, particularly with some plastics. This may also result in taint problems and possible effects on beer head retention. Unlike hypochlorite, iodine does not react with amines. Iodophors can be corrosive, therefore it is necessary to ensure that the correct concentration is used. In hand use, skin irritation can be a problem.

(c) Bromine

Bromine itself is not used as a brewery disinfectant (except for the treatment of recirculating water, described in section 10.15.1) because of handling difficulties and because it offers no advantages over chlorine. Bromine hydrolyses in water to form a weak acid, hypobromous acid (HOBr), which is the primary active anti-microbial agent.

10.13.2 Oxygen-releasing compounds

(a) Peracetic acid

Peracetic acid was first introduced as a disinfectant in 1955 (Hugo, 1991). The product supplied is an equilibrium mixture:

$$CH_3C(=O)O\text{—}OH + H_2O \longrightarrow CH_3C(=O)OH + H_2O_2$$

peracetic acid — acetic acid — hydrogen peroxide

As the equation above shows it is soluble in water; the breakdown products are harmless:

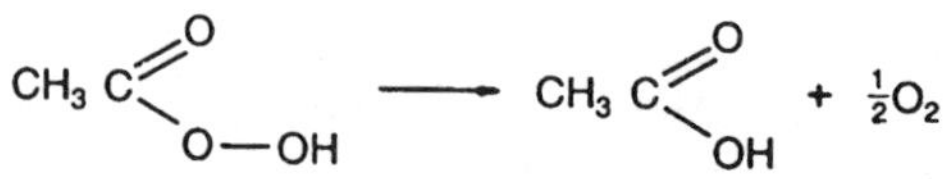

$$CH_3C(=O)O{-}OH \longrightarrow CH_3C(=O)OH + \tfrac{1}{2}O_2$$

There is sometimes concern about the production of oxygen from the breakdown of peracetic acid, but at the dilution used for disinfection this should not be a problem. Peracetic acid has a very unpleasant smell and in the concentrate is corrosive, highly reactive and strongly oxidizing. Because of these properties it is very unpleasant to handle and is therefore not recommended for manual use. Its main use is as a terminal disinfectant in CIP.

In use, solutions are not very stable and will react with organic materials. Peracetic acid may attack plant material such as rubber gaskets, and at higher concentrations pitting corrosion may be a problem in water with high levels of Cl^-.

Peracetic acid has a wide anti-microbial spectrum which includes viruses and bacterial spores, but resistant yeasts have been reported. Its activity is fast and is maintained at temperatures lower than ambient, which is particularly important in disinfecting bright beer tanks and conditioning vessels.

(b) Hydrogen peroxide

Hydrogen peroxide was introduced as a disinfectant in 1887 (Hugo, 1991). It is supplied in solution which has a tendency to decompose:

$$H_2O_2 \rightarrow H_2O + \tfrac{1}{2}O_2$$

Manual use of H_2O_2 is not recommended; its main use is in spray applications such as aseptic packaging. Hydrogen peroxide is bactericidal and fungicidal, but some bacteria and fungi are less sensitive because they produce the enzyme catalase which can destroy H_2O_2. Since it is slow acting, a long contact time and/or elevated temperatures are required for good anti-microbial activity.

10.14 NON-OXIDIZING DISINFECTANTS

10.14.1 Quaternary ammonium compounds

QACs were first introduced in 1916 (Hugo, 1991), and are probably the best-known cationic surface-active agents. Their general formula is

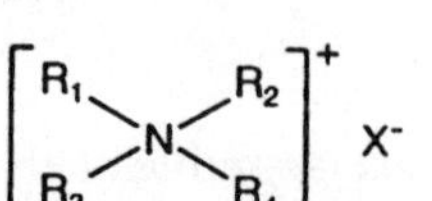

$$\left[N R_1 R_2 R_3 R_4 \right]^+ X^-$$

where X is usually a halide but is sometimes SO_4^{2-}. R_1, R_2, R_3 and R_4 may be a variety of alkyl or aryl groups.

QACs are generally poor detergents but good wetting agents allowing penetration of soil. In solution they ionize producing a cation, the substituted nitrogen part of the molecule, which provides the surface-active property. The length of the carbon chain in the group affects the disinfectant ability; usually C_8–C_{18} are the most effective. QACs which are effective disinfectants are generally too high-foaming for CIP applications. Low-foaming products generally have less anti-microbial activity and may be difficult to dissolve in water. Their main uses are in soak and manual cleaning at concentrations above 200 ppm active QAC. Optimum anti-microbial effect is at about neutral pH, but generally they show activity between pH 3.0 and 10.0. The use dilution is stable even at higher temperatures. They are less affected by organic soil than the halogens, but may be precipitated by tannins present in beer.

The cationic nature of these products means that they can absorb to surfaces and may be difficult to rinse off. QACs used in a brewery must be free rinsing or taint and head retention problems may arise.

The anti-microbial range of QACs is less than for the oxidizing disinfectants. They are generally less effective against Gram-negative bacteria than against Gram positive. There is also a possibility of resistant bacteria developing with prolonged use. QACs have limited activity against bacterial spores and almost no effect against viruses. To be effective against yeasts and moulds, a higher concentration is required. QACs tend to be expensive in use compared to the halogens.

10.14.2 Biguanides

Biguanides with anti-microbial activity were first reported in the 1950s (Hugo, 1971). The biguanides are derived from guanidine, a naturally occurring substance found in cereals and vegetables such as turnips:

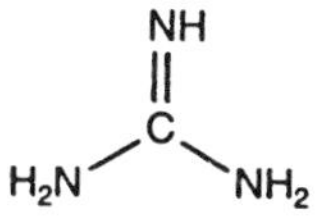

$$H_2N-C(=NH)-NH_2$$

guanidine

Biguanides are usually supplied as polymers and in the form of salts, mostly as the hydrochloride. Optimum activity lies between pH 3.0 and pH 9.0. Below pH 3.0 they are inactive against microorganisms. Above pH 9.0 they are precipitated, therefore care should be taken to ensure thorough rinsing after caustic detergent treatment. Biguanides are cationic in nature but are not regarded as surface active. If properly rinsed they will not affect beer head retention. They are also non-foaming and so are suitable for CIP.

Biguanides are also used in soak, manual cleaning and as tunnel pasteurizer biocides. Most biguanides have equal antibacterial activity against Gram-negative and Gram-positive bacteria. They are less effective against moulds and yeasts, especially at acid pH, and are ineffective against bacterial spores and viruses.

10.14.3 Amphoterics

Amphoterics have been in use as disinfectants since the early 1950s (Block, 1977). They are based on an amino acid, usually glycine, with substituents. Sometimes the term ampholyte is used because in solution they ionize to produce cations, anions or zwitterions depending on pH.

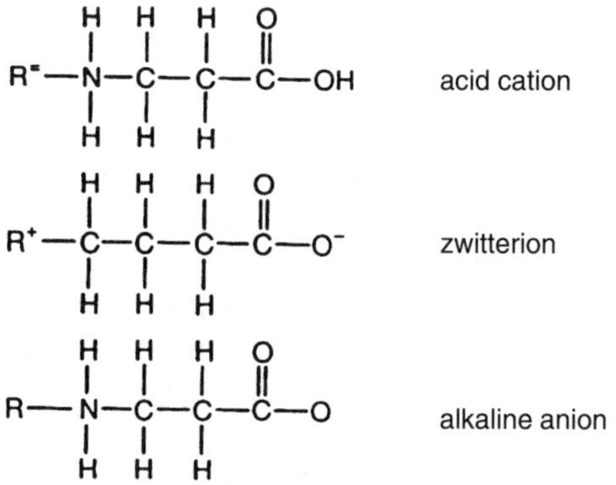

```
       H   H   H   O
       |   |   |   ||
R⁺—N—C—C—C—OH        acid cation
       |   |   |
       H   H   H

       H   H   H   O
       |   |   |   ||
R⁺—C—C—C—C—O⁻        zwitterion
       |   |   |
       H   H   H

       H   H   H   O
       |   |   |   ||
R—N—C—C—C—O          alkaline anion
       |   |   |
       H   H   H
```

Only certain amphoterics have both disinfecting and surface activity. The disinfecting ability appears to increase with the number of basic groups. Amphoterics have good wetting properties but cannot dissolve protein or beer stone. Although viscous liquids, they are freely soluble in water. Because of their surface activity, they tend to be too high-foaming for use in CIP.

Amphoteric disinfectants can be expensive to buy and use, but generally have low mammalian toxicity and low skin irritation, which can justify the cost. Their main applications are soak, spray and manual use. Optimum activity lies between pH 3.0 and 9.0, but depends on the formulation. They are equally effective against Gram-negative and Gram-positive bacteria, but are less effective against yeasts and moulds, and have little effect against viruses and bacterial spores. Properties such as soil tolerance and corrosion vary with the ampholyte, but corrosion is not normally a problem. The in-use solution, usually 1000 ppm, is stable.

10.14.4 Acid anionics

The active molecule in acid anionics varies considerably. There are two main types:

1. based on anionic surfactants combined with mineral acid

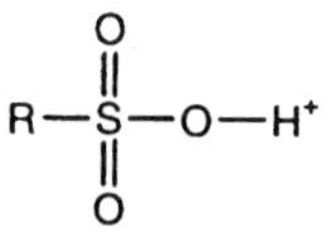

$$R{-}\overset{\displaystyle O}{\underset{\displaystyle O}{\overset{\|}{\underset{\|}{S}}}}{-}O{-}H^{+}$$

2. based on carboxylic acids, including fatty acids and derivatives

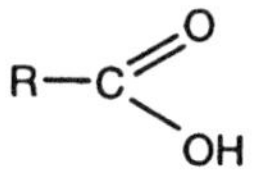

$$R{-}C\begin{matrix}{=}O\\ {-}OH\end{matrix}$$ carboxylic acid

Acid anionics tend to be formulated products with additions to assist activity and/or solubility. These properties will vary with the product. Acid anionics tend to have some detergency and wetting ability. The higher foaming products are unsuitable for CIP. Their general use is in spray applications. They are not suitable for hand use because of the low pH, 2, which is required for optimum anti-microbial activity.

In general, acid anionics show good anti-microbial activity against bacteria but are less effective against bacterial spores and viruses. Only some of the carboxylic acid types are active against yeasts and moulds.

10.15 WATER TREATMENT

Treatment of recirculating water in cooling towers and tunnel pasteurizers is a specialized area. In such situations, biocides are used to prevent slime developing through microbial growth. In tunnel pasteurizers, slime development blocks filters and spray bars, affecting the efficiency of pasteurization, and could result in problems with the product. Slime development in recirculating water is associated with bad odours, possible corrosion and potential health problems, e.g. legionellosis, caused by *Legionella* spp.

Any biocides used in tunnel pasteurizers must not affect can ink pigments or crown corks and must be compatible with other chemicals used, such as scale inhibitors and corrosion inhibitors. Some disinfectants already mentioned, such as QACs, biguanides and bromine, are used in recirculating water system. The properties of QACs and biguanides have already been discussed. The use of bromine is relatively new.

10.15.1 Bromine

Hypobromous and hypochlorous acids can be produced by dissolving bromo-chloro-dimethylhydantoin, supplied as a solid or powder, in water in a controlled manner through a brominator:

$$C_5H_6BrClN_2O_2 + 2H_2O \rightarrow C_5H_8N_2O_2 + HOBr + HOCl$$

The water flow through the brominator is regulated to give a residual level of bromine of 0.5 ppm available bromine. Bromine in this form is less corrosive than hypochlorite. Bromo-chloro-dimethylhydantoin provides a relatively safe way of handling bromine.

10.15.2 Aldehydes

This group of biocides contains highly reactive aldehyde groups, responsible for their anti-microbial properties. The main aldehydes for disinfection are formaldehyde and glutaraldehyde:

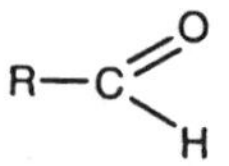

$R{-}C({=}O){-}H$

R = H formaldehyde

R = CHO–$(CH_2)_3$ glutaraldehyde

Aldehydes have an optimum pH range of alkaline to slightly acidic. For good anti-microbial activity, depending on the concentration, 15 min minimum will be needed, but frequently several hours' treatment is required.

Formaldehyde was first used as a disinfectant in 1886 (Hugo, 1991), but is rarely used as a disinfectant in breweries now because of its toxicity, by inhalation and skin contact, and its reported carcinogenicity. HCHO complexed in formaldehyde-release agents has provided a safe method for use as preservative or biocide.

Glutaraldehyde, currently used as a tunnel pasteurizer biocide, was introduced as a disinfectant in 1957 (Hugo, 1991). Glutaraldehyde is sometimes used in mixtures with QACs to increase their anti-microbial activity.

10.15.3 Isothiazolinones

Several isothiazolinones are used as biocides and preservatives. The most common is a mixture of 2-methyl-4-isothiazolin-3-one and 5-chloro-2-methyl-4-isothiazolin-3-one:

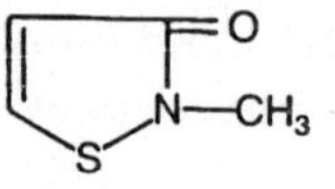

2-methyl-4-isothiazolin-3-one

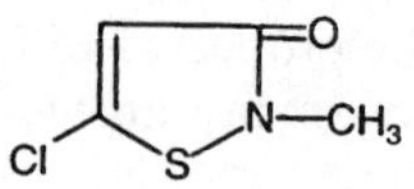

5-chloro-2-methyl-4-isothiazolin-3-one

These products are used in tunnel pasteurizers and cooling towers as anti-microbial agents and are added to dilute detergent tanks in CIP to prolong the useful life of the detergent.

Isothiazolinones tend to be bacteriostatic at very low concentrations, but are bactericidal after long contact of several hours. Their anti-microbial activity includes activity against sulphate-reducing bacteria, which if allowed to grow in tunnel pasteurizers can contribute to corrosion and produce bad odours and black slime.

10.15.4 BNPD

BNPD (2-bromo-2-nitropropane-1,3-diol) was originally developed as a preservative for cosmetic, pharmaceutical and toiletry products. It has now found an application in recirculation water treatment, although BNPD is only a slow-acting biocide.

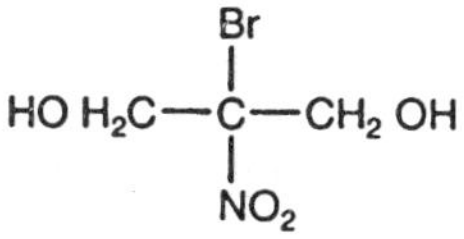

$$HOH_2C-\underset{NO_2}{\overset{Br}{\underset{|}{\overset{|}{C}}}}-CH_2OH$$

BNPD

Laboratory and field studies have shown BNPD to be effective against a wide range of microorganisms including *Legionella* spp. BNPD is readily soluble in water, but is more stable in acid conditions. At use dilutions it is non-toxic and the molecule is readily biodegradable.

10.16 STEAM

Steam is very effective for sterilizing equipment. It is effective in killing bacteria, moulds, yeast, bacterial spores and viruses. It has limited use in breweries, where in many areas chemical disinfection has replaced steam. Steam must be wet and air free. In breweries, steam sterilization can take up to 1.5 h to be effective.

Steam is expensive to produce, is dangerous to handle and bakes some types of soil onto surfaces, making it more difficult to remove. Heat-sensitive materials obviously cannot be steam sterilized.

10.17 SUMMARY

In the brewhouse, no terminal disinfectants are used. Cleaning is either manual or CIP, relying on heat and causticity to obtain a satisfactory level of cleanliness. From fermentation through to yeast recovery, cleaning and disinfection are achieved by CIP or manually. The products used here are detergents followed by disinfectants or combined detergent–sanitizer.

The main criteria for a CIP disinfectant are that it should not foam, be chemical and plant compatible, have broad anti-microbial activity at low temperature and concentration, and have no detrimental effect on the product. The disinfectants used in CIP are usually sodium hypochlorite, peracetic acid and iodophors.

The combined detergent–sanitizers are usually of the acid anionic type which are particularly effective at low temperatures. For manual cleaning, safety has to be a major consideration. For this reason, amphoterics are often used as the disinfectants. The other commonly used disinfectants are iodophors, chosen because of their broad anti-microbial spectrum and for their colour, which indicates whether there is any free I_2 left in solution. For cleaning conditioning vessels and bright beer tanks the disinfectant needs to be effective at low temperatures and not to have a detrimental effect on the finished product. Peracetic acid is often used in these areas.

The anti-microbial actives used in water treatment tend to be more aggressive chemicals. Here there is no cleaning stage, so the active compounds have to be able to cope with soil. Obviously chemical, plant and packaging compatibilities are major considerations, along with a broad anti-microbial activity.

REFERENCES

Anonymous (1972) *BASF handbook – the 'Trilon' Range of Sequestrants*, p. 10.
Anonymous (1976) *BS 5283: Glossary of Terms Relating to Disinfection*, British Standards Institute, London, p. 4.
Block, S.S. (1977) *Disinfection, Sterilisation and Preservation*, Lea & Febiger.
Elmore, D.G. (1980) *Proceedings of the Brewing Technology Conference, Harrogate*, 8.
Hobbs, G. and Wilson, G. S. (1942) *Journal of Hygiene*, **42,** 436.
Hugo, W.B. (1971) *Inhibitors and Destruction of the Microbial Cell*, Academic Press, London.
Hugo, W.B. (1991) *Journal of Applied Bacteriology*, **71,** 9.
Kelley, F.C. (1961) *Proceedings of the Royal Society of Medicine*, **54,** 831.
Sykes, G. (1972) *Disinfection and Sterilisation*, Spon, London.

Index

α-factor 44–5
Absidia 90, 106
Absidia corymbifera 93
Acetic acid bacteria 165–8, 185
  beer spoilage 168
Acetobacter 165–7, 205, 210
  characteristics differentiating species of 166
  general characteristics 165–6
  metabolism 165–6
  taxonomy 165
Acetolactate decarboxylase 68–71, 140, 142
Acetyl coenzyme A (acetyl CoA) 20
  fatty acid synthesis 27
  Lactobacillus 140
Acids
  cleaning/disinfecting 286–7
  hydrochloric 289
  nitric 286, 288
  phosphoric 286
  sulphanamic 286, 289
  see also individual acids
Acinetobacter 183
Acridine orange staining 30
Actidione agar 196–7
Adenosine triphosphate (ATP)
  bioluminescence 216–8
  glycolysis 17–19
  method for rapid viable cell counts 216
Aerococcus 129, 130
Alcaligenes 96, 183, 210
Alcohol dehydrogenase (ADH) 19
Aldehydes 298
Ales 33, 36, 39
Alternaria 84, 87
Alternaria alternata 87, 90, 99, 105
  on malt 99, 101
AMB-ID system 247
Amino acids 14
  uptake of isotopically labelled 26
  yeast nutrition 26
Aminocarboxylic acids 283
Amino-tris-(methylene phosphoric acid) (ATMP) 285
Ammonium ions 14
Amphibolic pathways 15
Amphoterics 296
Anabolic pathways 15
Anaerobic
  bacteria 180–83
  growth test 156
Anaplerotic pathways 15
Antigen–antibody reaction 222–4
API test kits 200, 260, 261
Arthrobacter globiformis 100
Ascomycetes 1–3
Aspergillus 84–115
  as storage fungi 89–96
Aspergillus candidus 102
Aspergillus flavus 94, 102
Aspergillus fumigatus 89, 91, 93, 94
Aspergillus glaucus 88, 90
  effect on beer 108
Aspergillus restrictus 91
Aspergillus versicolor 94, 102
Aureobasidium pullulans 101
Azotobacter chroococcum 103

Bacillus 210
  identification 155
  on malt 101
  nitrosamines 154

Bacillus (contd.)
in wort 154
Bacillus coagulans 154
Bacillus stearothermophilus 154
Barley, microflora of 83–96
Basidiomycetes 1, 6
Betabacterium 129
Betacoccus 128
Biguanides 295–6, 297
Binary fission 4
Biocides 297–9
Blastomycetes 2
Brettanomyces 6
as contaminants 196, 204–5
identification 199
Brewers' wort composition 14–15
Bromine 293, 297
Bromonitropropane diol (BNPD) 299
Budding
multilateral 4
polar 4

Candida 2, 6, 196, 205
on barley 86
in malting 99
Carbohydrate metabolism
acetic acid bacteria 165–7
during fermentation 17–26
enterobacteria 169–70
Gluconobacter 167
lactic acid bacteria 131–3
synthesis 24–26
Carnobacterium 130, 131
Catabolic pathways 15
Catabolite repression 25
Catalase test 155
Caustic soda 281
Cell
composition
analytical techniques 249–52
physical techniques 248
counts 28–9
DEFT 220–21
Chlorine 291–2
Chlorine dioxide 292
Chromosomes 46
karyotyping 72, 226–8
transfer 54–6
CIDS spectrum 248, 265
Citrobacter freundii 174
Cladosporium
on barley 87, 90
on malt 99, 100
Clamped homogeneous electric field electrophoresis (CHEF) 72, 226–8
Cleaning-in-place (CIP) systems 36, 275
detergents 286
methods 275–80
techniques 275–8
sequestrants 285
velocities 278
Colupulone 141
Conductance detection 21–4, 258–60
Continuous fermentation 39
Copper sulphate medium 195, 197
Corrosion profiles 288, 289
Crabtree effect 25
Cryptococcus 210
Cylindroconical fermentors 36, 38

Debaryomyces 2, 5, 199
Detergent
acid 275
alkali 275, 282
chemistry 281
CO_2-compatible 286
definition 272
near neutral 281
process 280
recircularization 278–80
recovery systems 278–89
Deuteromycetes 1
Diacetyl 68–71
Lactobacillus 241–2
Dimethl sulphide 172–4
Direct epifluorescent filter technique (DEFT) 30, 198, 220–21, 257–60
Disinfectant
definition 272
use 290–91
Disinfection 272, 290
Dispersants 285
DNA 46
random amplified polymorphic (RAPD) 230–31
recombinant 58, 61–71
transformation 58–9, 61
2 μm 46–8, 60, 61

Electrical detection, *see* Conductance detection
Electrophoresis
whole cell proteins 221–2, 246–7
Embden–Meyerhof–Parnas route (EMP) 17–23
in bacteria 169
Endospore-forming bacteria 154

Enterobacter aerogenes 68, 176
Enterobacter agglomerans, see Rhanella aquatalis
Enterobacter cloacae 171, 176
Enterobacteriaceae 169–77, 185
  beer spoilage by 170–71
  metabolism 169–70
  nitrosamines 177
Enterococcus 152
Enzyme-linked immunosorbent assay (ELISA) 198, 222–4, 244–6, 265
Enzyme tests for identification 263–4
Epicoccun 112
Epicoccun nigrum 112
Equilibrium relative humidity (ERH) 91
Erwinia herbicola 96
  in malting 98, 101
Ethylenediamine tetra-acetic acid (EDTA) 283–4
4-Ethyl guaicol 175
Eurotium
  on barley 86, 90, 93, 95
  and gushing 108
  in malting 99–102

Fatty acid methyl ester (FAME) profiles 249–52, 265
Fermentation 34–40
  attemperation of 36, 38
  carbohydrate metabolism during 17–24
  computer simulations 39
  continuous systems 39
  improved performance 48
  monitoring 39
  profile 39
  temperatures 39, 40
  time for 38–9
Ferulic acid 175–6
Filamentous fungi 86–8
Flavobacterium 96, 98, 171
  see also Obesumbacterium
Flocculation (*FLO*) genes 67
Flocculation of yeasts 67, 202–3
Flow cytometry 215, 248
Fluorescence microscopy 29–30, 219
Fourier transform infrared spectroscopy (FT-IR) 251–2
Freeze-drying 31
Fungi
  classification of 1
  filamentous 86–88
  seedling blight 103–5
  storage 88–96
Fusarium
  during malting 97
  germination stimulation 105
  gushing 107–8
  malt 106–7
  mycotoxins 116, 118

Gas
  chromatographic techniques 249–51
  production from glucose 156
Gene probe 224–6, 241–4
Gene-Trak system 243
Genetic information levels 238–41
Genetic techniques 50–73
Geotrichum candidum 97, 99–100
Germination stimulation 105
Gliocladium roseum 103
β-Glucanase 65–66
Gluconic acid 283–5
Gluconobacter 205
  general characteristics 167
  metabolism 167
Glutaraldehyde 298
Glycolytic sequence 17–23
Gram-negative bacteria 163–91
Gram-positive bacteria 127–62
  identification of genera 154
  test procedures 154–6
Gram stain 127–8, 163
Guanine + cytosine content (G + C%) 238–9
  generic description 238–9
  Gram positive bacteria 128

Hafnia 219
Hafnia alvei 169, 176
Hansenula 5
Health hazards 111–19
Helminthosporium 84–5
Heterofermentation 131–2
  test for 156
Hexose monophosphate pathway
  bacteria 166
  yeast 23
Homofermentation 131–2
Hops 141
Humulone 141
Hydrogen peroxide 293–4
Hydroxycarboxylic acid
Hydroxyethanol diphosphonic acid (HEDP) 285
Hypochlorite 291

Identification 237–8
  yeasts 9–10

Immunological techniques 193–4, 222–4, 244–6
Impedimetric techniques 211–13, 258–60
Iodine 292–3
Iodophores 292–3
Isothiazolinones 298–9

Killer yeasts 7, 67–8, 203–4
Klebsiella 169–71, 175–6
  beer spoilage by 171
Klebsiella pneumoniae 175
Klebsiella terrigena 69, 175
Kluyveromyces 4, 199

Lactic acid bacteria 128–53
Lactobacillus 134–47
  classification 134–8
  distribution in beer and breweries 138–9
  growth in beer 141
  identification of 144–7
    kits for 146–7
  isolation media 142–4
  phenotypic features of 145
  serological detection and identification 146
  species found in fermented beverages and related habitats 136
  spoilage of beer by 141–2
Lactobacillus acidophilus 247
Lactobacillus brevis 138
Lactobacillus casei 247
Lactobacillus coryneformis 138–9
Lactobacillus delbrueckii 138
Lactobacillus fermentum 138–9
Lactobacillus lindneri 138–9, 211
Lactobacillus plantarum 98
Lactococcus 152
Laser microprobe mass analyser (LAMMA) 257
Legionella 297, 299
Legionellosis 297
Leuconostoc 129–31, 151–2, 247
Leuconostoc oenos 152
Lipid metabolism 26–8
Luciferase reaction for ATP determination 216
Lysine agar 194–5
Lysis
  by lysostaphin and lysozyme 156

Malt
  analysis 105–7
  microflora 96–102
Malting
  effects of microorganisms on 102–9
Maltworkers' lung 112–14
Mass spectrometers 254–7
Mating type locus (*MAT*) 44–5
Megasphaera 164, 183
  culture process 186
Microbacterium imperiale 96
Microcalorimetry 214
Micrococcus 153–4
Micrococcus kristanae 153–4
Microcolony methods 218–19
Microflora
  barley 83–96
  effects on beer 107–10
  effects on distilled spirit 110–11
  malt 96–102
Micropolyspora faeni 112
Miniaturized methods for identification 260–62
Mitochondrial DNA 47
Morphological studies 257
Morphology and behaviour 257
  techniques for studying 257–64
Mould contamination assessment 119–20
Mucor 100
Mucor hiemalis 103
Mucor pusillus 89
Mycotoxins 114–19
  in beer 115–19

$NADH_2$ 16, 19, 20
Nicotinamide adenine dinucleotide (NAD^+) 19, 20, 23, 25
Nicotinamide adenine dinucleotide-phosphate ($NADP^+$) 23, 28
Nitrate reductase 177
Nitrite 177
Nitrogen
  metabolism (yeast) 26
  metabolism (lactic acid bacteria) 133
Nitrosamines 177
N-nitroso compounds 177
Nomenclature of yeasts 7–9
Non-fermentative bacteria 183

Obesumbacterium proteus 171–2
  culture media 185
  detection 230, 231

Pectinatus 180–81
  beer spoilage by 182–3
  detection 215

identification of 186
isolation of 186
taxonomy 181
Pectinatus cerevisiiphilus 181, 215
Pectinatus frisingensis 181
Pediococcus 147–52
classification 147–9
distribution in beer and breweries 149
effect on beer quality 150
growth in beer 150
identification of 151
isolation media 150–51
spoilage of beer 150
Pediococcus acidilactici 147
on malt 101
Pediococcus damnosus 147–51
detection 211, 215, 218, 230
Pediococcus inopinatus 149–51
Pediococcus pentosaceus 149–51
Penicillium
and gushing 108
in malting 100, 103
mycotoxins 114, 116
psychrotolerant 92–4
as storage fungi 88–9
Pentane dione 141
Peracetic acid 293
Phenolic off-flavour 52, 175–6
Phosphoenolpyruvate (PEP-dependent phosphotransferase) system (PTS) 131
Physarurn polycephalum 85
Pichia 3, 5, 204–5
Ploidy 46
Polar budding 4
Polymerase chain reaction (PCR) 198, 228–31, 242–3
Polyphosphate 283, 285
Post-harvest colonization 89–90
Proteins 244
examining techniques 244–7
Pseudomonas spp. 183
in malting 96
Pulsed-field gel electrophoresis (PFGE) 226–8
Pyrolysis 254–7
Pyrolysis gas chromatography (Py-GC) 254
Pyrolysis mass spectrometry (Py-MS) 254–7

Quaternary ammonium compounds (QACS) 289, 294–5

Raka-Ray agar 144
Rapid detection of microbial spoilage 209–31
Rapid identification of microorganisms 237–67
costs 265
Rare mating 53–6
Respiratory hazards 111–14
Rhanella aquatalis 172–4
beer spoilage 171, 173–4
culture 185
Rhizopus
effect on beer 109
on malt 99
Rhodotorula
on barley 86
on malt 99
RNA
double-stranded 47–8
ribosomal 135

Saccharomyces 4
Saccharomyces bayanus 195
Saccharomyces carlsbergensis 7, 34
Saccharomyces cerevisiae
acid washing 206
biochemistry and physiology 13–40, 49–50
flow cytometry 215
genetic features 44–8
hybridization 51–3
identification 9–10, 200
life cycle 44–6
rare mating 53–6
recombinant DNA methods 58–71
spheroplast fusion 56–8
strain typing 71–3, 201, 226–8
systematics 6–8
wild strains 197
Saccharomyces cerevisiae var. *ellipsoideus* 7
Saccharomyces diastaticus 62–3, 195
Saccharomyces ellipsoideus 7
Saccharomyces exiquus 195
Saccharomyces pastorianus 195
Saccharomyces uvarum 7, 34
Saccharomyces willianus 7
Sanitization 272
Sanitizers 281
Schizosaccharomyces 215
Seedling blight fungi 103–5
Selective media 195–7
Selenomonas lacticifex 181–2, 186
Sequestrants 282, 283–6
stoichiometric 283–5
threshold 285

Single chromosome transfer 54–6
Soaps 289
Spheroplast fusion 71–2
Spray
  high pressure 277
  low pressure 277
Sprayball 276
Staphylococcus 153–4
Steam 299
Sterilant, *see* Disinfectant
Storage
  bacteria 88
  fungi 88–96
  microflora 88–96
Streptobacterium 129
Streptococcus 129, 152–3
Streptomyces spp. 86
Sucrose agar 142
Surface active agents, *see* Surfactants
Surfactants 287–90
  classification 289

Tatumella 169
Tetracoccus 129
Thermoactinomyces vulgaris 86, 93
  and respiratory hazards 112
Thermobacterium 129
Torulaspora 4, 5
Torulopsis 86
Transformation 58–61
Transposable element, *Ty* 1, 71

Universal beer agar 142–4
Uridine diphosphate glucose (UDPG) 24

Vagococcus 152
VITEK identification system 262

Water-sensitivity in barley 102–3
Weissella 152, 155
Wild yeasts 193–208
  detection of 193–8
  effects in brewery 201–5
  elimination of 205–6
  identification of 198–201
  isolation media 194–7

Yeasts
  aerobic 204–5
  on barley 86
  biochemistry and physiology 13–42, 48–50
  bottom 34
  classification 1–6
  culture plant operation 32–4
  definition 1
  flocculation of 67
  genetic features 44–8
  genetic targets for brewing and distilling 48
  genetic techniques 50–61
  handling 30–32
  identification of 9–10
  improvement 48
  killer 67, 203–4
  life cycle 44–5
  mating types 45
  metabolism 15–28
  new products 73–4
  nomenclature 7–8
  nutrition 14–15
  propagation 28–34
  proteolytic activity 66–7
  pure cultures 30–32
  spores 3
  sporulation 2–6, 44–5
  systematics 1–11
  top 34
  viability of 30
  wild 193–200
  see also Saccharomyces cerevisiae

Zygomycetes 1
Zygosaccharomyces 4, 5
Zymomonas 177–80
  beer spoilage by 180
  detection of 185–6
  general characteristics 177
  metabolism 178–80
  taxonomy 178
Zymophilus 182
  beer spoilage by 182–3
  detection 186